AF761930

ANTIMICROBIAL ACTIVITY OF NANOPARTICLES

Advanced Topics in Biomaterials

ANTIMICROBIAL ACTIVITY OF NANOPARTICLES

Applications in Wound Healing and Infection Treatment

Edited by

GREGORY GUISBIERS

ELSEVIER

Elsevier
Radarweg 29, PO Box 211, 1000 AE Amsterdam, Netherlands
The Boulevard, Langford Lane, Kidlington, Oxford OX5 1GB, United Kingdom
50 Hampshire Street, 5th Floor, Cambridge, MA 02139, United States

Copyright © 2023 Elsevier Inc. All rights reserved.

No part of this publication may be reproduced or transmitted in any form or by any means, electronic or mechanical, including photocopying, recording, or any information storage and retrieval system, without permission in writing from the publisher. Details on how to seek permission, further information about the Publisher's permissions policies and our arrangements with organizations such as the Copyright Clearance Center and the Copyright Licensing Agency, can be found at our website: www.elsevier.com/permissions.

This book and the individual contributions contained in it are protected under copyright by the Publisher (other than as may be noted herein).

Notices
Knowledge and best practice in this field are constantly changing. As new research and experience broaden our understanding, changes in research methods, professional practices, or medical treatment may become necessary.

Practitioners and researchers must always rely on their own experience and knowledge in evaluating and using any information, methods, compounds, or experiments described herein. In using such information or methods they should be mindful of their own safety and the safety of others, including parties for whom they have a professional responsibility.

To the fullest extent of the law, neither the Publisher nor the authors, contributors, or editors, assume any liability for any injury and/or damage to persons or property as a matter of products liability, negligence or otherwise, or from any use or operation of any methods, products, instructions, or ideas contained in the material herein.

ISBN: 978-0-12-821637-8

For information on all Elsevier publications
visit our website at https://www.elsevier.com/books-and-journals

Publisher: Matthew Deans
Acquisitions Editor: Sabrina Webber
Editorial Project Manager: John Leonard
Production Project Manager: Surya Narayanan Jayachandran
Cover Designer: Greg Harris

Typeset by STRAIVE, India

Working together to grow libraries in developing countries
www.elsevier.com • www.bookaid.org

Contents

8. Toxicity of nanoparticles

Iliana E. Medina-Ramirez, Arturo Jimenez-Chavez, and Andrea De Vizcaya-Ruiz

Contributors

Brenda Jeanneth Acosta-Ruelas CONACYT, Coordinación para la Innovación y Aplicación de la Ciencia y la Tecnología, Universidad Autónoma de San Luis Potosí, San Luis Potosí, Mexico

Walid Azab Institute of Virology, Center for Infection Medicine, Freie Universitaet Berlin, Berlin, Germany

Hamed Barabadi Department of Pharmaceutical Biotechnology, School of Pharmacy, Shahid Beheshti University of Medical Sciences, Tehran, Iran

Aleksandra Benko Department of Chemical Engineering, Northeastern University, Boston, MA, United States; Faculty of Materials Science and Ceramics, AGH University of Science and Technology, Krakow, Poland

Clara Carreón-Álvarez Departamento de Ciencias Naturales y Exactas, Centro Universitario de los Valles, Universidad de Guadalajara, Ameca, Jalisco, Mexico

Jorge L. Cholula-Diaz School of Engineering and Sciences, Tecnologico de Monterrey, Monterrey, NL, Mexico

Andrea De Vizcaya-Ruiz Departamento de Toxicología. CINVESTAV-IPN, Mexico City, Mexico

José Miguel García-Martín Instituto de Micro y Nanotecnología, IMN-CNM, CSIC (CEI UAM+CSIC), Tres Cantos, Spain

Yves Huttel Instituto de Ciencia de Materiales de Madrid, ICMM-CSIC, Madrid, Spain

Arturo Jimenez-Chavez Departamento de Toxicología. CINVESTAV-IPN, Mexico City, Mexico

Harleen Kaur Institute of Technology, Estonian University of Life Science, Tartu, Estonia

Paulina Kazimierczak Independent Unit of Tissue Engineering and Regenerative Medicine, Medical University of Lublin, Lublin, Poland

Lidia Martínez Instituto de Ciencia de Materiales de Madrid, ICMM-CSIC, Madrid, Spain

David Medina-Cruz Department of Chemical Engineering, Northeastern University, Boston, MA, United States

Iliana E. Medina-Ramirez Departamento de Química, Universidad Autonoma de Aguascalientes, Aguascalientes, Mexico

Gabriela Navarro-Tovar CONACYT, Centro de Investigación en Ciencias de la Salud y Biomedicina, Universidad Autónoma de San Luis Potosí; CONACYT, Facultad de Ciencias Químicas, Universidad Autónoma de San Luis Potos, San Luis Potosí, Mexico

Catherine O'Connell Department of Chemical Engineering, Northeastern University, Boston, MA, United States

Elżbieta Pamuła Faculty of Materials Science and Ceramics, AGH University of Science and Technology, Krakow, Poland

Germán Plascencia-Villa Department of Neurosciences, Developmental and Regenerative Biology, The University of Texas at San Antonio (UTSA), San Antonio, TX, United States

Agata Przekora Independent Unit of Tissue Engineering and Regenerative Medicine, Medical University of Lublin, Lublin, Poland

Erwan Rauwel Institute of Technology, Estonian University of Life Science, Tartu, Estonia

Protima Rauwel Institute of Technology, Estonian University of Life Science, Tartu, Estonia

Katarzyna Reczyńska-Kolman Faculty of Materials Science and Ceramics, AGH University of Science and Technology, Krakow, Poland

José Luis Rodríguez-López Advanced Materials Department, Instituto Potosino de Investigación Científica y Tecnológica, A.C, 78216 San Luis Potosí, Mexico

Daniela Salado-Leza CONACYT, Facultad de Ciencias Químicas, Universidad Autónoma de San Luis Potos, San Luis Potosí, Mexico

Linh B. Truong Department of Chemical Engineering, Northeastern University, Boston, MA, United States

Thomas J. Webster Department of Chemical Engineering, Northeastern University, Boston, MA, United States

Sebastian Wilk Faculty of Materials Science and Ceramics, AGH University of Science and Technology, Krakow, Poland

Preface

When penicillin was discovered in September 1928 by Sir Alexander Fleming, a Scottish physician and microbiologist, it changed the face of medicine. This drug was able to offer a cure for some of the most nefarious diseases. However, due to the benefits of this new drug, doctors started prescribing it massively for all sorts of diseases. To make things worse, people also started to self-medicate themselves. This led to the spread of resistant bacteria; as more antibiotics are used, more strains adapt and survive.

According to the World Health Organization, the inappropriate use of antibiotics in animal husbandry is an underlying contributor to the emergence and spreading of antibiotic-resistant germs. It has been estimated that by 2050, antibiotic-resistant bugs could kill an estimated 10 million people each year.

Furthermore, the SARS-CoV-2 virus, which is profoundly affecting our daily life around the globe, demonstrated how disastrous the situation can be for a lack of preparation. Let me cite the Roman general Vegetius who said *"Igitur qui desiderat pacem, praeparet bellum"* meaning if you want peace, prepare for war! This is literally the purpose of this book to prepare students, scientists, and engineers to think and design new solutions to prevent future human infections.

Dr. G. Guisbiers

CHAPTER

1

Targeting the main responsible of human infections with nanoparticles

Germán Plascencia-Villa

Department of Neurosciences, Developmental and Regenerative Biology, The University of Texas at San Antonio (UTSA), San Antonio, TX, United States

1 Introduction

Infectious diseases are the main cause of illness and death worldwide, accounting for ~20% of global mortality [1]. Among the top 10 global causes of death reported by the World Health Organization, we find lower respiratory infections are the third cause with more than 3 million per year, diarrheal diseases are number 9th, and tuberculosis is number 10th, each one with around 2 million reported cases [2]. Remarkably, HIV/AIDS is no longer among the world's top 10 causes of death, but still, around 1–1.5 million people suffer from this viral infection. However, if we focus on the statistics for low-income countries, numbers indicate that lower respiratory infections and diarrheal diseases are the first and second causes of death, respectively, whereas HIV/AIDS figures as the fourth, malaria is the sixth, and tuberculosis is the seventh cause of death.

The development and widespread use of antibiotics and antivirals have been the main and sometimes exclusive therapeutic approach to fight these infections, especially for those microorganisms where a vaccine is not available. Nevertheless, infectious agents evolve constantly and surprisingly more rapidly than the treatments designed to fight them. Today, due to evolutionary pressure and overuse of antibiotics and other antimicrobial agents, many microorganisms have developed resistance, making it more difficult or even impossible to fight infections with our available drugs. According to the report CDC 2019 Antibiotic Resistance Threats in

Antimicrobial Activity of Nanoparticles
https://doi.org/10.1016/B978-0-12-821637-8.00001-8

Copyright © 2023 Elsevier Inc. All rights reserved.

the United States [3], more than 2.8 million antibiotic-resistant infections occur in the United States each year, and more than 35,000 people die as a result. There are at least 18 antibiotic-resistant bacteria and fungi, classified into three levels of concern to human health as urgent, serious, and concerning (Table 1; Fig. 1).

The development of new drugs to treat and prevent infectious diseases can take many years of research, including preclinical analysis, and clinical trials in phases I, II, and III that cost millions of dollars. Currently, there are around 42 new antibiotics in advanced development with the potential to treat serious bacterial infections, and 4 new drug applications were submitted to the FDA for review [4]. Data indicate that around 1 out of 5 infectious disease drugs that reach the clinical phase will advance through the process for FDA approval. Moreover, it is necessary to explore a novel drug class or alternative mechanisms of action to treat drug-resistant infections. Besides traditional antibiotics, antivirals, and antifungal drugs, the administration of vaccines, therapeutic or neutralizing antibodies, bacteriophages, and fecal microbiota transplants are used to treat infections. Remarkably, during the last decade, the use of functional nanoparticles has gained more interest as alternative active compounds to battle infectious agents, or as a complementary or synergistic treatment to current antimicrobials.

TABLE 1 Bacteria and fungi listed in the CDC 2019 threats report.

Urgent infectious threats	Serious infections threats	Concerning infectious threats
• Carbapenem-resistant *Acinetobacter* • *Candida auris* • *Clostridioides difficile* • Carbapenem-resistant Enterobacteriaceae • Drug-resistant *Neisseria gonorrhoeae*	• Drug-resistant *Campylobacter* • Drug-resistant *Candida* spp. • ESBL-producing Enterobacteriaceae • Vancomycin-resistant *Enterococci* (VRE) • Multidrug-resistant *Pseudomonas aeruginosa* • Multidrug-resistant nontyphoidal *Salmonella* • Multidrug-resistant nontyphoidal *Salmonella* serotype Typhi • Drug-resistant *Shigella* • Methicillin-resistant *Staphylococcus aureus* (MRSA) • Drug-resistant *Streptococcus pneumoniae* • Drug-resistant tuberculosis	• Erythromycin-resistant group A *Streptococcus* • Clindamycin-resistant group B *Streptococcus* • Azole-resistant *Aspergillus fumigatus* • Drug-resistant *Mycoplasma genitalium* • Drug-resistant *Bordetella pertussis*

Data adapted from CDC, Antibiotic Resistance Threats in the United States. https://www.cdc.gov/drugresistance/pdf/threats-report/2019-ar-threats-report-508.pdf, 2019, and do not include viruses (i.e., HIV, influenza, COVID) or parasites.

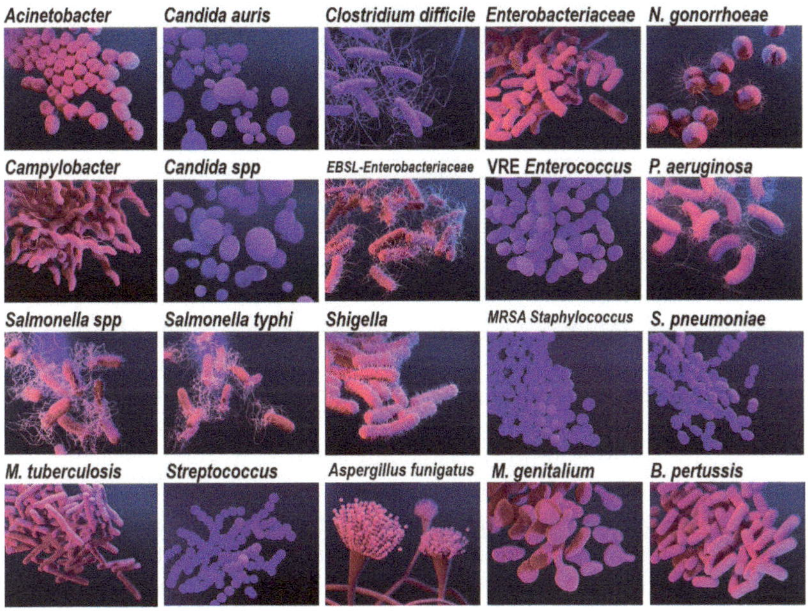

FIG. 1 Bacteria and fungi listed in the CDC 2019 threats report. *Images reproduced from CDC website CDC, Antibiotic Resistance Threats in the United States. https://www.cdc.gov/drugresistance/pdf/threats-report/2019-ar-threats-report-508.pdf, 2019.*

Nanoparticles are increasingly being utilized for clinical applications to treat infections, especially by multidrug-resistant microorganisms that cause high morbidity and mortality worldwide [5,6]. The structural, physical, and chemical properties of nanoparticles allow them to effectively interact with the cell wall and membranes of microorganisms, and upon internalization interfere with the metabolism and function of vital components promoting cell arrest and death. Moreover, nanoformulations can be administered in combination with antibiotics or antifungals, exerting a synergistic activity. Surface modification of nanoparticles with biomolecules may improve the pharmacokinetics and pharmacodynamics of the drugs, as well as improving the effectiveness through targeted delivery of bioactive drugs [7]. Furthermore, the use of nanoparticles in cleaning products or as a coating of surfaces can reduce microbial adhesion, proliferation, and biofilm formation through multiple antimicrobial properties of nanomaterials [8]. This application is particularly useful for medical devices, hospitals, self-cleaning surfaces, UV light-activated photocatalytic thin films, commercial applications, and consumer products to reduce potential infections. The versatility of nanoparticles and nanosystems makes them ideal for biodiagnosis of infectious organisms [9]. Many commercial diagnostic tests are based on nanoparticles and nanomaterials. These innovative applications of nanoparticles are particularly useful in rapid tests, point-of-case tests, and clinical tests, with the advantage to obtain results faster, easier, and at a lower cost than conventional diagnostic methods, and in some cases with higher sensitivity and specificity.

2 Bacterial infections

Bacterial infections are a major cause of chronic infections and mortality [10]. The use of antibiotics is the standard method of treatment for bacterial infections, with the advantages to be a cost-effective therapy, high effectiveness, spectrum of activity (broad spectrum, extended spectrum, and narrow spectrum), rapid response, and versatility for administration (oral, topical, intravenous, or intramuscular injections). The most common antibiotics such as sulfonamides, penicillin, erythromycin, methicillin, and ampicillin were discovered between 1930 and 1970. Derivatives from these compounds have been developed during the last decades, but the development of completely new antibiotics has virtually stopped, the cause of this could be related to the high cost for screening analysis and clinical trials, and also by the regulation of agencies. During the last decades, several bacterial strains have evolved resistance to common antibiotics, and in combination with the virtually null discovery and approval of new antibiotics have created a great problem for public health. Antibiotic resistance is directly linked to overuse of antibiotics, promoting genetic mutations or by direct transfer of genetic material (plasmids) that code for enzymes that confer specific resistance to antibiotics (β-lactamases, acetyltransferases, aminoglycoside modifying) [11]. For example, methicillin-resistant *Staphylococcus aureus* (MRSA) is an extremely difficult infection that can lead to sepsis and death (Fig. 2). It is particularly common in hospitals and healthcare facilities, approximately 5% of patients in US hospitals carry MRSA in their nose or on their skin, according to the CDC [12]. MRSA infections may evolve to skin infections, bloodstream infections, pneumonia, sepsis, and death. Here, nanostructured surfaces with ZnO, TiO_2, Ag, or Se nanoparticles can be effectively applied to prevent biofilm formation and growth of *Staphylococcus* in medical devices and clinical facilities [13].

Metals, as ions or nanoparticles, have been extensively investigated to treat bacterial infections as an alternative to antibiotics. The activity of metal nanoparticles can be classified as broad spectrum because they are not specific, impacting many strains of bacteria, and multiple biomolecules of the microorganisms. First, metal nanoparticles can interact with the cell wall; this charged outside layer attracts nanoparticles through electrostatics. Second, the bound nanoparticles will release ions that cause further disruption of the cell wall and internalization passing the membrane. Third, inside the cell body, these ions can alter the function of

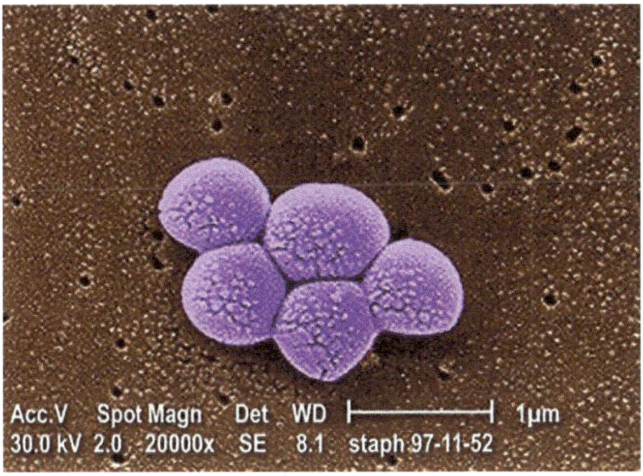

FIG. 2 Electron microscopy of methicillin-resistant *Staphylococcus aureus* (MRSA) bacteria. *Reproduced with permission from www.cdc.gov and Public Health Image Library.*

critical molecular components such as enzymes, proteins, and ribosomes that synthesize proteins, and even inhibition of DNA replication. Finally, the redox-active metal ions cause the overproduction of reactive oxygen species (ROS), lipid peroxides, and alteration of genetic material. Consequently, the accumulation of metal ions will cause metabolic failure, high oxidative stress, and cell death.

The design and formulation of metal nanoparticles targeting bacteria require considering several structural and physicochemical factors. In particular, size, shape, composition, and surface charge of nanoparticles to favor their interaction with the bacteria cell wall and then with biomolecules inside the bacteria. Metals in ionic form, ultra-small nanoclusters (0.5–3 nm), or small nanoparticles (3–50 nm) are more efficient in targeting bacteria in comparison with nanoparticles larger than 50 nm. The use of surfactants, ligands, or coatings during the synthesis of nanoparticles has a direct impact on the shape and surface charge of the nanosystems. In general, for bioapplications, it is preferred to obtain water-soluble, colloidal-stable, and partially charged nanoparticles. Several metals have demonstrated antibacterial properties, including silver (Ag), copper (Cu), iron (Fe), zinc oxide (ZnO), selenium (Se), gold (Au), quantum dots (CdSe, SeTe), titanium oxide (TiO_2), aluminum (Al), cerium oxide (CeO_2), and tungsten carbide (WC).

Moreover, metal nanoparticles can be employed as antiseptic or bioactive nanostructured surfaces. Surfaces are coated or sprayed with a solution of nanoparticles creating a protecting layer to inhibit the growth of microorganisms, particularly those bacteria that produce biofilms. Applications of nanoparticles against bacteria are versatile, including potable water filters, water treatment, clothing, medical devices, washing solutions, food containers, antibacterial creams, lotions, ointments, deodorants, self-cleaning glass, and air purifiers [14].

3 Viral infections

Viruses are nanometric infectious agents. In general, a virus is made of genetic material (DNA or RNA) enclosed in a protein capsid, and in the case of enveloped viruses, an additional outmost layer made of lipids and glycoproteins. Viruses can cause different types of serious diseases, including respiratory viral diseases (coronavirus, influenza, common cold, respiratory syncytial virus, severe acute respiratory syndrome); gastrointestinal viral diseases (rotavirus, norovirus, astrovirus, adenovirus); exanthematous (skin rash) viral infections (measles, rubella, varicella-chickenpox or shingles, roseola, smallpox, chikungunya, zika); hepatic viral infections (hepatitis A, B, C, D, and E); cutaneous viral infections (herpes, papillomavirus); hemorrhagic viral infections (ebola, lassa, dengue, yellow fever, Marburg, Crimean-Congo); neurological viral infections (poliovirus, viral meningitis, viral encephalitis, rabies); and sexual viral infections (HIV, papillomavirus, herpes, cytomegalovirus) (Fig. 3).

The illnesses caused by viral infections are of different degrees of harm, some of them just causing mild symptoms and the body heal on their own or with over-the-counter medications, allowing the immune system to fight the viruses, but others severely compromise the function of the organs or tissues infected. Furthermore, some viral infections may trigger the development of cancer, immunodeficiency, inflammation, hemorrhage, or death. By its nature, viruses are spread from person to person, through contaminated objects, contaminated food and water, or through biofluids exchange. Antibiotics do not work against viral infections.

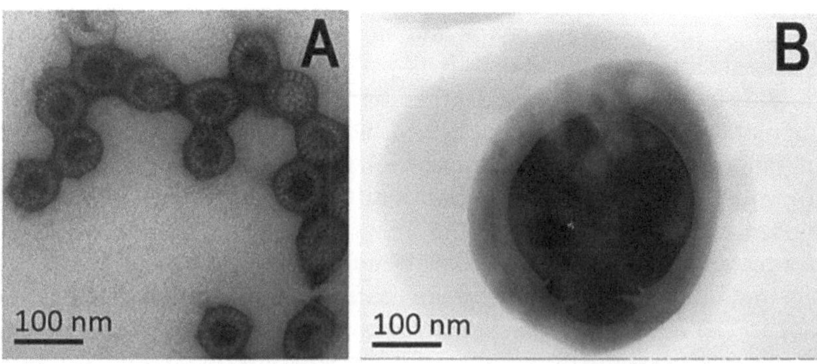

FIG. 3 Viral infectious agents. (A) Rotavirus. (B) Varicella-zoster virus (chickenpox).

Vaccination is by far the most efficient strategy to prevent viral diseases, training the immune system to efficiently recognize and remove viral particles before they can cause any systemic damage.

The drug treatments for viral infections are usually virus-specific, targeting specific molecular components of the virus, its cellular receptor, or host cells. Antiretroviral drugs are divided into nucleoside reverse transcriptase inhibitors (NRTIs), nucleotide reverse transcriptase inhibitors (NtRTIs), nonnucleoside reverse transcriptase inhibitors (NRTTIs), general antiviral drugs, nucleoside analogs, nucleotide analogs, antisense drugs, pyrimidines, immune modulators, inhibitors of polymerase and proteases, inhibitors of fusion and transport, and therapeutic antibodies [15]. In most cases it is necessary to administer combinations of agents to achieve synergistic inhibition of viruses, to delay resistance, and to decrease dosages of toxic antiviral drugs [16].

Nanoparticles function against viral infections at different levels, being broad-spectrum antiviral substances (Fig. 4). Activities of antiviral nanoparticles are divided into blocking or inhibition of cell entry, modification of viral proteins, inhibition of viral replication, sanitation of surfaces, and as nanotools for the diagnosis of viral infections. Au nanoparticles coated with biomolecules mimicking heparan sulfate proteoglycan have been employed to inhibit virus-cell interactions, serving as virucidal or virustatic agents against herpes simplex virus (HSV), human papillomavirus, respiratory syncytial virus (RSV), dengue virus, and lentivirus [17]. AuNPs are promising diagnostic and therapeutic tools against HIV/AIDS, including the delivery of antiretroviral drug raltegravir to inhibit HIV replication within $CD8^+$ cells [18,19]. Moreover, gold nanoparticles served as a vehicle for small interfering RNA to inhibit dengue virus infections [20]. AuNP of 7–8 nm coated with gallic acid showed efficiency to inhibit in vitro infection of herpes simplex virus; the antiviral activity was related to inhibition of viral attachment and internalization into Vero cells [21]. Silver nanoparticles are known for their antimicrobial properties. AgNP (1–10 nm) demonstrated antiviral activity against HIV by

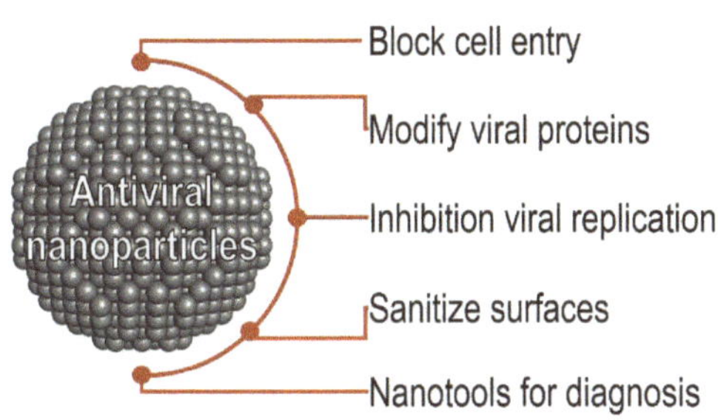

FIG. 4 Antiviral nanoparticles.

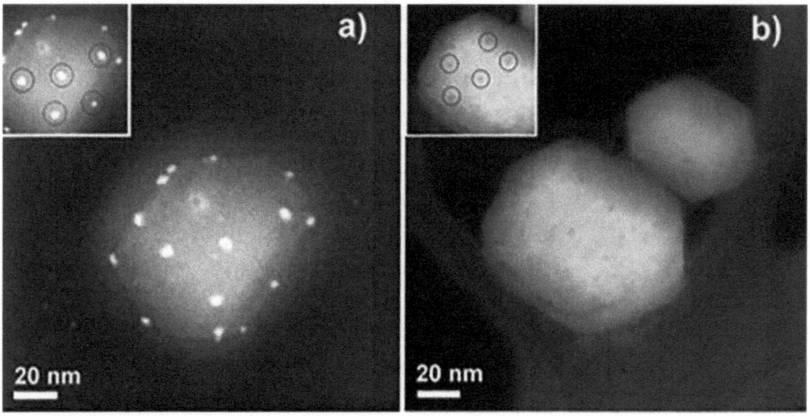

FIG. 5 HIV-1 virus with silver nanoparticles. (A) STEM imaging of HIV-1 virus with BSA-Ag nanoparticles. (B) Control HIV-1 virus without nanoparticles. *Reproduced from J.L. Elechiguerra, et al., Interaction of silver nanoparticles with HIV-1, J. Nanobiotechnol. 3(1) (2005) 6.*

modification of the viral surface glycoprotein gp120, blocking the interaction with the cellular receptor CD4 (Fig. 5) [22,23]. Furthermore, AgNP have also proven to be active against hepatitis B, herpes simplex virus, influenza (H1N1) syncytial virus, and monkeypox virus [24,25].

Therapeutic drugs can be delivered using nanoparticles as targeted delivery of antiviral agents, not only with metal-based nanoparticles but also within liposomes, micelles, microspheres, or dendrimers [26]. Polymer-based nanoformulations have been tested to deliver antiviral drugs (efavirenz, acyclovir, lamivudine, nevirapine, zidovudine, lopinavir, ritonavir, elvitegravir) against HIV, herpes virus, hepatitis B, VZV, and chickenpox [27]. Even carbon-based compounds, such as carbon nanotubes, C-dots, fullerenes, and graphene, have demonstrated antiviral applications [28]. Functionalized graphene oxide inactivated and inhibited attachment of respiratory syncytial virus [29]. Polymeric nanogels inhibited interactions of viral glycoproteins with cellular receptors, efficiently blocking infections of HSV [30]. Metal and metal oxide nanoparticles (CuO, SiO_2, TiO_2, ZnO, and CeO_2) have shown potential as broad-spectrum antiviral compounds against HIV, hepatitis B, influenza (H3N2 and H1N1) and herpes virus, dengue virus, and vesicular stomatitis virus [31,32]. Different nanodiagnostic tools are based on metal and metal oxide nanoparticles through colorimetric reactions, absorbance-based assays, immunoassays with conjugated antibodies, and chips [31]. Quantum dots nanocrystals were used to deliver saquinavir to treat HIV [33]. Remarkably, silver and copper nanoparticles are proposed as antiviral surface coating to prevent spread of infections [34]. These nanoparticles can be attached to surfaces such as fabrics, plastics, metals, and concretes as bioactive antiviral agents.

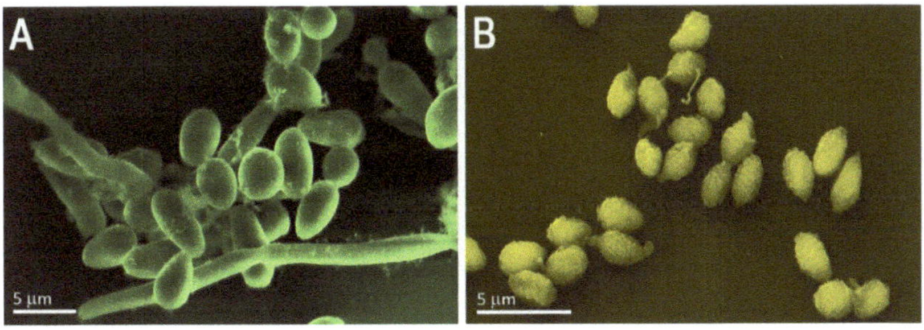

FIG. 6 Electron microscopy of fungi. (A) *Candida albicans*. (B) *Aspergillus niger*.

4 Fungal infections

Fungal infections are very common and a global public health problem. Although some fungi cause mild skin diseases, rashes, asthma, and allergies, other fungal diseases are serious, affecting the lungs (pneumonia), skin, bloodstream, and even developing meningitis [35]. Fungal diseases affect particularly patients with a compromised immune system, such as those with HIV/AIDS, cancer, and the elderly. The most common fungal diseases are fungal nail infections (onychomycosis) called *athlete's foot*, caused by different types of fungus including dermatophytes and *Fusarium*. Dermatophytosis, commonly known as ringworm, is a skin infection caused by more than 40 types of fungus, including *Trichophyton*, *Microsporum*, and *Epidermophyton*. Candidiasis is an infection by a yeast named *Candida*, the infections are common in mucosal areas of the mouth, throat, esophagus, gut, and vagina (Fig. 6A). Fungal infections can affect people who live in or acquired during travel to certain areas, such as *Blastomycosis* (soil in parts of the United States and Canada); *Cryptococcus gattii* (tropical and subtropical areas); *Paracoccidioidomycosis* (Central and South America); *Coccidioidomycosis* (Valley Fever) in southwest United States and parts of Mexico, Central, and South America; and *Histoplasma*, close to bird or bat droppings. As mentioned, fungal infections are opportunistic, affecting patients with weakened immune systems. In this group of fungi, we find *Aspergillus* spp. (mold) (Fig. 6B), *Candida auris* (multidrug-resistant fungus in healthcare facilities), invasive candidiasis in hospitalized patients, *Pneumocystis jirovecii* (pneumonia), *Cryptococcus neoformans* (meningitis), *Mucormycosis* (mold), and *Talaromyces* (present in areas of Southeast Asia, China, and India). Finally, some fungi can cause eye infections (*Fusarium*, *Aspergillus*, and *Candida*), sporotrichosis (rose gardener's disease by fungus *Sporothrix*), and mycetoma caused by fungi found in soil and water (*Eumycetoma*).

Treatments to fight fungal infections are similar to those used for bacterial infections, active compounds being applied topically, orally, or by injections that interfere with the cell membrane or metabolism to inhibit the growth of fungus (fungistatic), or kill parasitic fungi and spores (fungicides). Antifungal agents such as terbinafine and azoles are prescribed for onychomycosis. Creams, lotions, and powders with griseofulvin, terbinafine, itraconazole, and fluconazole are commonly used to treat skin areas affected by ringworm, whereas for candidiasis the antifungal treatment includes fluconazole, miconazole, clotrimazole, nystatin, or flucytosine. The overuse of antifungal drugs and fungicides in agriculture has caused some strains of fungi to develop resistance. Around 7% of all Candida are resistant to fluconazole, including *Candida albicans, Candida glabrata,* and *Candida parapsilosis,* but certain types of fungi, such as *Candida auris,* became resistant to all drug types available, causing severe invasive fungal infections [36,37]. Similarly, the fungus *Aspergillus fumigatus* also has developed antifungal resistance, and the drugs voriconazole and azole drugs are no longer useful to fight this infection [38,39].

Nanotechnology approaches are under development for the treatment of fungal infections on skin, including infections by *Candida, Trichophyton, Epidermophyton,* and Microsporum species, and to inhibit molds (*Scopulariopsis brevicaulis, Aspergillus* sp., *Acremonium* sp., *Fusarium* sp., *Hendersonula toluroidea,* and *Scytalidium hyalinum*) [40]. Colloidal silver nanoparticles (5nm) showed fungicidal activity against *Candida albicans* and *Candida glabrata* at a concentration of 0.4–3.3μg/mL, reducing cell viability and biofilm formation for up to 90% [41]. The antifungal activity of silver nanoparticles (25nm) was assessed for pathogenic *Candida albicans.* The inhibitory effect was observed at a low concentration of 0.21mg/L and even improved to 0.05mg/L with the addition of sodium dodecyl sulfate as a surfactant [42]. In comparison, ionic silver ($AgNO_3$) inhibited fungi at 1mg/L. In the same direction, Ag-Fe_3O_4 and Ag-Fe_2O_3 nanoparticles exhibited significant antifungal activities against several *Candida* species, confirming the utility of Ag-based nanoparticles in biomedicine and as a disinfectant agent [43]. The antifungal mechanism of AgNP is related to the production of ROS and the formation of hydroxyl radicals ($^{\bullet}OH$) that alter mitochondrial functions and induce cell death through apoptosis [44]. Ag, Au, and Au-Ag nanoparticles of 5nm demonstrated high antifungal activity against *Candida parapsilosis, Candida krusei, Candida glabrata, Candida guilliermondii,* and *Candida albicans,* with minimum inhibitory concentrations less than 0.7ppm in all cases [45]. Gold nanoparticles have demonstrated antifungal activity against *Candida* and *Trichophyton rubrum,* a dermatophytic fungus [46]. The antifungal property of silver is also effective against *Aspergillus niger* and *Aspergillus terreus* [47]. Some strains of Candida are resistant to drugs, in this case, AuNP served as an efficient nanodelivery system of indolicidin, to fight fluconazole-resistant *Candida albicans* isolated from patients with burn infections [48].

Copper oxide (CuO) nanoparticles (30 ± 2 nm) showed antifungal activity against two strains of *Aspergillus niger* and *Aspergillus flavus*, the concentrations of 500 ppm showed 81%–83% inhibition, whereas at 1000 pm was of 85%–86% [49]. Copper nanoparticles (4–6 nm) demonstrated significant antifungal activity against pathogenic fungi (*Fusarium oxysporum*, *Phoma destructiva*, and *Curvularia lunata*) that affect agriculture, as well as a disinfectant in poultry and animal husbandry [50]. Zinc oxide (ZnO) and titanium oxide (TiO_2) nanoparticles are used in treatments against fungal infections by *Aspergillus niger*, *Trichophyton*, *Fonsecaea*, *Aspergillus flavus*, *Rhizopus oryzae*, *Fusarium*, *Ramichloridium schilzeri*, and *Cladosporium* [51]; these nanomaterials showed similar antifungal activity as miconazole that is commonly used to treat mycosis. Moreover, nanoparticle formulations with MgO nanoparticles demonstrated its effectiveness against soilborne fungal phytopathogens (*Phytophthora nicotianae* and *Thielaviopsis basicola*) with 50%–62% of efficiency [52].

5 Emerging infectious diseases and pathogens

Every day new infectious agents emerge, and these transmissible organisms are classified as *emerging infectious diseases*, this definition includes newly appeared in a population or have existed but are rapidly increasing in incidence or geographic range [53]. The National Institute of Allergy and Infectious Diseases (NIAID) from the National Institutes of Health (NIH) classifies the emerging infectious diseases in category A, B, or C by priority or biodefense research efforts (Table 2; Fig. 7). The infectious agents in Category A are organisms or biological agents that present the highest risk to national security and public health. These microorganisms are quickly disseminated or transmitted from person to person, resulting in high mortality rates, having a significant public health impact, causing panic and social disruption, and requiring immediate action for public health responses. The infectious agents in Category B are the second-highest priority; these biological agents are moderately easy to disseminate, cause moderate morbidity rates and low mortality rates, but require specific diagnostics and disease surveillance. Finally, the pathogens in Category C include organisms that could be engineered for mass dissemination as potential biohazards, because of their high availability, ease of production and dissemination, and presenting a potential for high morbidity and mortality rates and major health impact [53].

Several of the infectious agents listed in Table 2 have a treatment or vaccines with different degrees of effectiveness. However, it is evident the vast diversity of biological agents with the potential to infect and harm humans, making it impossible to develop specific antimicrobials or vaccines targeting each one of them. Here, broad-spectrum or multifunctional

TABLE 2 Categories of emerging infectious pathogens.

Category A priority	Category B priority	Category C priority
• *Bacillus anthracis* (anthrax) • *Clostridium botulinum* toxin (botulism) • *Yersinia pestis* (plague) • Variola major (smallpox) and other related pox viruses • *Francisella tularensis* (tularemia) • Viral hemorrhagic fevers: Arenaviruses (Junin, Machupo, Guanarito, Lassa, Lujo); Bunyaviruses (Hantaviruses causing Hanta Pulmonary syndrome, Rift Valley Fever, Crimean Congo Hemorrhagic Fever); Flaviviruses (Dengue); Filoviruses (Ebola and Marburg viruses)	• *Burkholderia pseudomallei* (melioidosis) • *Coxiella burnetii* (Q fever) • *Brucella* spp. (brucellosis) • *Burkholderia psittaci* (Psittacosis) • Ricin toxin (*Ricinus communis*) • Epsilon toxin (*Clostridium perfringens*) • *Staphylococcus* enterotoxin B (SEB) • Typhus fever (*Rickettsia prowazekii*) • Food and waterborne pathogens: Bacteria (Diarrheagenic *Escherichia coli*, Pathogenic vibrios, *Shigella* spp., *Salmonella, Listeria monocytogenes, Campylobacter jejuni, Yersinia enterocolitica*); Viruses (Caliciviruses, Hepatitis A); Protozoa (*Cryptosporidium parvum, Cyclospora cayetanensis, Giardia lamblia, Entamoeba histolytica, Toxoplasma gondii, Naegleria fowleri, Balamuthia mandrillaris*); Fungi (*Microsporidia*) • Mosquito-borne viruses: West Nile virus (WNV),	• Nipah and Hendra viruses • Hantaviruses • Tickborne hemorrhagic fever viruses: Bunyaviruses (severe fever with thrombocytopenia syndrome virus, heartland virus); Flaviviruses (Omsk hemorrhagic fever virus, Alkhurma virus, Kyasanur Forest virus) • Tickborne encephalitis complex flaviviruses (Tickborne encephalitis viruses, European subtype, Far Eastern subtype, Siberian subtype, Powassan/Deer Tick virus) • Tuberculosis, including drug-resistant TB • Influenza virus • Rickettsias • Rabies virus • Prions • *Coccidioides* spp. • Severe acute respiratory syndrome-associated coronavirus (SARS-CoV), MERC-CoV, and other highly pathogenic human coronaviruses • Multidrug-resistant bacteria • Multidrug-resistant fungi

Data obtained from NIAID with content last reviewed on July 26, 2018 NIAID, NIAID Emerging Infectious Diseases/Pathogens. https://www.niaid.nih.gov/research/emerging-infectious-diseases-pathogens, 2018.

compounds such as nanoparticles and nanomaterials can be employed or combined with available treatments.

Severe acute respiratory syndrome coronavirus 2 (SARS-CoV-2) is the infectious agent of coronavirus disease 2019 (COVID-19) (Fig. 8). Since its outbreak in Wuhan, China in December 2019, SARS-CoV-2 has infected

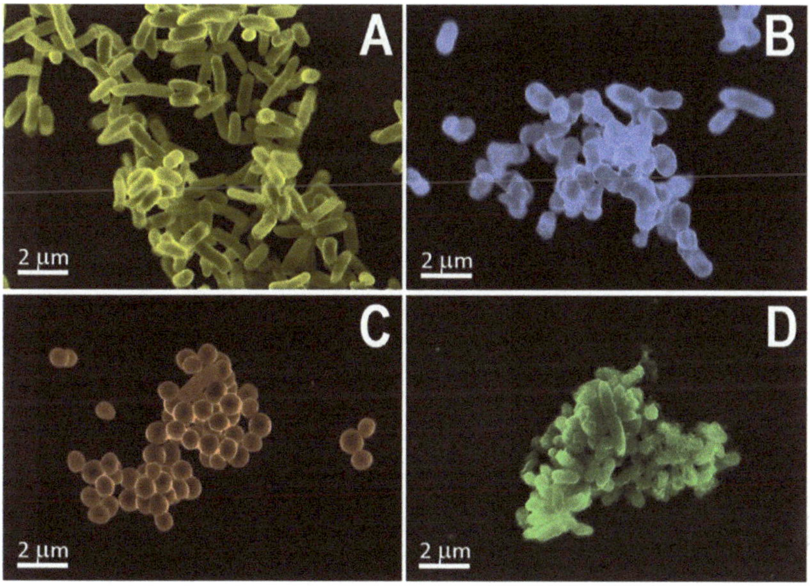

FIG. 7 Emerging infectious pathogens. (A) Enterobacteriaceae *Escherichia coli*. (B) *Acinetobacter baumannii*. (C) *Staphylococcus aureus*. (D) *Pseudomonas aeruginosa*.

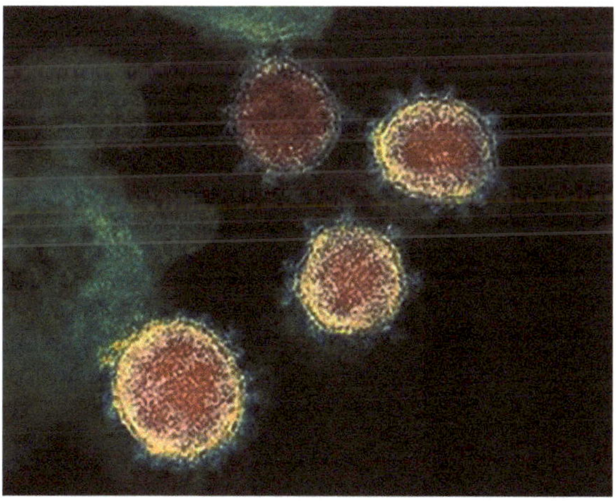

FIG. 8 Transmission electron microscopy of SARS-CoV-2, the virus that causes COVID-19. Sample isolated from a patient in the United States. Virus particles are emerging from the surface of cells cultured in the lab. The spikes on the outer edge of the virus particles give coronaviruses their name, crown-like. *Reproduced from NIH, Novel Coronavirus Structure Reveals Targets for Vaccines and Treatments. https://www.nih.gov/news-events/nih-research-matters/novel-coronavirus-structure-reveals-targets-vaccines-treatments, 2020.*

more than 514 million people globally (confirmed cases), with more than 6.25 million deaths as reported by WHO (May-2022) (https://covid19.who.int). SARS-CoV-2 has high infectivity, spreading through close contact via droplets and aerosols, especially in enclosed areas with poor ventilation [54]. This situation is exacerbated by the susceptibility of unexposed populations to a new type of virus. After almost a year, vaccines were available (Oxford-AstraZeneca, Pfizer-BioNTech, and Moderna), but there are no effective drug treatments, several therapeutic compounds have been submitted for emergency use to the FDA, in particular, bamlanivimab, a therapeutic monoclonal antibody developed by Eli Lilly [55].

The coronavirus pandemic encouraged the creative use of nanotechnology to battle this infectious agent through the development and use of functional nanoscale materials, drug delivery systems (liposomes), vaccines, nanoformulations, and nanodiagnostic tools [56]. Chloroquine and its derivate, hydroxychloroquine (an antimalaria drug) was formulated as nanomedicine and assessed as a potential treatment of COVID-19. This drug has been used in nanomedicine to study nanoparticle interactions and uptake in cells [57]. Chloroquine exerts direct antiviral effects and immunomodulatory effects, inhibiting replication of several viruses including flaviviruses, retroviruses, and coronaviruses (strain HCoV43 and SARS) [58–60]. The mechanisms of action of (hydroxy)chloroquine are not well understood, and clinical trials will confirm the utility of this drug as prophylactic and/or therapy against COVID-19. Dexamethasone nanoformulations are administered to improve inflammatory and immune responses of antiviral treatments and management of COVID-19 complications, including cytokine storm, edema, and fibrosis [61]. Nanotechnology-based antimicrobial and antiviral formulations are proposed to reduce SARS-CoV-2 dissemination [62]. These nano-based antimicrobial and antiviral formulations are proposed for disinfecting air, surfaces, and personal protective equipment (facial respirators). The bioactive nanoparticles tested in viral disinfectants are silver nanoparticles, graphene, titanium dioxide, Ag-zeolites, and copper nanoparticles. The mechanism of action is through releasing of free radicals, interaction and inhibition of viral glycoproteins, prevent binding and cell entry of viruses, overproduction of ROS, and viral membrane dissolution [62,63]. Silver nanoclusters/silica composite were sputtered coated on face masks, showing virucidal effects against coronavirus SARS-CoV-2 [64]. Effective diagnosis is critical to identify whether patients are infected or already developed antibodies against SARS-CoV-2. Several nanotechnology-based technologies have been developed for viral detection. Colloidal gold nanoparticles are employed in a rapid IgM-IgG antibody test for SARS-CoV-2 infections [65]. This rapid test cassette received FDA emergency authorization approval to detect COVID-19 IgG/IgM in whole blood/serum/plasma using a 10-min lateral flow immunoassay

(Healgen Scientific LLC) [66]. Plasmonic Au nanoparticles conjugated with antisense oligonucleotides (N gene nucleocapsid phosphoprotein) are used for a 10-min colorimetric assay to detect coronaviruses [67]. A dual-functional plasmonic gold biosensor combining plasmonic photothermal effect and localized surface plasmon resonance sensing was developed for the clinical diagnosis of SARS-CoV-2 [68]. This platform showed high sensitivity and selectivity toward coronavirus with a limit of detection of 0.22 pM. A graphene-based nanosensor with SiO_2/Si substrate was fabricated for detection of SARS-COV-2 in nasopharyngeal swab specimens by field-effected transistor-based biosensing [69]. This nanosensor demonstrated high sensitivity by direct detection of SARS-COV-2 spike protein at concentrations of 1 fg/mL in PBS buffer and 100 fg/mL in clinical transport medium, requiring no sample processing or labeling. Over the last couple of years of the pandemic, several nanotechnology-based solutions have been proposed; it is expected that with the accumulation of research evidence and data, these approaches will be improved to tackle COVID-19 and other viral infections, and even applied for the formulation of innovative vaccines to address infectious diseases impacting global health [70].

References

[1] R. Lozano, et al., Global and regional mortality from 235 causes of death for 20 age groups in 1990 and 2010: a systematic analysis for the global burden of disease study 2010, Lancet 380 (9859) (2012) 2095–2128.

[2] WHO, The Top 10 Causes of Death, 2018. https://www.who.int/news-room/fact-sheets/detail/the-top-10-causes-of-death.

[3] CDC, Antibiotic Resistance Threats in the United States, 2019, 2019. https://www.cdc.gov/drugresistance/pdf/threats-report/2019-ar-threats-report-508.pdf.

[4] P. Trusts, Antibiotics Currently in Global Clinical Development, 2020. https://www.pewtrusts.org/en/research-and-analysis/data-visualizations/2014/antibiotics-currently-in-clinical-development.

[5] N.-Y. Lee, W.-C. Ko, P.-R. Hsueh, Nanoparticles in the treatment of infections caused by multidrug-resistant organisms, Front. Pharmacol. 10 (1153) (2019).

[6] B.A. Aderibigbe, Metal-based nanoparticles for the treatment of infectious diseases, Molecules (Basel, Switz.) 22 (8) (2017) 1370.

[7] V.A. Aparna, R. Biswas, R. Jayakumar, Chapter 12 - Targeted nanoparticles for treating infectious diseases, in: A.R. Unnithan, et al. (Eds.), Biomimetic Nanoengineered Materials for Advanced Drug Delivery, Elsevier, 2019, pp. 169–185.

[8] E. Taylor, T.J. Webster, Reducing infections through nanotechnology and nanoparticles, Int. J. Nanomedicine 6 (2011) 1463–1473.

[9] C.I. Colino, C.G. Millán, J.M. Lanao, Nanoparticles for signaling in biodiagnosis and treatment of infectious diseases, Int. J. Mol. Sci. 19 (6) (2018) 1627.

[10] E. Sánchez-López, et al., Metal-based nanoparticles as antimicrobial agents: an overview, Nanomaterials (Basel, Switz.) 10 (2) (2020) 292.

[11] J.M.A. Munita, C.A. Arias, Mechanisms of antibiotic resistance, in: Virulence Mechanisms of Bacterial Pathogens, 2016, pp. 481–511.

[12] CDC, CDC. What is MRSA? 2020. https://www.cdc.gov/mrsa/community/index.html.
[13] Q. Wang, T.J. Webster, Nanostructured selenium for preventing biofilm formation on polycarbonate medical devices, J. Biomed. Mater. Res. A 100A (12) (2012) 3205–3210.
[14] K.M. Varier, et al., Nanoparticles: antimicrobial applications and its prospects, in: M. Naushad, S. Rajendran, F. Gracia (Eds.), Advanced Nanostructured Materials for Environmental Remediation, Springer International Publishing, Cham, 2019, pp. 321–355.
[15] S. Jain, et al., Occurrence of antiviral drugs in environment and their removal: a review, Water Air Soil Pollut. 224 (1410) 2013.
[16] M. Van Westreenen, C.A.B. Boucher, Chapter 1 - Classes of antiviral drugs, in: C.A.B. Boucher, et al. (Eds.), Practical Guidelines in Antiviral Therapy, Elsevier, Amsterdam, 2002, pp. 1–12.
[17] V. Cagno, et al., Broad-spectrum non-toxic antiviral nanoparticles with a virucidal inhibition mechanism, Nat. Mater. 17 (2) (2018) 195–203.
[18] C. Garrido, et al., Gold nanoparticles to improve HIV drug delivery, Future Med. Chem. 7 (9) (2015) 1097–1107.
[19] S. Gulati, et al., Functionalized gold nanoparticles: promising and efficient diagnostic and therapeutic tools for HIV/AIDS, RSC Med. Chem. 11 (11) (2020) 1252–1266.
[20] A.M. Paul, et al., Delivery of antiviral small interfering RNA with gold nanoparticles inhibits dengue virus infection in vitro, J. Gen. Virol. 95 (Pt 8) (2014) 1712–1722.
[21] A. Halder, et al., Highly monodispersed gold nanoparticles synthesis and inhibition of herpes simplex virus infections, Mater. Sci. Eng. C Mater. Biol. Appl. 89 (2018) 413–421.
[22] S. Galdiero, et al., Silver nanoparticles as potential antiviral agents, Molecules 16 (10) (2011) 8894–8918.
[23] J.L. Elechiguerra, et al., Interaction of silver nanoparticles with HIV-1, J. Nanobiotechnol. 3 (1) (2005) 6.
[24] D. Baram-Pinto, et al., Inhibition of herpes simplex virus type 1 infection by silver nanoparticles capped with mercaptoethane sulfonate, Bioconjug. Chem. 20 (8) (2009) 1497–1502.
[25] D.X. Xiang, et al., Inhibitory effects of silver nanoparticles on H1N1 influenza a virus in vitro, J. Virol. Methods 178 (1-2) (2011) 137–142.
[26] M. Milovanovic, et al., Nanoparticles in antiviral therapy, in: Antimicrobial Nanoarchitectonics, Elsevier, 2017, pp. 383–410.
[27] M. Chakravarty, A. Vora, Nanotechnology-based antiviral therapeutics, Drug Deliv. Transl. Res. (2020) 1–40.
[28] P. Innocenzi, L. Stagi, Carbon-based antiviral nanomaterials: graphene, C-dots, and fullerenes. A perspective, Chem. Sci. 11 (26) (2020) 6606–6622.
[29] X.X. Yang, et al., Synergistic antiviral effect of curcumin functionalized graphene oxide against respiratory syncytial virus infection, Nanoscale 9 (41) (2017) 16086–16092.
[30] P. Dey, et al., Multivalent flexible nanogels exhibit broad-spectrum antiviral activity by blocking virus entry, ACS Nano 12 (7) (2018) 6429–6442.
[31] T. Yadavalli, D. Shukla, Role of metal and metal oxide nanoparticles as diagnostic and therapeutic tools for highly prevalent viral infections, Nanomedicine 13 (1) (2017) 219–230.
[32] T.E. Antoine, et al., Intravaginal zinc oxide tetrapod nanoparticles as novel Immunoprotective agents against genital herpes, J. Immunol. 196 (11) (2016) 4566–4575.
[33] K.T. Yong, et al., Preparation of quantum dot/drug nanoparticle formulations for traceable targeted delivery and therapy, Theranostics 2 (7) (2012) 681–694.
[34] Anti-viral surface coating to prevent the spread of COVID-19, Focus Powder Coat. 2020 (7) (2020) 5.
[35] CDC, Fungal Diseases, 2020. https://www.cdc.gov/fungal/index.html.

[36] B. Ostrowsky, et al., Candida auris isolates resistant to three classes of antifungal medications - New York, 2019, MMWR Morb. Mortal. Wkly Rep. 69 (1) (2020) 6–9.
[37] M. Toda, S.R. Williams, E.L. Berkow, Population-based active surveillance for culture-confirmed candidemia — four sites, United States, 2012–2016, MMWR Suveill. Summ. 68 (SS-8) (2019) 1–15.
[38] P.P. Lestrade, et al., Voriconazole resistance and mortality in invasive aspergillosis: a multicenter retrospective cohort study, Clin. Infect. Dis. 68 (9) (2019) 1463–1471.
[39] S.F. Hurst, et al., Isolation of azole-resistant aspergillus fumigatus from the environment in the South-Eastern USA, J. Antimicrob. Chemother. 72 (9) (2017) 2443–2446.
[40] M. Rai, et al., Chapter 11 - Nanotechnology for the treatment of fungal infections on human skin, in: K. Kon, M. Rai (Eds.), The Microbiology of Skin, Soft Tissue, Bone and Joint Infections, Academic Press, 2017, pp. 169–184.
[41] D.R. Monteiro, et al., Silver colloidal nanoparticles: antifungal effect against adhered cells and biofilms of Candida albicans and Candida glabrata, Biofouling 27 (7) (2011) 711–719.
[42] A. Panácek, et al., Antifungal activity of silver nanoparticles against Candida spp, Biomaterials 30 (31) (2009) 6333–6340.
[43] R. Prucek, et al., The targeted antibacterial and antifungal properties of magnetic nanocomposite of iron oxide and silver nanoparticles, Biomaterials 32 (21) (2011) 4704–4713.
[44] K.-J. Kim, et al., Antifungal activity and mode of action of silver nano-particles on Candida albicans, Biometals 22 (2) (2009) 235–242.
[45] J.A. Gutiérrez, et al., High antifungal activity against Candida species of monometallic and bimetallic nanoparticles synthesized in nanoreactors, ACS Biomater Sci. Eng. 4 (2) (2018) 647–653.
[46] R. Anbarasu, et al., Synthesis of Evolvulus alsinoides derived gold nanoparticles for medical applications, Int. J. Adv. Sci. Res. 2 (1) (2016) 38–44.
[47] K. Alananbeh, W. Refaee, Z. Al-Qodah, Antifungal effect of silver nanoparticles on selected fungi isolated from raw and waste water, Indian J. Pharm. Sci. 79 (2017).
[48] H. Rahimi, et al., Antifungal effects of indolicidin-conjugated gold nanoparticles against fluconazole-resistant strains of Candida albicans isolated from patients with burn infection, Int. J. Nanomedicine 14 (2019) 5323–5338.
[49] D. Devipriya, S.M. Roopan, Cissus quadrangularis mediated ecofriendly synthesis of copper oxide nanoparticles and its antifungal studies against aspergillus Niger, aspergillus flavus, Mater. Sci. Eng. C Mater. Biol. Appl. 80 (2017) 38–44.
[50] P. Kanhed, et al., In vitro antifungal efficacy of copper nanoparticles against selected crop pathogenic fungi, Mater. Lett. 115 (2014) 13–17.
[51] S.A. George, et al., A comparative study of the anti-fungal activity of zinc oxide and titanium dioxide Nano and Bulk particles with anti-Fungals against fungi isolated from infected skin and dandruff flakes, Res. Rev.: J. Microbiol. Biotechnol. 3 (2014) 23–30.
[52] J. Chen, et al., Comparative study on the fungicidal activity of metallic MgO nanoparticles and macroscale MgO against Soilborne fungal phytopathogens, Front. Microbiol. 11 (365) (2020).
[53] NIAID, NIAID Emerging Infectious Diseases/Pathogens, 2018. https://www.niaid.nih.gov/research/emerging-infectious-diseases-pathogens.
[54] N.A. Alwan, et al., Scientific consensus on the COVID-19 pandemic: we need to act now, Lancet 396 (10260) (2020) e71–e72.
[55] Administration, U.S.F.A.D, Coronavirus (COVID-19) Update: FDA Authorizes Monoclonal Antibody for Treatment of COVID-19, 2020. https://www.fda.gov/news-events/press-announcements/coronavirus-covid-19-update-fda-authorizes-monoclonal-antibody-treatment-covid-19.
[56] Nanotechnology versus coronavirus, Nat. Nanotechnol. 15 (8) (2020) 617.

[57] T.Y. Hu, M. Frieman, J. Wolfram, Insights from nanomedicine into chloroquine efficacy against COVID-19, Nat. Nanotechnol. 15 (4) (2020) 247–249.
[58] A. Savarino, et al., Effects of chloroquine on viral infections: an old drug against today's diseases, Lancet Infect. Dis. 3 (11) (2003) 722–727.
[59] E. Keyaerts, et al., Antiviral activity of chloroquine against human coronavirus OC43 infection in newborn mice, Antimicrob. Agents Chemother. 53 (8) (2009) 3416.
[60] M.J. Vincent, et al., Chloroquine is a potent inhibitor of SARS coronavirus infection and spread, Virol. J. 2 (1) (2005) 69.
[61] T. Lammers, et al., Dexamethasone nanomedicines for COVID-19, Nat. Nanotechnol. 15 (8) (2020) 622–624.
[62] S. Talebian, et al., Nanotechnology-based disinfectants and sensors for SARS-CoV-2, Nat. Nanotechnol. 15 (8) (2020) 618–621.
[63] M.C. Sportelli, M. Izzi, E.A. Kukushkina, S.I. Hossain, R.A. Picca, N. Ditaranto, N. Cioffi, Can nanotechnology and materials science help the fight against SARS-CoV-2? Nanomaterials 10 (4) (2020) 802.
[64] C. Balagna, et al., Virucidal effect against coronavirus SARS-CoV-2 of a silver nanocluster/silica composite sputtered coating, Open Ceram. 1 (2020) 100006.
[65] Z. Li, et al., Development and clinical application of a rapid IgM-IgG combined antibody test for SARS-CoV-2 infection diagnosis, J. Med. Virol. 92 (9) (2020) 1518–1524.
[66] F. Healgen, COVID-19 IgG/IgM Rapid Test Cassette (Whole Blood/Serum/Plasma), 2020. https://www.fda.gov/media/138438/download.
[67] P. Moitra, et al., Selective naked-eye detection of SARS-CoV-2 mediated by N gene targeted antisense oligonucleotide capped plasmonic nanoparticles, ACS Nano 14 (6) (2020) 7617–7627.
[68] G. Qiu, et al., Dual-functional plasmonic photothermal biosensors for highly accurate severe acute respiratory syndrome coronavirus 2 detection, ACS Nano 14 (5) (2020) 5268–5277.
[69] G. Seo, et al., Rapid detection of COVID-19 causative virus (SARS-CoV-2) in human nasopharyngeal swab specimens using field-effect transistor-based biosensor, ACS Nano 14 (4) (2020) 5135–5142.
[70] C.N. Fries, et al., Advances in nanomaterial vaccine strategies to address infectious diseases impacting global health, Nat. Nanotechnol. 16 (1) (2020) 1–14.

CHAPTER

2

Nanomaterials to aid wound healing and infection control

Aleksandra Benko[a,b], Katarzyna Reczyńska-Kolman[b], David Medina-Cruz[a], Jorge L. Cholula-Diaz[c], Catherine O'Connell[a], Linh B. Truong[a], Lidia Martínez[d], Paulina Kazimierczak[e], Agata Przekora[e], Sebastian Wilk[b], Hamed Barabadi[f], Yves Huttel[d], José Miguel García-Martín[g], Elżbieta Pamuła[b], and Thomas J. Webster[a]

[a]Department of Chemical Engineering, Northeastern University, Boston, MA, United States, [b]Faculty of Materials Science and Ceramics, AGH University of Science and Technology, Krakow, Poland, [c]School of Engineering and Sciences, Tecnologico de Monterrey, Monterrey, NL, Mexico, [d]Instituto de Ciencia de Materiales de Madrid, ICMM-CSIC, Madrid, Spain, [e]Independent Unit of Tissue Engineering and Regenerative Medicine, Medical University of Lublin, Lublin, Poland, [f]Department of Pharmaceutical Biotechnology, School of Pharmacy, Shahid Beheshti University of Medical Sciences, Tehran, Iran, [g]Instituto de Micro y Nanotecnología, IMN-CNM, CSIC (CEI UAM+CSIC), Tres Cantos, Spain

1 Introduction

1.1 Wound healing

Wound healing constitutes a dynamic and evolving process in which different elements, such as soluble molecules, blood cells, extracellular matrix (ECM) proteins, and parenchymal cells, help the organism to replace an

Antimicrobial Activity of Nanoparticles
https://doi.org/10.1016/B978-0-12-821637-8.00002-X

Copyright © 2023 Elsevier Inc. All rights reserved.

injured or destroyed tissue. This process has four distinctive phases: hemostasis, inflammation, proliferation, and maturation [1].

In the first phase, hemostasis occurs as soon as the injury occurs within a particular tissue. Its objective is to stop bleeding. In this phase, the blood clotting system is activated, leading to the formation of a cellular dam that blocks blood drainage. Then, platelets come into contact with collagen, triggering their activation and aggregation. Immediately afterwards, the enzyme thrombin initiates the formation of a fibrin mesh, which strengthens the clumps of platelets into a stable clot, successfully stopping the bleeding [2].

During the inflammatory phase, neutrophils, a type of white blood cells, enter the wound to destroy any signs of microbial infections, remove debris and fragments of necrotic tissues. These cells often reach their peak population between 24 and 48h after injury, which is then reduced significantly after 3 days. Once the white blood cells have completed this task, macrophages arrive to continue clearing debris and secrete growth factors and proteins that attract other immune system cells to the wound to facilitate tissue repair. This phase often lasts 4–6 days [3].

Once the wound is cleaned out, it enters the third phase: proliferation, in which the healing system aims to fill and cover the open wound. This phase is subsequently divided into three subphases: (a) filling the wound and the formation of new blood vessels around the injury; (b) contraction of the wound margins; and (c) covering the wound, also known as epithelialization. This phase often lasts anywhere from four to 24days, depending on the size and severity of the damage [4].

During the last phase, maturation, the new tissue slowly gains strength and flexibility. Here, collagen fibers reorganize, the tissue remodels and matures, resulting in an overall increase in tensile strength. The maturation phase varies greatly from wound to wound, often lasting anywhere from 21days to 2 years, depending on the tissue and the type of injury [5].

The healing process is remarkably complex and often susceptible to interruption due to local and systemic factors, including moisture, infection, and maceration, as well as other factors, such as age or nutritional habits [5]. Thus, microbial infections become more prevalent due to a wide variety of outcomes and factors that might impact the wound healing process.

1.2 Bacterial infection control in wound healing

The wound healing process is connected to an increased risk of bacterial invasion because the wound is more exposed to external environmental factors than regular tissue, and its defense systems are depleted.

Moreover, factors such as arterial insufficiency, venous hypertension, or the presence of systemic diseases, such as diabetes mellitus and rheumatoid arthritis, can increase the appearance of microbial infections during the different phases of wound healing [6]. Once invasive bacterial strains arrive at the wound, they will reach critical levels of bacterial cells per gram within varioustimelines, consequently triggering the formation of a wound that is not completely healed. Nonhealing or chronic wounds are typically characterized by prolonged inflammation, defective reepithelialization, and impaired matrix remodeling, although different abnormalities may well be present in individual patients.

In addition, in all chronic wounds, there is an interaction between the patient and the bacteria present in the wound. This interaction ranges from contamination through colonization of local infection to a final stage, in which the infection spreads, leading to cellulitis or septicemia. Once bacteria reach the wound, they can start delaying the wound healing process via toxin secretion, either directly from the viable cells (exotoxins) or as a result of cell lysis (endotoxins). These toxins tend to cause local necrosis and disrupt the delicate balance of critical mediators such as cytokines and proteases, necessary for healing progression [7].

Skin wounds are especially susceptible to the colonization of bacteria since they have lost many of the protective defense mechanisms of intact skin. The absence of an intact epithelium provides a favorable environment for bacteria. The successful closure of wounds, such as burns and pressure ulcers, along with the survival of skin grafts and flap procedures, have all been shown to be dependent on maintaining a bacterial level below 10^5 cells per gram of tissue [8]. The number of microbial-host interactions determines whether the colonizing organisms invade the tissues, which will result in wound infection. This, in turn, could lead to an increased number of bacteria per gram of tissue. The virulence and pathogenicity of the organism will affect the ability of the host to mount an effective immune response [9].

1.3 Nanomedicine for bacterial infection control

Many approaches for wound healing currently fail to deliver specific outcomes, such as wound closure, and control of fluid loss while exhibiting attractive properties such as durability, elasticity, and histocompatibility. Furthermore, the field of chronic wounds is in dire need of robust, more efficacious treatments to address a dysfunctional healing process at multiple cellular levels.

The need to overcome such challenges has led to an increasing number of products emerging from the application of nanotechnology to the science of wound healing. Nanomaterials (NMs), confined to the nanoscale (1–100nm in at least one dimension), have been shown to possess great potential for many biomedical applications, with wound healing becoming an interesting target for nanotechnological research. The major advantage of nanomaterials over their bulk counterparts is the versatility and tunability of the nanomaterial's physicochemical properties, such as hydrophobicity, charge, or morphology [10]. Furthermore, the high surface area-to-volume ratio endows nanostructures with unique features [11].

Most of these advantages can be successfully implemented in the wound healing process. For example, NMs have a high probability of interacting with biological targets, thus enhancing their penetration into the wound site. As a result, nanoparticles have the ability to deliver a sustained and controlled release of therapeutics, which results in an accelerated healing process [12,13].

While some NMs are currently under clinical investigation, most nanotechnology-based platforms are in research and development. Current nanoscale strategies can be divided into carriers, drug related, and scaffold platforms—all of them based on the use of one or more nanomaterials. This book chapter, however, will be entirely focused on the use of nanoparticles (NPs) as therapeutic agents. Since the scientific community has been releasing an increasing number of innovative nanotherapies associated to wound healing, some of which are currently under clinical investigation, this chapter will be centered around those. Here, most emphasis will be placed on recent contributions toward infection control in wound healing. Therefore, in this book chapter, the authors will discuss the various types of NPs (metallic, carbon based, polymeric, lipid, and other) and their applications in controlling infection and improving the efficiency of the wound healing process.

2 Metal nanoparticles for infection control in wound healing

2.1 Monometallic noble metal nanoparticles (Au, Ag, Pd, Pt)

Metals have been used since ancient times as antibacterial agents. Among them, silver (Ag) has been one of the most commonly used due to its antimicrobial efficiency [14]. At the nanoscale, metallic nanoparticles (MNPs) are a subject of interest due to their antibacterial activity related to their small size and high surface area that improves their ability to react with multiple biomolecules in targeted microorganisms. The bactericidal pathways evoked by MNPs are dependent on their core material, shape, size, and surface functionalization [15]. MNPs constitute a type of NPs against which bacteria can hardly develop resistance because they target several cellular pathways at a time. This is one of the main reasons why these nanosystems can be a great substitute for conventional antibiotics to treat antibiotic-resistant bacterial infections [16].

Even though the mechanisms that MNPs evoke to fight bacteria have not yet been fully elucidated, their activity is known to be related to at least one of the following factors: (a) inhibition of cell membrane synthesis; (b) disruption of energy transduction; and (c) production of toxic ROS, such as superoxide radicals (O^-), hydroxyl radicals (·OH), hydrogen peroxide (H_2O_2), or singlet oxygen (O_2). ROS-related antimicrobial inhibition alters bacterial membrane permeability, photocatalysis, enzyme inhibition, and reduction of DNA production [17].

Aside from their intrinsic antimicrobial characteristics, MNPs can serve as effective delivery vectors for therapeutic agents. As such, the use of MNPs as nanocarriers for drugs and antibiotics helps to enhance the activity of pharmaceuticals against resistant microbes. The surface chemistry of MNPs is critical to modulate their interaction with bacteria, improving their broad-spectrum of activity, while simultaneously reducing their toxicity against mammalian cells. Therefore, appropriately functionalized nanomaterials provide a powerful tool to penetrate bacterial cell membranes [15].

Additionally, there are approaches that combine MNPs with external stimuli. For instance, it is possible to use photodynamic therapy (PDT) or generate ROS upon irradiation or photothermal therapy (PTT), producing heat upon irradiation, an effect which is related to their localized surface plasmon resonance (LSPR), leading to the further physical disruption of bacteria [18]. Combining MNPs with PTT and PDT are just some exemplary strategies for combating bacterial infections [19].

The most studied MNPs are those made of silver and gold, due to their unique antibacterial activity and reduced skin penetration properties [18]. As such, silver NPs (AgNPs) possess an intrinsic broad-spectrum of antibacterial activity and have already been applied in the clinics. However, they also represent a threat due to their toxicity toward human cells [20]. In this sense, other MNPs like those made of Au, Pd, or Pt are less toxic and are being explored as promising viable candidates for safer NP-based antibiotics [20,21]. Moreover, all of these MNPs present a high LSPR, which makes them especially well-suited for using them in combination with PTT or PDT. In the following subsections, we will detail the characteristics and applications of each of these MNPs for infection control and wound healing.

2.1.1 *Silver nanoparticles (AgNPs)*

Since the advent of nanomedicine, AgNPs have been widely studied for their antimicrobial action, including tests against multidrug-resistant bacteria [14]. The main antibacterial mechanism of AgNPs is assumed to be based on the formation of sulfuric bonds with either bacterial cell membrane proteins or thiol groups of various enzymes, especially those involved in the respiratory chain, hence leading to apoptosis [21]. Similarly, AgNPs can interfere with the sulfurous and phosphorous bonds in DNA, disrupting its synthesis during cell division, thus inhibiting bacterial multiplication. Ag ions released from NPs can damage the cell wall and all cellular components, leading to apoptosis and bactericidal activity [22]. They activate oxygen and produce reactive oxygen ions and hydroxyl radicals, which can hinder bacterial proliferation or destroy the cells [16].

Unfortunately, along with excellent antibacterial efficiency comes strong cytotoxic activity [23]. Generally, the relatively high toxic effects on human cells and low yield for penetration through bacterial biofilms are the two main obstacles for the use of AgNPs in biomedical applications [24]. To overcome this, smaller doses of AgNPs are combined with traditional antibacterial agents or dressings, showing an enhanced efficiency in treating the infected wounds, while still being biocompatible [17]. It has been proven that AgNPs can modulate antiinflammatory cytokine release, promoting rapid wound closure without increasing scarring [11]. Additionally, by inducing the differentiation of myofibroblasts from normal fibroblasts, AgNPs promote wound contractility, thus accelerating the healing process. AgNPs can also stimulate epidermal reepithelialization through the proliferation and relocation of keratinocytes [25].

2.1.2 *Gold nanoparticles (AuNPs)*

AuNPs are widely used in the field of wound healing due to their biocompatibility which is higher than any other metal NPs, combined with excellent antimicrobial activity. These properties can be further boosted by functionalization (with biomolecules, antibiotics, pathogen-specific antibodies, or photosensitizing molecules), which is relatively easy to complete [16]. The antimicrobial mechanism of AuNPs is related to a strong electrostatic attraction, accumulation at cell surfaces, and alteration of the membrane potential inhibiting ATP enzyme synthesis [16,25,26]. AuNPs are also known to exhibit bactericidal activity in multidrug-resistant bacteria via a ROS-independent manner, increasing the oxidative stress in bacteria. In addition, they can bind to bacterial DNA, blocking the double helix from uncoiling during replication or transcription, thus, exerting bactericidal and bacteriostatic properties [17].

Regarding the wound healing applications, AuNPs represent an interesting option due to their chemical stability and capacity for absorbing near-infrared light, while, at the same time, being relatively easy to synthesize [22]. They have inherent antioxidant and antibacterial activities that aid the wound healing process through inflammatory and hemostasis phases [25,27]. They are able to promote healing and inhibit microbial colonization in burns while being transdermally active, as demonstrated by recent ex vivo permeation studies [28].

2.1.3 *Palladium nanoparticles (PdNPs)*

Palladium nanoparticles (PdNPs) are known to have remarkable catalytic properties, mainly due to the high Pd affinity to hydrogen [29]. Although the biomedical use of PdNPs is not as popular as that of Ag- or AuNPs, there already are some reviews on this matter [30]. PdNPs present antibacterial activity by damaging the cell membrane and causing an imbalance in the metabolism system through ROS production [31]. Recently, Fang et al. found that Pd nanocrystals display facet-dependent antibacterial efficiency. {100}-faceted Pd cubes had higher enzyme-like activities and killed Gram-positive bacteria more effectively [20]. On the contrary, Gram-negative bacteria were more efficiently killed by the {111}-faceted Pd octahedrons, with the main mechanism identified as their penetration into bacterial membranes. In a different study, smaller PdNPs were found to be more effective toward killing Gram-positive than Gram-negative bacteria [32]. In addition, these MNPs can inhibit the formation of bacterial biofilms, caused by human pathogens, on a synthetic scaffold material [33].

2.1.4 *Platinum nanoparticles (PtNPs)*

The use of platinum nanoparticles (PtNPs) in bacterial infections for wound applications is far less widespread than that of the previously described nanosystems. It is, however, starting to attract increasing attention because of some interesting intrinsic properties. Pt is a precious metal that is unlikely to cause allergies and does not demonstrate genotoxicity, thus having a high potential for clinical applications. However, although the growth inhibition of *Escherichia coli* by Pt^{2+} has been known since the 1960s, the use of PtNPs in wound healing applications has been rarely assessed.

The bactericidal effect of PtNPs is thought to be induced by chemical interactions, in which free radicals, produced from the antioxidant response of PtNPs, damage the cell membrane of the bacteria, leading to outflow of the cytoplasm and invasion of NPs into the cell. This is due to the functional potential of PtNPs to decompose proteins and their strong effectiveness against lipopolysaccharides (LPS), which are constituents of the cell wall of Gram-negative bacteria. Therefore, PtNPs have a stronger antibacterial effect against Gram-negative bacteria, although they have also been reported to show antibacterial activity against Gram-positive microorganisms [23]. Extensive evaluation of PtNPs' bactericidal effect against different bacteria, with respect to their synthesis method, was recently published in a review by Jeyaraj et al. [34].

It is also known that the antibacterial properties of PtNPs can be further boosted by combining them with NIR irradiation (PTT and PDT). It was found that spatiotemporally light-activated dual-valent platinum nanobiotics (which combine photothermal therapy and chemotherapy) could effectively induce the rapid death and long-term inhibition of both Gram-positive and Gram-negative bacteria. Once exposed to NIR, the PtNPs showed a strong hyperthermal effect due to the excellent photothermal conversion ability of the Pt^0 core, together with the simultaneous and rapid release of chemotherapeutic Pt^{2+} ions, and creation of ROS. Thus, PtNPs precisely synchronizedchemo/photothermal/photodynamic tri-mode therapies [35].

To summarize, up to now, AgNPs have been the most widely studied NPs to control bacterial infections, due to their excellent antibacterial properties. However, their strong cytotoxicity has stemmed from the need to find an alternative in other noble metals like Au, with significantly lower toxicity due to their chemical stability. However, it has been shown that AuNPs need to be combined with proper molecules to show antimicrobial activity. In this sense, Pd- and PtNPs are promising novel alternatives, as they have intrinsic bactericidal properties and possess a strong catalytic activity that can favor many different antimicrobial effects. All of these MNPs also present an intense LSPR, which make them ideal candidates for employing combination therapies, like PTT or PDT. Table 1 presents some representative examples of the use of these MNPs against bacteria.

2.2 Bimetallic nanoparticles

Merging two metals is an effective way to combine different properties into one nanosystem. The physicochemical properties of the obtained NPs depend on their composition, dimensions, and morphology. Hence, the selection of controlled synthesis methodologies is a worthwhile area of research for the production of multimetallic formulations, specifically those combining two elements [41]. As such, scientists have managed to successfully control and eradicate bacterial infections by using bimetallic nanomaterials with finely tuned surface modifications, controlled composition, and improved uniformity of particle size. In this section, a short review of current work using bimetallic MNPs for the control of bacterial infections and wound healing is presented. Some of the examples of studies in this field are presented in Table 2.

TABLE 1 Representative examples of monometallic nanomaterials for the control of bacterial infections in wound healing applications.

System		Synthesis method	NP size (nm)	Antimicrobial activity	Reference
AgNPs	AgNPs + Simvastatin	Fungus-mediated synthesis (*Fusarium oxysporum*)	78	MSSA MRSA EC	[14]
	AgNPs + carboxyl betaine groups	Reduction of $AgNO_3$ in the presence of carboxyl betaine groups	2–12	SA EC	[24]
	AgNPs + gallic acid	One-pot method + hydrogel	43 ± 20	SA EC	[19]
	AgNPs + PVA	PVA hydrogel + immersion in $AgNO_3$	45–65	SA PA	[36]
AuNPs	PVP/PVA-capped AuNPs + irradiation	$HAuCl_4$ + boiling sodium citrate dihydrate + PVA or PVA	15 ± 3	EC BC	[15]
	AuNPs + *Solanum nigrum* leaves (SNL)	$HAuCl_4$ + SNL extract	3–33	EC	[21]
	AuNPs + *Cyperus rotundus*	Aqueous extract of *Cyperus rotundus* root + gold chloride under different conditions	<30	SP EC BS SA	[37]
	AuNPs + laser	Coprecipitation through the reduction of $HAuCl_4$	25	*Corynebacterium pseudotuberculosis*	[18]
PdNPs	PdNPs on PPy/rGO nanocomposite	Reduction of $PdCl_2$ by $NaBH_4$ in PPy/rGO	2–5	BS PA EC KP	[33]
	Pd nanocrystals with the {111} and {100} surfaces	Hydrothermal method	10	EC SA	[20]
	PdNPs + *Diospyros kaki*	Biosynthesis with an extract of *D. kaki* leaves	50–120	EC SA	[38]
	PdNPs + glucosamine	Glucosamine and $PdCl_2$ solution	4–9	EC	[31]

Continued

TABLE 1 Representative examples of monometallic nanomaterials for the control of bacterial infections in wound healing applications—cont'd

System		Synthesis method	NP size (nm)	Antimicrobial activity	Reference
PtNPs	PtNPs	Pt irradiation with infrared pulsed laser in liquid	2–19	EF SM *Porphyromonas gingivalis*	[23]
	PtNPs + NIR light	H_2PtCl_6-induced coordination-crosslinking nanogels with pAA_{100}-b-$pNIPAM_{200}$-b-pAA_{100} (PNA) (Pt (IV) nanogels)	1.6–130	EC MRSA	[35]
	PtNPs + doxycycline	Reduction method with doxycycline drug as a reducing and capping agent	10–20	SA SP EC ST	[39]
	Peptide-capped PtNPs	H_2PtCl_6 + pectin on reduction with $NaBH_4$	2–6	EC	[40]

BC, *Bacillus cereus;* BS, *Bacillus subtilis;* EC, *Escherichia coli;* EF, *Enterococcus faecalis;* KP, *Klebsiella pneumoniae;* MRSA, methicillin-resistant *Staphylococcus aureus;* MSSA, methicillin-susceptible *Staphylococcus aureus;* PA, *Pseudomonas aeruginosa;* SA, *Staphylococcus aureus;* SM, *Streptococcus mutants;* SP, *Salmonella paratyphi;* SP, *Staphylococcus pyogenes;* ST, *Salmonella typhimurium.*

As previously stated in Section 2.1.1, AgNPs are the most studied materials for fighting multidrug-resistant bacteria because of their high bacteria-killing efficiency. However, their high toxicity toward healthy cells limits their biomedical applicability [47]. One potential approach to decrease this negative effect is to form an alloy with other more cytocompatible metals, such as Au, Pd, or iron (Fe).

An interesting approach was suggested in a study by Li et al., who indicated the urgency to develop new antibacterial materials to confront multidrug-resistant bacteria. The authors proposed a novel method to synthesize AuNPs inlaid onto larger AgNPs, using an egg white protein. This inlaid structure was confirmed by EDX mapping, showing that Ag existed in the interior, and Au was distributed across the entire particle. TEM analysis was performed in order to determine the size and the space lattice of the particles. The sizes of the Au and AgNPs were 5 nm and 10 nm, respectively. The space lattice of Au-Ag was 0.221 nm and 0.211 nm, respectively, whereas the values for the single NPs were 0.235

TABLE 2 Representative examples of bimetallic nanomaterials for the control of bacterial infections in wound healing applications.

System	Synthesis method	NP size (nm)	Antimicrobial activity	Reference
AgPd NPs	Coreduction method	20	MRSA PA	[42]
Ag@Fe core-shell NPs	Coreduction method	5–40	SA EC	[43]
Au-Ag	Hydrothermal coreduction method	9.4 ± 2.8	MRSA MDREC	[44]
Au-Ag	Hydrothermal coreduction method	5–10	SA EC	[45]
Au-Ag, Pt-Ag, Pd-Ag	Galvanic replacement reactions	61.7 ± 16.9 to 66.9 ± 10.8	SA EC	[46]

EC, *Escherichia coli;* MDREC, multidrug-resistant *Escherichia coli;* MRSA, methicillin-resistant *Staphylococcus aureus;* PA, *Pseudomonas aeruginosa;* SA, *Staphylococcus aureus.*

(Au) and 0.236 nm (Ag). The space lattice change confirmed the formation of a Au-AgNP alloy (see Fig. 1A). The antibacterial activity of the Au-AgNPs was evaluated on *Staphylococcus aureus* and *Escherichia coli* by using the inhibition ring test (Kirby-Bauer method). The inhibition zone for Au-AgNPs against *Escherichia coli* and *Staphylococcus aureus* was 36.4 mm and 35.3 mm, respectively. In comparison, the inhibition zone for monometallic AgNPs for the same bacteria was 24.5 mm and 23.8 mm, respectively. This increase in inhibition supports the hypothesis that the bimetallic Au-AgNPs have higher antimicrobial activity. Furthermore, the authors examined the wound healing activity of Au-AgNPs in rabbits for 13 days. Four circular skin wounds were carved in rabbits in order to compare their reactions against Au-AgNPs, AgNPs, a model approach, and a blank control group. One hundred microliters of Au-AgNPs and AgNPs at its MIC were placed onto the respective wounds once per day. Then, 150 μL of a *Staphylococcus aureus* suspension (1.0×10^6 CFU/mL) was added to the wounds with Au-AgNPs, AgNPs, and the model approach. On day 3, the wound treated with Au-AgNPs began to heal with a 35% contraction, whereas the model showed inflammation and the blank showed edema. As the days passed, the wound started to contract for all the tested animals. On day 8, the contraction level was 64% for Ag-AuNPs, 43% for Ag NPs, 35% for the blank, and 6% for the model approach. At the end of the test, the wound with Au-AgNPs possessed complete reepithelialization (see Fig. 1B). This experiment allowed the authors to confirm that the high antibacterial activity and negligible

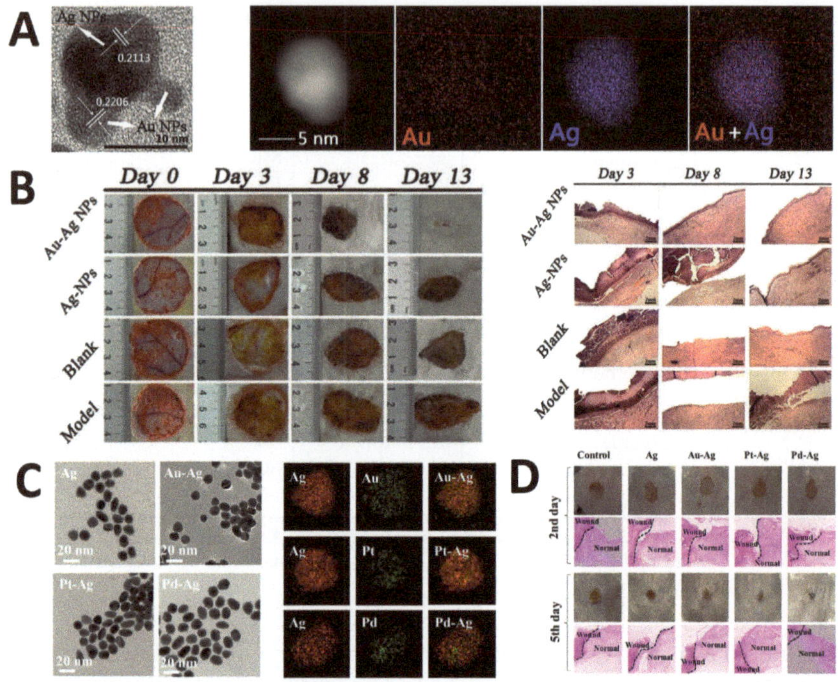

FIG. 1 (A) EDX mapping and TEM analysis of Au-Ag NPs; (B) Wound evaluation with Au-Ag NPs; (C) TEM analysis with energy-dispersive X-ray spectroscopy (EDX) confirmed the formation of Au-Ag, Pt-Ag, and Pd-Ag alloys; (D) Wound healing performance in mice skin infected with *Escherichia coli* and treated with phosphate-buffered saline (PBS) as the control, Ag, Au-Ag, Pt-Ag, and Pd-Ag NPs. *(A and B) Reproduced from Q. Li, F. Lu, H. Ye, K. Yu, B. Lu, R. Bao, Y. Xiao, F. Dai, G. Lan, Silver inlaid with gold nanoparticles: enhanced antibacterial ability coupled with the ability to visualize antibacterial efficacy, ACS Sustain. Chem. Eng. 6 (2018) 9813–9821. https://doi.org/10.1021/acssuschemeng.8b00931. (C and D) Reproduced from Y. Chang, Y. Cheng, Y. Feng, K. Li, H. Jian, H. Zhang, Upshift of the d band center toward the fermi level for promoting silver ion release, bacteria inactivation, and wound healing of alloy silver nanoparticles, ACS Appl. Mater. Interfaces. 11 (2019) 12224–12231. https://doi.org/10.1021/acsami.8b21768. Both are used with permission from the ACS, copyrights 2018 and 2019, respectively.*

cell cytotoxicity from the bimetallic alloys facilitated wound healing, as compared to the monometallic NPs [45].

Another solution to improving the antibacterial properties of AgNPs, while reducing their cytotoxicity, was proposed by Chang et al. In this paper, it was suggested that the antibacterial effect of AgNPs is attributed to releasing silver ions (Ag^{+}) after their surface oxidation. The electronic motion during oxygen absorption on the Ag surface (coupling between Ag 4d and O 2p) creates separate bonding and antibonding states. The former is fully filled, whereas the electron filling of antibonding states depends on their energy states. As antibonding states are above the d states, an increase of the d band center (energy) generates antibonding states with higher energy, leading to less filling and stronger bond formation. In order to upshift the d band to increase Ag-O bond formation, the authors proposed the synthesis of Au-Ag, Pt-Ag, and Pd-Ag bimetallic alloy NPs, due to their active electronic properties. TEM analysis with EDX confirmed the formation of the alloy, showing that the noble metals and Ag were homogeneously mixed in the NPs (see Fig. 1C). Moreover, the hydrodynamic diameters obtained by DLS ranged from 61.7 ± 16.9 nm to 66.9 ± 10.8 nm. The band energy shift, calculated by the density function theory (DFT), found a d band center of −4.12, −4.11, −3.95, and −3.94 eV for Ag, Au-Ag, Pt-Ag, and Pd-Ag NPs, respectively. These results showed that the chemical composition with the highest probability of increasing antimicrobial and wound healing activity was Pd-Ag, due to the highest upshift of the d band. Then, the ICP-OES technique was used to assess Ag^+ release. It was found that all the systems released Ag^+ within 2h, until 10h. In addition, the system with the most potent Ag^+ release ability was Pd-Ag, with Ag^+ release of 1.5 μg/mL in 2h. The authors used Gram-positive *Staphylococcus aureus* bacteria and Gram-negative *Escherichia coli* bacteria to assess antimicrobial activity. They treated the bacteria with 12.5 μg/mL of mono and bimetallic alloy NPs. Using SEM imaging, they noticed that the untreated *Escherichia coli* bacteria were smooth and rod shaped; however, when the bacteria were treated with the NPs, the cell membrane exhibited partial or complete lysis. Similar results were obtained with *Staphylococcus aureus*. Although cell walls could still be recognized, abnormal morphologies appeared, meaning that the cell wall had been destroyed. The morphology change was attributed to the interaction between Ag^+ and the cellular membrane. When Ag^+ are released, they bind to membrane proteins; this attachment increases membrane permeability that leads to the flow of different components between intracellular and extracellular fluids. This leads to cell death. Another source of cell damage is ROS. ROS oxidize the polyunsaturated fatty acids in the cell membrane, leading to lipid peroxidation. Cytotoxicity was evaluated with human keratinocytes (HaCat) cells and mouse fibroblast (NIH 3T3) cells by MTS assays, and a high viability (>90%) of cells was confirmed at 12.5 μg/mL. The reason for the high viability of cells is that Ag nanomaterials interact strongly with negatively charged membranes such as bacteria, while the cholesterol components of mammalian cells provide higher membrane stabilization [48]. Finally, the wound healing performance was assessed in mice skin wounds infected with *Escherichia coli* and treated with phosphate-buffered saline (PBS) as the control, Ag, Au-Ag, Pt-Ag, and Pd-Ag NPs (see Fig. 1D). The appearance of scars was evident on day 2 in all wounds; hematoxylin and eosin images revealed the boundary between normal and wound tissue. On day 5, the smallest scars were observed in the wound treated with Pd-AgNPs. Moreover, the wound treated with this nanosystem was the first to return to normal skin on day 12. It was concluded that the Pd-AgNPs showed the best antibacterial and wound healing performance due to the largest shift of the d band [46].

Interestingly, bimetallic NPs based on potent Ag ions can be synthesized using a green chemistry approach, as suggested by Sharma et al. [49]. For example, natural extracts from fruits, plant leaves, roots, seeds, and similar organisms can be used as both reducing and stabilizing agents in the synthesis of NPs. The biocompatible characteristics of these nanomaterials enable them to be safely applied in biomedicine. For instance, Sivamaruthi et al. used a *Terminalia chebula* fruit extract to synthesize bimetallic AgPd NPs alloys [42]. The formation of nanocrystals was confirmed qualitatively by a color change, from yellow to dark brown. The face-centered cubic (FCC) structure of the AgPd NPs was analyzed by XRD analysis. The Ag and Pd signals in the EDX analysis revealed the uniform mixing of both metals that proves the alloy structure of the bimetallic NPs. They concluded that the particle size was around 20nm, as determined by DLS, XRD, and HRTEM imaging. The latter also showed a spherically shaped and uniform size of the AgPd NPs. The authors analyzed the antimicrobial activity of the AgPd NPs against Gram-positive bacteria MRSA and the Gram-negative bacteria *Pseudomonas aeruginosa* by the Kirby-Bauer method. They noticed a zone of inhibition for MRSA and *Pseudomonas aeruginosa* at 14 and 16mm, respectively, at a AgPd NP concentration of 30μg/mL. The minimum inhibitory concentration (MIC) for MRSA was 160μg/mL, whereas for *Pseudomonas aeruginosa* it was 320μg/mL. Moreover, the minimum bactericidal concentration (MBC), the concentration of AgPd NPs at which bacteria are killed, wasa 320μg/mL for MRSA and 640μg/mL for *Pseudomonas aeruginosa.* While the molecular mechanisms for the antibacterial effect of the NPs were not identified, two possible mechanisms were suggested. The first one was a size-related mechanism. As it has been reported, particles with sizes less than 25nm are toxic to bacterial cells [50,51], as they can interact with the cell membrane, altering its fluidity and releasing the intracellular content that leads to cell death. The second mechanism was ROS related, leading to oxidative stress. The cytotoxicity of AgPd NPs was evaluated in human peripheral blood mononuclear cells (PBMC) by a trypan blue exclusion assay. The results did not show a significant difference between cells treated with AgPd NPs and vancomycin (the control) after a 24-h exposure. Therefore, it was validated that the AgPd NPs did not trigger a cytotoxic effect against normal human cells [42].

The diversity of natural sources around the world has given rise to numerous studies about their potential use as reducing and capping agents for the synthesis of NPs. Al-Asfar studied the effective activity of date flesh as a natural reducing agent for the synthesis of core-shell

Ag@Fe NPs. The palm date fruit is the main food resource in North Africa, the Middle East, and Asia. The main chemical constituents of date flesh are amino acids, alkaloids, polyphenols, flavonoids, and carbohydrates. The latter confirms that 70%–80% of the total constituents are mainly responsible for the reducing ability of date flesh, while the rest of the compounds are attributed to colloid stability. The structure of the obtained NPs was found to be dependent on the reduction potential of metal ions. The ions with higher reduction potential are first reduced and create a core, whereas the other metal ions, with lower reduction potentials, are deposited as a shell [52]. In the article, the Ag^+ ions ($Ag^+/Ag^0 = +0.788\,V$) were first reduced by β-D-glucopyranose and β-D-fructopyranose. Then, the Fe^{3+} ions ($Fe^{3+}/Fe^0 = -0.04\,V$) were reduced on the surface of Ag^0 clusters in underpotential deposition, resulting in core-shell NPs as the final product. The structure of core-shell particles was confirmed by UV-visible (UV-vis) spectroscopy. The characteristic absorption band of AgNPs at an approx. 450 nm wavelength was not visible in the absorbance spectrum of Ag@FeNPs, suggesting the core-shell formation. The obtained particles had an irregular shape, i.e., triangular polyhedral nanodisks and spheres, with sizes ranging from 5 to 40 nm.

In order to assess the antimicrobial activity of Ag@Fe against *Staphylococcus aureus* and *Escherichia coli*, the cells were exposed to different concentrations of Ag@Fe NPs (5, 10, 15, and 20 μg/mL). The results demonstrated that antimicrobial activity was directly dependent on the concentration. The zone of inhibition of *Staphylococcus aureus* and *Escherichia coli* was 15 and 8 mm, respectively, and in *Escherichia coli*, it was 8 mm at 5 μg/mL. At 20 μg/mL, the zone of inhibition was 25 mm and 20 mm for *Staphylococcus aureus* and *Escherichia coli*, respectively. Moreover, the MIC was 60 μg/mL for *Staphylococcus aureus* and 80 μg/mL for *Escherichia coli*. The authors related the inhibiting action of Ag@Fe to the electrostatic interaction between Ag@Fe and DNA, cellular proteins, and -SH groups of cysteine. The latter is of higher importance because sulfur (S) is more susceptible to gaining electrons from oxidizing species, causing protein denaturation and inhibiting respiration processes that finally cause cell death [43].

One disadvantage of green chemistry approaches for the production of nanomaterials is the lack of reproducibility due to the use of natural extracts [53]. Nevertheless, there are some green chemical processes in the literature that overcome this issue, such as the one proposed by Lomelí-Marroquín et al. [44]. The authors developed a novel synthesis method of Au-AgNPs using starch as both reducing and stabilizing agent. Starch is a hydrophilic and biocompatible biopolymer composed of amylose. During the synthesis method, at a high temperature and basic conditions, glucose molecules are released. Glucose is a reducing sugar which is able to reduce the metal ions. As it was previously mentioned, AgNPs have a relevant cytotoxic effect; hence, the authors proposed an Au-Ag alloy in order to enhance cytocompatibility. UV-vis spectroscopy was used as a primary characterization technique to confirm the synthesis of bimetallic NPs. Monometallic NPs had a single absorption band related to the surface plasmon resonance (SPR), at wavelengths of 406 nm for AgNPs and 521 nm for AuNPs. The result for the bimetallic NPs was an intermediate band at 463 nm. Furthermore, the reproducibility was assessed by analyzing six sample batches that were independently synthesized. No significant changes in the maximum absorption wavelength (462 ± 1 nm) were observed, confirming the high reproducibility of the proposed method.

The colloidal stability of the suspension was identified as an important factor in the biomedical application and, as such, was investigated in the study, using zeta potential measurements (ζ-potential). The result was a stable aqueous dispersion with an absolute ζ-potential above 30 mV. Finally, according to the TEM analysis, XRD diffraction pattern, and DLS quantification, the determined NP size was 9.7 ± 3.2 nm. The antimicrobial performances of the Au-AgNPs and AuNPs were evaluated on MRSA and multidrug-resistant *Escherichia coli* (MDR-EC) bacterial strains. The authors noticed that the activity of AuNPs was insignificant at lower concentrations (1–10 μg/mL) for MDR-EC strains, whereas, for MRSA, no significant antibacterial effect was observed over the total concentration range (1–20 μg/mL). However, for the Au-AgNPs, the results were remarkably different. For MDR-EC, the inhibition was noticeable at lower concentrations (1–5 μg/mL), whereas, for MRSA, the effect was found at concentrations between 10 and 20 μg/mL. The MIC values for the AuNPs and Au-AgNPs against MDR-EC were 7.02 ± 1.31 μg/mL and 4.92 ± 0.81 μg/mL, respectively, whereas for MRSA they were 12.29 ± 3.77 and 6.95 ± 1.70, respectively. The authors suggested that the enhanced effect of bimetallic NPs is related to ROS production and particle size. Cytotoxicity of both systems was analyzed using HDF cells. The concentration range was 1–20 μg/mL for periods between 1 and 3 days. The authors found a slight delay in cell proliferation for Ag-AuNPs, as compared to cells treated with AuNPs. However, the Ag-AuNPs were concluded to be noncytotoxic as over 60% of the cells grew and proliferated [44].

To summarize, bimetallic NPs are effective nanomaterials in inhibiting the cell proliferation of antibiotic-resistant bacteria. In all of the reviewed works, the type of bacteria cells with the highest resistance was *Staphylococcus aureus*. This characteristic could be attributed to the membrane structure of Gram-positive bacteria [54]. Scientists are running out of time to solve the current health issues raised by multidrug-resistant bacteria. Therefore, research in the development of bimetallic formulations must continue in an effort to find the best nanosystems that void the viability of antibiotic-resistant bacteria, while being biocompatible and able to enhance the process of wound healing.

3 Nonmetal nanoparticles

3.1 Carbon-based nanomaterials

There are five types of carbon-based nanomaterials (CNMs) that might find applications as advanced wound healing materials: carbon nanofibers (CNFs), carbon nanotubes (CNTs), fullerenes, carbon dots (CDs), graphene, and graphene oxide (GO). When used in biomedical applications, CNMs can elicit an antimicrobial effect through methods such as: (a) utilization of intrinsic physicochemical properties [55,56], (b) by serving as cargo carriers of antimicrobial molecules (selectively delivering the cargo inside the cells) [57], (c) by serving as inducers for photothermal (PTT) [58] or photodynamic therapies (PDT) (due to effective production of reactive oxygen species (ROS) and/or singlet oxygen (SO) [59]), (d) by enabling in vivo targeted imaging of infections [60], and (e) by creating surfaces that inhibit the adhesion of pathogens (through electronic properties [61,62], surface chemistry [63], or specific surface morphology [64,65]). A graphical representation of antimicrobial effects elucidated by CNMs is given in Fig. 2. The CNMs can be directly delivered into the infected site, loosely embedded in the material for their successive release into the desired place upon certain stimuli (pH, heat, etc.), used as matrix modifiers of polymeric dressings, or form a coating on the substrate of choice. Following is a list of CNMs with attributed antimicrobial strategies and some exemplary applications in various wound healing solutions, along with other important information (Table 3).

Materials with diameters below 10nm have been reported to cause physical damage to various pathogens by targeting cellular membrane/bacterial walls, causing so-called membrane stress. In this matter, graphene, CNFs, and CNTs have been referred to as nanoblades/nanoknifes or nanodarts [65], which can cut the cellular membrane/wall, destroying its integrity. Meanwhile, fullerenes, NDs, and CDs disrupt the membrane/wall by attaching to it and interacting with its structural components. All of these mechanisms lead to bacteriostatic and bactericidal effects [55]. These effects are more pronounced for materials with good dispersibility in polar liquids, i.e., most typically oxidized, and those with smaller diameters/planar sizes.

Other destructive mechanisms are insertion into the cell membrane/wall (which causes lipid extraction and wrapping of the cells [68,74]) or wrapping of the ion/gas exchange channels (which alters the osmotic

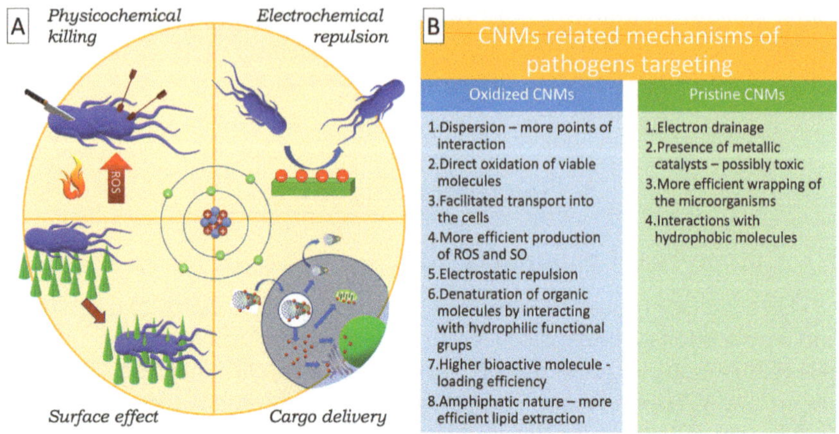

FIG. 2 (A) Schematic of CNM-induced antimicrobial effects; (B) CNM-related pathogen targeting mechanisms in their dispersed state depending on the oxidation state.

pressure and isolates the cells from external stimuli). When CNMs possess a significantly positive surface charge, they may drain the electrons from the negatively charged membranes of bacteria, affecting their viability and damaging membrane integrity. On the other hand, materials with a negative surface charge may be inhibitory toward cell adhesion by electrostatic repulsion. Insertion, wrapping, and electron transfer are more efficient for nonfunctionalized materials with a large share of delocalized π electrons (graphene, CNTs, fullerenes), while electrostatic repulsion should be more efficient on oxidized surfaces. A list of microorganism-killing effects, depending on the oxidation state of CNMs, is given in Fig. 2.

Upon entering the cells, CNMs can interact with viable molecules (such as peptides or lipids) and participate in the production of ROS and SO, both in the presence or absence of light as an external stimulus. ROS and SO production have been found to be more efficient in the presence of functional groups (such as hydroxyl or amine) introduced into the structure of CNMs. Oxidized materials can also cause damage to viable molecules of pathogens by directly oxidizing them [55]. All of these mechanisms are known to kill microorganisms effectively.

The physical killing of pathogens can be even more efficient by using an external stimulus in the form of light irradiation, which enhances ROS and SO production and generates local heat [71,79]. As such, PDT and PTT have been proven to kill even the most drug-resistant microorganisms effectively, and be able to sensitize or target the microorganism inside treatment-resistant biofilms. By targeting a wide range of biomolecules

TABLE 3 Representative examples of carbon-based nanomaterials for the control of bacterial infections in wound healing applications.

	CNMs size	Antimicrobial activity	In vivo model	Reference
CDs (derived from wheat); *Antibacterial mechanism*: pH-sensitive and fluorescent drug delivery systems for amoxicillin	2.5–8.5nm	SA EC	None	[57]
CDs (derived from Vit C); *Antibacterial mechanism*: PDT	~5nm	SA BS	None	[66]
C60 fullerene functionalized with one methylpyrrolidinium group; *Antibacterial mechanism*: PDT	<1nm (theoretical)	MRSA EC PA	Mice with wounds down to the panniculus carnosus, infected with MRSA	[67]
SWCNTs, covered with mesoporous silica, pores loaded with silver; *Antibacterial mechanism*: through ROS generation, wrapping and piercing the membrane	~20nm	MRSA MDREC	Rats, full-thickness skin defect on the back, infected with MRSA	[68]
MWCNTs, inside functionalized chitosan cryogel. Injectable material for stopping the hemorrhage of deep noncompressible wounds; *Antibacterial mechanism*: PTT, inhibitory surface	10–20nm diameter, 10–30μm length	SA EC	Mice, full-thickness skin defect model, mouse liver injury model, mouse-tail amputation model, rabbit noncompressible liver hemorrhage model	[69]
rGO coated with Ag/AgCl nanoparticles; *Antibacterial mechanism*: PDT	Not specified	SA EC	Mice with second-degree burn wounds, not infected	[70]

Continued

TABLE 3 Representative examples of carbon-based nanomaterials for the control of bacterial infections in wound healing applications—cont'd

	CNMs size	Antimicrobial activity	In vivo model	Reference
rGO—iron oxide nanoparticles; *Antibacterial mechanism*: PTT and PDT	Not specified	MRSA	Mice with subcutaneous skin infections of MRSA	[71]
Kapton HN foil, covered with Au nanoholes, and **rGO**; *Antibacterial mechanism*: inhibitory surface, PTT	Not specified	EF SE EC	Mice with subcutaneous skin infections of *Staphylococcus epidermidis*	[58]
rGO embedded in chitosan/silk fibroin scaffold; *Antibacterial mechanism*: inhibitory surface, specific core chemistry	Not specified	None	Rats, full-thickness skin defect on back	[72]
rGO embedded in PLA fibers; *Antibacterial mechanism*: inhibitory surface, PTT, drug delivery system for ampicillin or cefepime	Not specified	SA SE EC	Mice with superficial skin wounds, infected with *Staphylococcus aureus*	[73]
GO; *Antibacterial mechanism*: wrapping of the cells	830 ± 50 nm	SA PA	None	[74]
GO modified with PHMG, embedded in the PVA/chitosan matrix; *Antibacterial mechanism*: inhibitory surface, specific core chemistry	Not specified	SA EC	Mice, full-thickness skin defect, infected with a mix of *Escherichia coli* and *Staphylococcus aureus*	[75]
GO embedded in chitosan fibers; *Antibacterial mechanism*: inhibitory surface	68–105 nm (fiber diameter)	Not specified	Rat, deep wounds through panniculus carnosus, not infected	[76]

TABLE 3 Representative examples of carbon-based nanomaterials for the control of bacterial infections in wound healing applications—cont'd

	CNMs size	Antimicrobial activity	In vivo model	Reference
GO–collagen conjugates; *Antibacterial mechanism*: inhibitory surface, drug delivery system for curcumin	60–600 nm (fiber diameter)	SA	Rats, full-thickness abdominal skin defect, not infected	[77]
GO on the surface of chitosan/PLLA core-shell nanofibers; *Antibacterial mechanism*: inhibitory surface	~240 nm (fiber diameter)	SA EC	Rats, full-thickness abdominal skin defect, not infected	[78]

BS, *Bacillus subtilis;* EC, *Escherichia coli;* EF, *Enterococcus faecalis;* MRSA, methicillin-resistant *Staphylococcus aureus;* PA, *Pseudomonas aeruginosa;* SA, *Staphylococcus aureus;* SE, *Staphylococcus epidermidis.*

at a one time, PDT and PTT are regarded broad-spectrum antimicrobial therapies. These strategies are considered to bear a low risk for drug resistance development in microorganisms [59]. Upon internalization, PT and PD effects from carbon dots [66], fullerenes [67], CNTs [68], or rGO [70,71] provide excellent bacteria-killing efficiency and good biocompatibility, leading to successful recovery in animal models of infected wounds, even in case of infection with multidrug-resistant bacteria.

Furthermore, CNMs are known to enter cells fairly easily via passive diffusion [80] or through various adenosine triphosphate (ATP)-dependent uptake mechanisms (dynamin dependent): pinocytosis, phagocytosis, or endocytosis [81]. While the first one is spontaneous and nonselective, the rest of them can be selective by specific surface modification, enabling targeting of the desired microorganism and delivering of the selected cargo, thus, improving the biocompatibility toward healthy cells.

This cargo can either be a metal/metal oxide particle, which may be used to boost ROS-dependent damage [55], or a bioactive molecule, such as an antibiotic. In the latter case, CNMs can sensitize otherwise treatment-resistant pathogens, reduce the effective dose, and provide a more desired sustained release kinetics, improving killing efficiency and reducing the risk of developing a new resistance. Furthermore, pH and temperature-sensitive smart delivery systems can be obtained, which release their cargo in the acidic environment or at elevated temperatures of chronic wounds. In more advanced strategies, a combination of more than one cargo as well as a targeting molecule is used for fabricating multifunctional CNM-based carriers [55,59,79]. In a recent study by John et al., CDs were used as smart, pH-sensitive antibiotic delivery systems that enabled real-time fluorescence observation of the pathogens. Such systems may be used to monitor infection progress and evaluate the efficiency of a particular treatment. It was found that the CDs were easily internalized and performed as effective bactericidal materials while being safe to normal cells [57].

In different approaches, GO or rGO were loaded with antibiotics [73], natural polyphenols [77], growth factors [72], or synthetic bactericidal compounds [75] and embedded in a polymer matrix to obtain cargo eluting scaffolds for efficient wound closure and infection control. Such materials represent drug-releasing kinetics favorable for effective infection eradication reducing the risk of pathogens developing immunity. Zhu et al. fabricated a SWCNT-silica, silver eluting nanocomposite. Silica enabled good dispersion of SWCNTs, enabling killing mechanisms through their intrinsic properties—by wrapping the bacteria and piercing their membrane. Material efficiency was further boosted through silver NPs released inside the cells causing ROS damage. An in vivo study in rats proved that this was an effective strategy to disinfect wounds at day 3, followed by a complete recovery with low scarification and no systemic toxicity observed [68]. Similar efficiency was also reported for rGO/AgNPs nanocomposites [70].

Despite the obvious advantage of using CNMs as cargo carriers, this application is not popular for fighting infections because the CNMs have intrinsic antimicrobial properties insufficient for most applications. Thus, instead of being used as cargo carriers, the CNMs are suggested as sensitizers. In an important study by Mortimer et al., the authors decided to test the impact of low concentrations (below MIC) of various CNMs (MWCNTs, graphene, carbon black) and boron nitride on bacteria metabolism and life cycle, with the goal of identifying what is happening with the pathogens below obviously detrimental doses. An analysis of transcriptomic responses in *Pseudomonas aeruginosa PG201* revealed that while all of the NPs tested affected bacterial metabolism, the most significant effect (with an altered regulation of 111 genes) was observed for the MWCNTs, followed by graphene (44 genes). Remarkably, it was found that the MWCNTs impacted the processes responsible for developing antibiotic immunity, reversing them, and sensitizing the cells toward traditional treatment strategies. Unfortunately, the origin of significantly stronger responses evoked by CNTs as compared to other NMs was not evaluated, and biocompatibility studies were not undertaken. Still, this study provides an important guideline for the future employment of CNMs in fighting various types of antibiotic-resistant pathogens [82].

CNTs and GOs are two types of CNMs that are commonly applied as antimicrobial coatings. When the coatings are hydrophobic, they can be inhibitory toward the adhesion of bacteria and other pathogens and even cause their death [83]. Additionally, specific alignment of CNMs has been reported to affect microorganisms by piercing their membranes. This effect has been observed for vertically aligned CNT forests or GO laying on a surface in such a way that its sharp edges are exposed [84,85]. Surface electrochemical properties are also important dictators of a pathogen's fate upon contacting the material. In their proof-of-concept studies [61,62], Kotarba's team investigated the impact of electrochemical properties of graphene surfaces on the adhesion of various bacterial strains. They found that the adhesion of Gram-negative *Escherichia coli* with a highly negative zeta potential surface was significantly reduced on materials with lower amounts of oxygen-based functional groups and lower work function values. By introducing larger amounts of oxygen (and increasing the work function), this bacteriostatic effect was logarithmically reduced. Meanwhile, *Staphylococcus aureus, Staphylococcus epidermidis,* and *Pseudomonas aeruginosa,* which had the highest zeta potential values among all of the pathogen studied, adhered readily to all of the tested surfaces, which slightly improved adhesion on oxygen plasma-treated materials, linearly [61,62]. Similar results were observed recently [86] where materials (EDP-deposited layers of MWCNTs lying flat on the substrate) with a higher work function were more readily colonized by both Gram-positive and Gram-negative bacteria. Meanwhile, a work function reduction created materials inhibitory toward the MDR-EC and MRSA. On the contrary, Yiu et al. found that Gram-negative *Escherichia coli* adhesion and proliferation were not affected by the presence of oxygen-based species, but Gram-positive *Bacillus subtilis* was. In this case, the different reactions of bacteria should be attributed to different surface morphologies, which affect the work function values. This parameter has a decisive impact on cellular adhesion, but was not analyzed in the cited study.

In a study by Silva et al., MWCNTs were codeposited with polypyrrole on the surface of a polyurethane sponge to obtain an antibacterial, piezoresistive material for wound healing and the detection of movement. In their preliminary evaluation, high biofunctionality and extremely good bacteriostatic properties were proven for this new class of smart biomaterials [56]. However, as no biocompatibility tests with cells were performed, it is hard to judge the actual applicability of this material in wound healing applications. In vivo, GO and rGO have been used as coatings for dressing materials with an ability to eradicate pathogens from infected wounds in mice [58] and speed up wound closure in full-thickness abdominal skin defects in rats [14].

On an additional note, it is well known that CNMs can be typically incorporated into a polymer matrix to improve its structural stability [76,77] and to grant materials with additional functionalities. These can be: an altered nanotopography, surface chemistry, or swelling capability; the possibility of employing PTT or PDT [58]; and the possibility of fabricating smart wound treating materials. In the latter, temperature, pH, or electrical conductivity measurements and/or electrical stimulation enable the real-time evaluation of the healing process and/or stimulate a more efficient healing wounds [87]. In this application, the most commonly used CNMs are GO and rGO [72,73,75–77], which were reported to support healing and cellular infiltration, also with the aid of electrical stimulation [72], while maintaining proper wound homeostasis and removing pathogens from infected wounds.

The availability of numerous pathogen killing mechanisms offered by CNMs places them among the top choices of NPs used in effective infection control for the successful healing of wounds. Certainly, the most interesting performance comes from their intrinsic properties(morphological, chemical, and electrochemical) which enable PTT, PDT, drug delivery, and physical disruption of cells. Among all the challenges to be faced, one of the most significant is the selective targeting of pathogens without affecting normal healthy cell functions within the wounds. This can be achieved by specific modifications and introducing smart responsive materials (pH, temperature, magnetic field, light), which can be triggered by the infected wound environment or an external stimulus. It seems that these targets lie within reach of CNMs, with their multiple binding options and physicochemical properties that enable the fabrication of multifunctional nanocomposites.

3.2 Polymeric nanoparticles

Polymers are macromolecules consisting of multiple repeating units (monomers), which can differ in structure. Polymers differ in origin (e.g., natural, semisynthetic, and synthetic) [88], structure and composition of the monomer chain (e.g., linear, branched, or cross-linked and homopolymers or copolymers) [89,90], or behavior in physiological conditions (e.g., nondegradable, degradable, resorbable) [91]. Well-established synthesis methods enabling the formation of numerous materials with diverse properties make polymers particularly useful for manufacturing drug delivery systems (DDS) for different medical purposes [92].

Among all polymeric DDS, nanoparticulate drug carriers are the most versatile and widely used. By definition, "nanoparticle" refers to a particle with diameters below 100 nm. However, it is agreed that due to the diverse nature of polymeric nanoparticles, many of the carriers exceeding 100 nm (but usually not bigger than 500 nm) are still described as nanoparticles as well [88]. Based on their structure and properties, polymeric NPs can be divided into solid nanospheres, core-shell NPs, micelles, dendrimers, and polymersomes [88,93]. Polymeric NPs differ in size (a few nm to 500 nm), shape (e.g., spherical or rod like), and drug loading capacity, depending on their type, administration mode, manufacturing method, and processing parameters [93]. Some examples of the application of polymeric NPs in DDS can be found in Table 4.

The creation of such a vast selection of NP types is possible due to the development of advanced methods of polymer nanoparticle preparation. Polymeric nanoparticles or nanospheres can be fabricated using emulsification methods [93]. Layer-by-layer deposition of polyelectrolytes allows for the fabrication of multifunctional NPs [100]. Polymeric micelles or polymersomes form upon the self-assembly of amphiphilic polymers [9]. Polymeric NPs or dendrimers can also be synthesized from monomers (using emulsion or interfacial polymerization) [101]. Besides having different sizes and forms, these NPs offer several mechanisms for drug release [102]. DDS based on degradable or resorbable polymers release their cargo upon bulk degradation or surface erosion. Degradation rate, and thus drug release kinetics, depends on polymer type/composition and can be adjusted for a specific purpose [4]. Drugs may also be released from nondegradable polymeric NPs due to simple diffusion, swelling, desorption from the surface, or after exposure to external stimuli (e.g., light, temperature, or mechanical cues) [102,103].

Chitosan, a natural amino polysaccharide obtained via deacetylation of chitin, is one of the most commonly used materials for the fabrication of drug delivery carriers, antimicrobial wound dressings, and tissue engineering scaffolds [104,105]. Besides being biocompatible and biodegradable, chitosan can be easily modified and functionalized [104]. In terms of antimicrobial drug delivery systems, chitosan is particularly interesting, as it possesses intrinsic antimicrobial properties. At a low pH (below 6.5, thus similar to a biofilm pH), most amino groups in chitosan are protonated, so that the positively charged chitosan molecules can bind to the negatively charged bacterial walls, leading to the disruption of cell membranes [105]. Rhamnolipid-loaded chitosan nanoparticles synthesized by Marangon et al. showed supreme antibacterial efficacy against *Staphylococcus aureus* and *Staphylococcus epidermidis*, which was attributed to increased binding of chitosan nanoparticles to the bacterial surface, as compared to chitosan or rhamnolipid alone [94]. Similar results were obtained by Alqahtani et al., who prepared diclofenac-loaded NPs based on low or high molecular weight chitosan [106]. The inclusion of diclofenac in chitosan NPs significantly improved its antimicrobial activity against *Staphylococcus aureus* and *Bacillus subtilis*. It was also observed that antimicrobial activity increased with an increase in chitosan molecular

TABLE 4 Representative examples of polymeric nanoparticles for the control of bacterial infections in wound healing applications.

System	Synthesis/ manufacturing method	NP size (nm)	Antimicrobial activity	Reference
Chitosan nanoparticles loaded with rhamnolipid	Ionic gelation with sodium tripolyphosphate crosslinking	287.8 ± 6.2	SA SE	[94]
PLGA nanoparticles loaded with ciprofloxacin	Double emulsion solvent evaporation	202.9 ± 4.6	EF	[95]
Polyanhydride nanoparticles loaded with doxycycline and rifampicin	Antisolvent nanoencapsulation	Ranging from 162.8 ± 52.0 to 326.8 ± 117.8	EC	[96]
PHB nanoparticles loaded with epirubicin	Nanoprecipitation	Ranging from 152.3 ± 0.6 to 160.6 ± 1.4	EC SA	[97]
PEG-PCL micelles loaded with ciprofloxacin and modified with vancomycin	Solvent evaporation	76.9	PA	[98]
Polymersomes based on hyaluronic acid-oleylamine copolymers loaded with vancomycin	Ultrasonication	Ranging from 201.4 ± 3.25 to 360.9 ± 5.84	MRSA	[99]

EC, *Escherichia coli*; EF, *E. faecalis*; MRSA, methicillin-resistant *Staphylococcus aureus*; PA, *Pseudomonas aeruginosa*; SA, *Staphylococcus aureus*; SE, *Staphylococcus epidermidis*.

weight. MIC90 against *Staphylococcus aureus* and *Bacillus subtilis* was almost two times lower for high molecular weight chitosan in comparison to low molecular weight chitosan diclofenac-loaded nanoparticles, while drug-loaded efficacy was similar for both types of chitosan (29.3 ± 4.3% for low molecular weight chitosan and 31.1 ± 3.1% for high molecular weight chitosan). Similarly, Cao et al. fabricated chitosan NPs loaded with triclosan and coated them with in situ cross-linked phosphorylcholine (PMPC) [107]. PMPC-modified nanoparticles showed improved penetration and accumulation in biofilms (*Staphylococcus aureus*). The acidic environment of a *Staphylococcus aureus* biofilm caused chitosan swelling, which triggered triclosan release. Most importantly, the dose of triclosan required for biofilm eradication was significantly lower for triclosan-loaded chitosan NPs in comparison to triclosan alone. Alternatively, chitosan is also commonly used as a component of composite drug delivery systems, e.g., with silver NPs [108], alginate [109], or synthetic polymers [110].

Other types of naturally occurring polymers used in DDS are polyhydroxyalkanoates (PHAs). They have attracted much attention as possible drug delivery carriers since they are biocompatible, biodegradable, and sustainable (can be produced via microbial fermentation). Among over 150 different PHAs, poly-3-hydroxybutyrate (PHB) and poly-3-hydroxybutyrate-*co*-3-hydroxyvalerate (PHBV) are the most intensively studied materials [97,111]. Perveen et al. prepared PHB and PHBV nanoparticles, coated with polyethylene glycol, and loaded with epirubicin. Drug encapsulation efficacy was higher for PHB-based nanoparticles. However, in both cases, the release of the drug was sustained for 8 days. Encapsulation of the drug in a PHB or PHBV matrix increased its antibacterial efficacy against *Escherichia coli* and MRSA [97].

Although natural polymers offer a broad range of advantages as matrix materials for the fabrication of antimicrobial nanoparticles, it should be remembered that since they are obtained from natural sources, their properties may vary from batch to batch. Thus, synthetic polymers have been widely investigated, being more reliable, with fully controlled structures and properties. The synthesis of polymers with different compositions and properties allows for the fabrication of various nanoparticles, including solid nanospheres, micelles, dendrimers, or polymersomes [105].

Poly (D, L-lactic-*co*-glycolic acid) (PLGA) was approved by the Food and Drug Administration (FDA) and by the European Medicines Agency (EMA) and has already been used for the fabrication of various drug delivery systems and implants [112]. PLGA is nontoxic, biocompatible, and its hydrolytic degradation leads to controlled drug release kinetics, depending on the LA-to-GA monomer ratio. Being a hydrophobic polymer, PLGA is commonly used for the fabrication of drug-loaded nanospheres or nanocapsules. Arafa et al. prepared ciprofloxacin-loaded PLGA nanoparticles and modified their surfaces with chitosan. Chitosan changed the surface charge of the nanoparticles (from −0.026 mV for uncoated nanoparticles to +28.5 mV for chitosan-coated ones), slightly slowing down ciprofloxacin release, but significantly increasing antibacterial efficacy of the nanoparticles against *Enterococcus faecalis* [95]. Hill et al. used a double emulsification method to obtain tobramycin-loaded

PLGA NPs. A sustained release of tobramycin effectively suppressed the proliferation of *Pseudomonas aeruginosa* [113]. Similar PLGA and PLGA-polyethyleneimine (PLGA-PEI) NPs, loaded with clindamycin, were fabricated by Hasan et al. Both types of nanoparticles effectively decreased the viability of MRSA within 24 h of incubation. Moreover, in vivo tests proved that clindamycin-loaded nanoparticles significantly decreased the wound bacterial burden, which allowed for faster wound closure (decreased bacteria viability and accelerated healing were more prominent in the case of PLGA-PEI-based NPs).

Other degradable polymers utilized for manufacturing antimicrobial DDS systems are polyanhydrides [114]. Degradation time, cellular uptake, and drug loading capacity of polyanhydrides strongly depend on their copolymer composition. Thus, they may be easily modified to meet specific requirements [115]. The most popular polyanhydrides are based on sebacic acid, 1,3-bis(4-carboxyphenoxy)-propane, and maleic anhydride [93]. Lueth et al. encapsulated doxycycline and rifampicin in nanoparticles based on copolymers of sebacic acid, 1,6-bis(-*p*-carboxyphenoxy)hexane, and 1,8-bis(*p*-carboxyphenoxy)-3,6-dioxaoctane. Rifampicin-loaded nanoparticles were able to effectively eradicate *Brucella melitensis* from macrophage cultures within 48 h of treatment. A prolonged release of antibiotics was also observed in vivo in BALB/c mice, which allowed for a sevenfold decrease in antibiotic dosing [96].

Polycaprolactone (PCL) is a semicrystalline, hydrophobic aliphatic polyester. However, as its degradation time is longer in comparison to PLGA (even up to a few years in physiological conditions), it is more suitable for long-term drug delivery. Thus, PCL is rarely used for the fabrication of nanoparticles loaded with antimicrobials [93]. It has, however, been utilized for the fabrication of polymeric micelles, designed to deliver hydrophobic drugs. In general, micelles are composed of amphiphilic block copolymers that self-organize into core-shell structures (hydrophobic parts gather inside, while hydrophilic chains are exposed on the outside) [105]. In a majority of micelles, hydrophilic parts are from poly(ethylene glycol) (PEG), which provides stealth properties and improves drug pharmacokinetics and pharmacodynamics [93]. Guo et al. prepared lipase-sensitive, hypocrellin A-loaded micelles based on methoxy-PEG and PCL copolymers [116]. PCL degradation occurs upon contact with lipases secreted by bacteria, leading to drug release. The antimicrobial efficacy of hypocrellin A was further increased via light irradiation due to the triggered production of ROS. Similar PEG-PCL micelles, this time loaded with ciprofloxacin, were also fabricated by Chen et al. [98]. In addition, vancomycin was conjugated to the surface of the micelle, allowing for bacteria targeting. Upon a decrease in environmental pH (e.g., in close proximity to bacteria), vancomycin detached from the micelles (cleavage of hydrazine bonds), and the hydrophobic/hydrophilic balance became disrupted, thus leading to variations in micelle size and exposure of PCL segments to lipases for degradation. As a consequence of the PCL degradation, ciprofloxacin embedded inside of the micelles was released. In vivo studies provided evidence of a higher survival rate of *Pseudomonas aeruginosa*-infected mice treated with the micelle, in comparison to the ones treated with free ciprofloxacin and vancomycin. Other studies utilized PLA [117], polyurethanes [118], inulin [119], or poly(oligoethylene glycol methyl ether methacrylate) [120] as hydrophobic blocks in the micelles, while hydrophilic PEG might be replaced with, e.g., poly(4-vinylpyridine) [120].

Due to the presence of a hydrophobic core, micelles are not suitable for encapsulating hydrophilic drugs. Thus, a novel group of nanoparticles was developed—polymersomes. Polymersomes are composed of a hydrophilic core and hydrophobic shell, surrounded by a hydrophilic periphery. On the contrary to micelles, polymersomes can be loaded with both hydrophilic and hydrophobic drugs [105]. Walvekar et al. synthesized copolymers based on hydrophilic hyaluronic acid and hydrophobic oleylamine and used them for manufacturing polymersomes loaded with vancomycin. The minimal inhibitory concentration against MRSA was fourfold lower than for free vancomycin. It was also observed that polymersome encapsulated drugs exhibited a stronger impact on the MRSA membrane, in comparison to vancomycin alone [99].

Alternatively, Lanzilotto et al. demonstrated the fabrication of polymersomes based on poly(2-methyloxazoline)-*block*-poly(dimethylsiloxane)-*block*-poly(2-methyloxazoline) containing a water-soluble photosensitizer (porphyrin TPyCP). The antimicrobial efficacy of such polymersomes against *Escherichia coli* was based on the generation of a singlet oxygen in the hydrophilic compartment and its diffusion through the hydrophobic shell, causing oxidative stress in bacteria, consequently leading to their death [121].

Considering a broad range of different materials and the relative ease of their functionalization, it is not surprising that polymers are commonly used for the fabrication of nanocarriers. They are used not only as antimicrobial agents but also in cancer treatment [122,123] or gene delivery [124]. Well-established methods for polymer synthesis and modification, availability, and ease of processing allow for the manufacturing of countless polymeric carriers, differing in size, shape, and drug loading capacity. Tunable surface properties enable the attachment of drugs, signaling molecules, or targeting ligands, which may increase the binding of such nanocarriers to bacteria, increase biofilm penetration, and improve drug bioavailability [105]. Although it may seem like polymeric NPs are ideal drug carriers, they have their limitations as well. Polymers of natural origin may vary from batch to batch, resulting in altered properties of the final products. On the other hand, synthetic polymers often require catalysts or organic solvents, which, if not completely removed, may be toxic. Drug-loaded nanoparticles made of degradable polymers may be unstable in long-term storage. Nevertheless, polymeric NPs are some of the most versatile drug delivery systems. They can be used alone as a suspension, can be embedded in electrospun fibers, or finally can be attached to the surface of scaffolds or implants. Additionally, polymeric NPs can be the foundation for the development of various hybrid drug delivery systems.

3.3 Lipid-based nanoparticles

Lipid-based nanoparticles (LNPs), such as solid lipid nanoparticles (SLNs) and liposomes, have emerged as a novel approach for wound healing and infection control. The topical administration of LNPs for wound healing purposes prevents microbial contamination and skin water loss by evaporation, thus enhancing the fluidization of the stratum corneum lipid packing, resulting in a rise in drug permeability. Fig. 3A shows a schematic representation of wound healing using LNPs [125]. SLNs are composed of a solid lipid matrix (e.g., waxes, triglycerides, steroids, fatty acids, and partial glycerides), surfactants, and occasionally cosurfactants in the range of 50–1000nm [126,127].

Some of the advantageous factors of SLNs for drug delivery systems involve biodegradability of the components, lack of organic solvents in the preparation process, drug protection from environmental damage, etc. [126–128]. In addition, SLNs can entrap hydrophilic/hydrophobic drugs [129,130], as well as peptides/proteins [131], and nucleic acids [132], which make SLNs quite effective in DDS. Table 5 details the various SLNs and their synthesis methods and antimicrobial activity.

Infection control is essential for wound care [146]. Using SLNs for wound healing and infection control is a novel approach. Gad et al. formulated chamomile oil (CM)-loaded SLNs with a particle size of 542.1 ± 27.51 nm. Interestingly, the topical administration of CM-SLNs showed significant wound closure and healing activity when compared to free CM and SLNs, as evaluated through an in vivo study involving 40 rats (Fig. 3B) [147]. For example, Arantes et al. formulated retinoic acid-loaded SLNs (ATRA-SLNs) surrounded by a chitosan film for wound healing and tested their performance in an in vivo study. The findings showed that chitosan films containing ATRA-SLNs enhanced the closure of wounds in diabetic mice in comparison to a control of chitosan films without ATRA-SLNs (Fig. 3C) [133].

Similarly, Khezri et al. produced a formulated gel containing *Mentha pulegium* essential oil (MPO) loaded into nanostructured lipid carriers

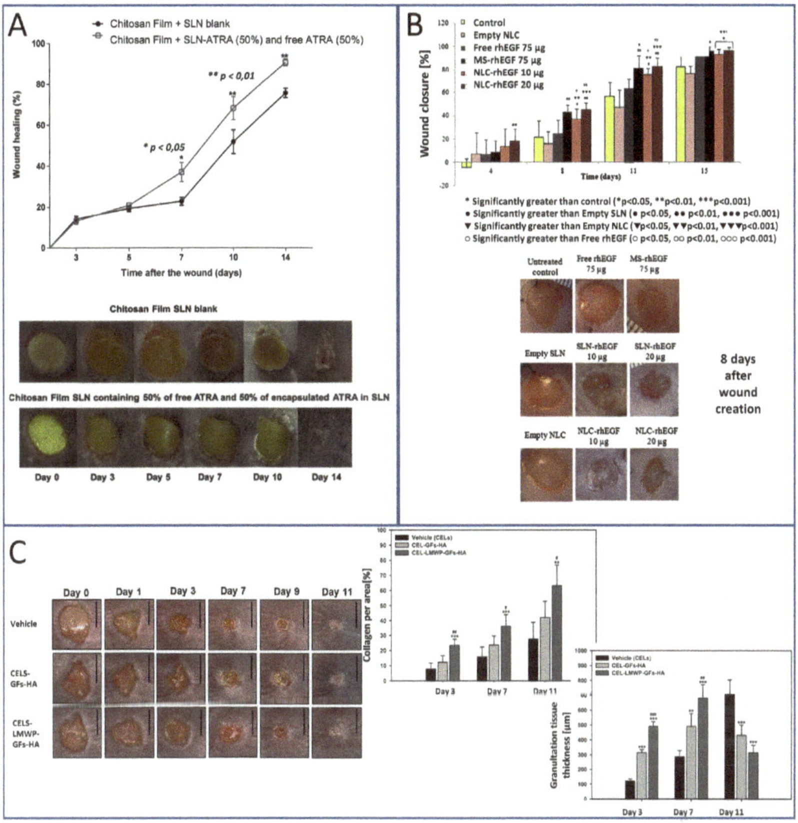

FIG. 3 Examples of lipid NPs used in wound healing. (A) Chitosan films containing SLN-ATRA (50%) and free ATRA (50%), 0.1% of total ATRA concentration. =Graphical representation and images of the wound healing process; (B) comparison of SLN and NLC (rhEGF encapsulated) efficiency in closing the wound in db/db mice; (C) Effect of cationic elastic liposomes comprising low molecular weight protamine (LMWP)-conjugated growth factors and hyaluronic acid (HA) on wound area reduction in the full thickness excisional diabetic mouse wound model, presented with granulation tissue thicknesses and collagen deposition in the wound area 3, 7, and 11 days after treatment. *(A) Reproduced from V.T. Arantes, A.A.G. Faraco, F.B. Ferreira, C.A. Oliveira, E. Martins-Santos, P. Cassini-Vieira, L.S. Barcelos, L.A.M. Ferreira, G.A.C. Goulart, Retinoic acid-loaded solid lipid nanoparticles surrounded by chitosan film support diabetic wound healing in in vivo study, Colloids Surf. B Biointerfaces 188 (2020) 110749. https://doi.org/10.1016/j.colsurfb.2019.110749. (B) Reproduced from G. Gainza, M. Pastor, J.J. Aguirre, S. Villullas, J.L. Pedraz, R.M. Hernandez, M. Igartua, A novel strategy for the treatment of chronic wounds based on the topical administration of rhEGF-loaded lipid nanoparticles: in vitro bioactivity and in vivo effectiveness in healing-impaired db/db mice, J. Control. Release 185 (2014) 51–61. https://doi.org/10.1016/j.jconrel.2014.04.032. (C) Reproduced from J.U. Choi, S.W. Lee, R. Pangeni, Y. Byun, I.S. Yoon, J.W. Park, Preparation and in vivo evaluation of cationic elastic liposomes comprising highly skin-permeable growth factors combined with hyaluronic acid for enhanced diabetic wound-healing therapy, Acta Biomater. 57 (2017) 197–215. https://doi.org/10.1016/j.actbio.2017.04.034. All are used with permission from the Elsevier, copyrights 2020, 2014, and 2017, respectively.*

TABLE 5 Representative examples of lipid-based nanomaterials for the control of bacterial infections in wound healing applications.

System	Synthesis method	NP size (nm)	Antimicrobial activity	Animal model	Reference
Chitosan films loaded with SLN encapsulating all-trans retinoic acid (ATRA)	Hot melting, and drop-wise adding into oil phase Compritol + cholesterol + polyoxyl 20 ethyl ether	83–194 (depending on composition)	Not performed	Mice with induced diabetes and full skin thickness wounds	[133]
SLN loaded with a silver complex of clotrimazole	Emulsification-ultrasonication-based method: Compritol 888 ATO	124	SA MRSA	Not tested	[134]
SLN and NLC encapsulated with rhEGF	Emulsification-ultrasonication-based method. SLN: Precirol ATO 5 NLC: Precirol ATO 5 +Poloxamer + Miglyol, hot melted	332 (SLN) and 348 (NLC)	Not tested	Full-thickness excision wound model in mice	[135].
Nanostructured lipid carriers (NLC), loaded with *Mentha pulegium* essential oil (MPO)	Hot melting, and drop-wise adding into oil phase MPO +Miglyol + Poloxamer –	40–250	SP SE SA PA EC LM BA ST	Mice, wounds infected with *Staphylococcus aureus* and *Pseudomonas aeruginosa*	[136]
pH-responsive NLCs, stearic and oleic acid derived, loaded with vancomycin	Two branched lipids (liquid and solid) were synthesized and mixed together	225	SA MRSA	Mice MRSA skin infection model	[137]

Madecassoside liposomes, modified with PEG-PCL-PEG (PECE) copolymer	Double emulsion method: PECE + egg yolk and cholesterol as oil phase, MA in PBS as the water phase	213.4	Not tested	Rats with second-degree burn wounds	[138]
Cationic elastic liposomes encapsulated with HA and LMWP bonded to growth factors	Double emulsion with microfluidization method: HPC + DOTAP + cholesterol as oil, LMWP-GF + Tween 80as water phase	107	Not tested	Mice with induced diabetes and full skin thickness wounds	[139]
Liposomes with PS or PEG bilayers at the surface, encapsulated with dexamethasone phosphate	Film hydration method: DPPC + cholesterol + PEG-DSPE/DOPS, solvent removed and liposomes hydrated in DexP containing solutions	100–150	Not tested	Not tested	[140]
Liposomes loaded with curcumin	Film hydration method: cholesterol – phosphatidylcholine + curcumin, solvent removed, hydrated with PBS	189–395	Not tested	Rats with full-thickness excision wounds	[141]
Liposomes loaded with bFGF	pH gradient method: Soybean lecithin + cholesterol, solvent removed, hydrated in citric acid followed by bFGF solution	120	Not tested	Rats with second-degree burn wounds	[142]
Liposomes loaded with polyphenols from *Salvia* and *Rosmarinus*	Film hydration method: P90G + cholesterol + essential oils	200	EC KP PM EC PA SA	Not tested	[143]

Continued

TABLE 5 Representative examples of lipid-based nanomaterials for the control of bacterial infections in wound healing applications—cont'd

System	Synthesis method	NP size (nm)	Antimicrobial activity	Animal model	Reference
Liposomes loaded with gallic acids and coated with lactoferrin	Film hydration method: Soybean lecithin + cholesterol, Tween 80 and vit. E, solvent removed, hydrated in water with GA	153nm	EC SA	Not tested	[144]
Cationic liposomes loaded with ticarcillin	Film hydration method followed by filtration and centrifugation: Egg lecithin + cholesterol, solvent removed, hydrated in ticarcillin aqueous solution	96	PA	Mice with second-degree burn wounds, infected with *Pseudomonas aeruginosa*	[145]

BA, *Bacillus anthracis*; EC, *Escherichia coli*; EF, *Enterococcus faecalis*; KP, *Klebsiella pneumoniae*; LM, *Listeria monocytogenes*; MRSA, methicillin-resistant *Staphylococcus aureus*; PA, *Pseudomonas aeruginosa*; PM, *Proteus mirabilis*; SA, *Staphylococcus aureus*; SM, *Streptococcus mutants*; SP, *Staphylococcus pneumoniae*; SP, *Staphylococcus pyogenes*; ST, *Salmonella typhimurium*.

(MPO-NLCs) and evaluated in vivo wound healing and in vitro antibacterial activity. The findings proved the in vivo wound healing activity of MPO-NLCs in BALB/c mice as well as significant antibacterial activity against Gram-negative *Escherichia coli* and *Pseudomonas aeruginosa* and also Gram-positive *Staphylococcus epidermidis, Listeria monocytogenes,* and *Staphylococcus aureus,* as compared to free MPO ($P < .01$) [136]. Gainza et al. used nanostructured lipid carriers (NLC) and SLNs as carriers for the delivery of recombinant human epidermal growth factor (rhEGF) for chronic wound treatment using topical administration. As shown in Fig. 3B, both the formulated rhEGF-loaded NLC and rhEGF-loaded SLNs showed an in vivo effectiveness in healing-impaired db/db mice by improving wound closure and reepithelialization grade [135].

In a different set of examples, Osman et al. formulated vancomycin (VCM)-loaded nanostructured lipid carriers (NLCs) and evaluated their in vitro and in vivo antibacterial efficacy compared to bare VCM. The findings showed significant in vitro antibacterial activity of VCM-loaded NLCs against methicillin-susceptible and methicillin-resistant *Staphylococcus aureus* compared to bare VCM. In addition, the in vivo study using a mouse model with MRSA revealed a 37-fold decrease ($P < .05$) in bacterial populations treated with VCM-loaded NLCs compared to that of bare VCM [137]. Kalhapure et al. formulated a silver complex of clotrimazole encapsulated into SLNs for antibacterial studies. The findings showed that the minimum inhibitory concentrations (MIC) of clotrimazole and clotrimazole-silver were 31.25 g/mL and 9.76 g/mL against *Staphylococcus aureus,* and 31.25 g/mL and 15.62 g/mL against MRSA at the end of 18 h, respectively. Clotrimazole SLNs exhibited MIC values of 104 g/mL and 208 g/mL against both *Staphylococcus aureus* and MRSA at the end of 18 h, respectively. Clotrimazole-silver SLNs had a MIC value of 52 g/mL up to 54 h against both strains. Hence, modification of clotrimazole with silver boosted its antibacterial properties [134].

Some of the advantages of liposomes for drug delivery systems involve the ability to carry both lipophilic and hydrophilic drugs and protect them from degradation. Furthermore, the liposomes are biocompatible, biodegradable, and are characterized by a low immunogenicity when they are comprised of natural lipids. Additionally, liposomes can improve bactericidal efficacy by increasing antibiotic concentrations at the site of infection, leading to drug uptake improvement [148]. Studies that include the design of successful liposome-based drug delivery vehicles for wound healing and infection control have been performed. For example, Liu et al. formulated surface-modified madecassoside liposomes with an average diameter of 213.43 ± 4.68 nm. Madecassoside has considerable therapeutic effects for wound healing. The surface of the formulated liposomes was modified by temperature-responsive copolymers to maintain sufficient adhesion to the wound surface. The findings revealed that surface-modified liposomes caused superior wound healing effects in second-degree burn experiments using a rat model, when compared to unmodified liposomes. The authors confirmed that the formulated liposomes have excellent efficacy for transporting drugs through the epidermal barrier, owing to their tissue compatibility and affinity [138].

Another example appears from Choi et al., where cationic elastic liposomes, comprising of highly skin-permeable growth factors, and combined with hyaluronic acid were formulated. The obtained NPs had an average diameter of 107 ± 0.757 nm and induced a significant wound healing efficacy in a diabetic mouse wound model (Fig. 3A). Moreover, histology of the wound site revealed reepithelialization and production of a collagen matrix in the treated skin [139]. In addition, Gauthier et al. formulated dexamethasone-loaded liposomes with modified surfaces, composed of polyethylene glycol and phosphatidylserine. The findings revealed that phosphatidylserine-modified liposomes had a faster uptake than polyethylene glycol-modified liposomes. According to the 3D skin equivalent model, locally administered liposomes were preferentially phagocytosed by macrophages, while the drug-free liposomes were largely inert in terms of influencing macrophage polarization and function. The authors also confirmed that phosphatidylserine enhanced dexamethasone delivery to macrophages and induced a phenotype suitable to support chronic wound healing [140].

Choudhary et al. formulated curcumin-loaded liposomes in the range of 189–395 nm for wound healing applications in a rat skin model. Fig. 3B shows that wound closure for curcumin-loaded liposomes was faster than free curcumin, free drug liposomes, and normal saline, especially after 10 and 14 days of treatment. The authors also confirmed skin localization of the curcumin-loaded liposomes using rhodamine 123 and reported effective permeation of rhodamine 123 labeled curcumin-loaded liposomes in the skin up to 1125 μm at 5 h [141]. Xiang et al. formulated basic fibroblast growth factor (bFGF)-encapsulated liposomes with an average diameter of 120 nm. The formulated bFGF-liposomes at a concentration of 60 IU/mL exhibited the fastest tissue collagen generation as well as the earliest and highest tumor growth factor (TGF)-β1 amount and dermal cell proliferation (proliferating cell nuclear antigen), indicating promising prospects for their application in wound healing therapy [142].

On the other hand, Risaliti et al. formulated liposomes loaded with *Salvia triloba* and *Rosmarinus officinalis* essential oils. Both liposomes were found to have an average diameter of about 200 nm. The findings revealed that the liposomes loaded with *Salvia triloba* and *R. officinalis* essential oils exhibited significant antibacterial activity against *K. pneumoniae* through a disk diffusion assay, with a zone of inhibition (ZOI) of 22.6 ± 0.6 nm and 18.5 ± 0.6 mm, respectively, while the *Salvia triloba* and *R. officinalis* essential oils showed lower antibacterial activity against *K. pneumoniae* with a

ZOI of 19.0 ± 1.0 mm and 12.7 ± 0.6 mm, respectively [143]. Zhang et al. produced gallic acid liposomes (GA-LIP) and gallic acid liposomes decorated with lactoferrin (LF-GA-LIP). The MIC values of both formulated liposomes were found to be 0.25 μg/mL. Furthermore, the ZOI values for LF-GA-LIP were found to be 12.0 ± 1.2 mm and 10.2 ± 2.3 mm against *Escherichia coli* and *Staphylococcus aureus*, respectively, while the ZOI values for GA-LIP were found to be 12.0 ± 1.2 mm and 10.2 ± 2.3 mm against *Escherichia coli* and *Staphylococcus aureus*, respectively. These results indicate higher antibacterial activity of LF-GA-LIP than GA-LIP against *Escherichia coli* and *Staphylococcus aureus* [144]. Moreover, Gharib et al. formulated ticarcillin-loaded nanoliposomes with positive, negative, and neutral surface charges. The MIC of free, cationic, neutral, and anionic nanoliposomal forms of ticarcillin against *Pseudomonas aeruginosa* was found to be 24 μg/mL, 3 μg/mL, 6 μg/mL, and 48 μg/mL, respectively. Additionally, the in vivo study revealed that treatment with ticarcillin-loaded nanoliposomes with positive, neutral, and negative surface charges led to almost 100%, 60%, and 20% survival rates in BALB/c mice, respectively. The authors suggested that the electrostatic interaction between the cationic nanoliposome-loaded ticarcillin and the outer membrane lipopolysaccharides of *Pseudomonas aeruginosa* could improve drug entry inside the bacteria [145].

4 Conclusion

Hard-to-heal wounds, especially those connected with microbial infections, have becomeserious challenges in current medicine. The evolution and widespread appearance of drug-resistant bacterial strains are significantly contributing to the severity of this issue. Where traditional therapies fail, nanomaterials offer promise to deliver effective solutions to improve healing efficacy and fight such infections. NMs can be used as cargo carriers, surface, or in-mass modifiers for microbial killing/repulsion, or as a means to apply PDT/PTT. Some of the most commonly suggested materials include metallic nanoparticles, bimetallic nanoparticles, carbon nanomaterials, polymeric or lipidic nanoparticles. Some examples of the effective use of these are described in the presented review.

MNPs, such as those made of Au or Ag, have been effectively used as antimicrobial agents. The bacterial activity of metal nanoparticles depends on material shape, size, and surface functionalization [15]. The mechanisms that MNPs use to inhibit bacterial growth are: disrupted energy transduction and production of ROS. ROS can alter membrane permeability, photocatalysis, and can inhibit proteins, which can lead to DNA damage [17]. Although these mechanisms are very effective, MNPs do present cytotoxic effects toward healthy cells that must be overcome in the close future for their successful implementation in the biomedical field. Moreover, bimetallic NPs have also been used as therapies for bacterial infections. The creation of these structures allows for the integration of multiple properties. For example, AgNPs have been combined with other cytocompatible metals to create a more biocompatible nanoparticle. For even better biocompatibility, green synthesis methods have been employed. Still, more research needs to be done concerning monometallic and bimetallic NPs in order to establish them as an easy choice for the formulation of nanostructured drugs.

This review also discussed three types of nonmetal nanomaterials and their biomedical applications. While carbon-based nanomaterials can enter cells via passive diffusion and are characterized by several bacteria killing mechanisms which have been proven to be highly efficient, their use may be controversial due to large discrepancies regarding their cytotoxicity. We believe that these originate from an insufficient characterization of the CNMs employed in the studies, which can vary in size, shape, and chemical composition. Meanwhile, polymeric NPs are commonly employed due to their versatility, which allows for the generation of a wide range of nanostructures that can easily vary in their shape, size, and drug loading capacity. These factors are dependent on their type, administration route, manufacturing method, and processing parameters, all of which greatly impact their antimicrobial features [93]. On the other hand, lipid-based nanomaterials seem ideal for wound healing because they prevent skin water loss by evaporation, are biodegradable, and are able to protect the drugs from environmental damage. Lipid nanoparticles have been tested against various bacterial strains and have shown high antimicrobial efficacy in combination with various bioactive molecules (drug, polyphenols, etc.). Thus, they are very promising treatments for open wounds and abrasions.

To sum up, recent years have seen some promising results when it comes to the application of NMs in wound healing and infection control. Careful selection of the type of NM, based on their desired role and mechanism of action, may lead to a significant progress in this field, offering unparalleled solutions.

Acknowledgments

A.B. would like to acknowledge financial support from the National Science Center, Poland, under grant no. UMO-2017/24/C/ST8/00400. K.R-K. and E.P. would like to acknowledge financial support from the National Science Center, Poland, under grants no. 2018/29/N/ST5/01543 and 2019/35/B/ST5/01103, the Program "Excellence Initiative-Research University" for the AGH University of Science and Technology, Kraków, Poland and the subsidy of the Ministry of Education and Science for the AGH University of Science and Technology in Kraków (Project No 16.16.160.557). A.P. and P.K. would like to acknowledge the National Science Centre (NCN) in Poland for support (grant no. UMO-2018/31/B/ST8/00945). S.W. would like to acknowledge financial support from the National Centre for Research and Development under grant no. LIDER/7/0020/L-11/19/NCBR/2020.

References

[1] M. Rodrigues, N. Kosaric, C.A. Bonham, G.C. Gurtner, Wound healing: a cellular perspective, Physiol. Rev. 99 (2019) 665–706, https://doi.org/10.1152/physrev.00067.2017.

[2] P. Martin, Wound healing—aiming for perfect skin regeneration, Science (80-) 276 (1997) 75–81, https://doi.org/10.1126/science.276.5309.75.

[3] Y. Yamaguchi, K. Yoshikawa, Cutaneous wound healing: an update, J. Dermatol. 28 (2001) 521–534, https://doi.org/10.1111/j.1346-8138.2001.tb00025.x.

[4] N.N. Nissen, P.J. Polverini, A.E. Koch, M.V. Volin, R.L. Gamelli, L.A. DiPietro, Vascular endothelial growth factor mediates angiogenic activity during the proliferative phase of wound healing, Am. J. Pathol. 152 (1998) 1445–1452. (Accessed 22 June 2020).

[5] M.B. Witte, A. Barbul, General principles of wound healing, Surg. Clin. North Am. 77 (1997) 509–528, https://doi.org/10.1016/S0039-6109(05)70566-1.

[6] K. Penhallow, A review of studies that examine the impact of infection on the normal wound-healing process, J. Wound Care 14 (2005) 123–126, https://doi.org/10.12968/jowc.2005.14.3.26747.

[7] I.G. Rojas, D.A. Padgett, J.F. Sheridan, P.T. Marucha, Stress-induced susceptibility to bacterial infection during cutaneous wound healing, Brain Behav. Immun. 16 (2002) 74–84, https://doi.org/10.1006/brbi.2000.0619.

[8] P.G. Bowler, B.I. Duerden, D.G. Armstrong, Wound microbiology and associated approaches to wound management, Clin. Microbiol. Rev. 14 (2001) 244–269, https://doi.org/10.1128/CMR.14.2.244-269.2001.

[9] M.C. Robson, Wound infection: a failure of wound healing caused by an imbalance of bacteria, Surg. Clin. North Am. 77 (1997) 637–650, https://doi.org/10.1016/S0039-6109(05)70572-7.

[10] S. Sandhiya, S.A. Dkhar, A. Surendiran, Emerging trends of nanomedicine—an overview, Fundam. Clin. Pharmacol. 23 (2009) 263–269, https://doi.org/10.1111/j.1472-8206.2009.00692.x.

[11] S. Hamdan, I. Pastar, S. Drakulich, E. Dikici, M. Tomic-Canic, S. Deo, S. Daunert, Nanotechnology-driven therapeutic interventions in wound healing: potential uses and applications, ACS Cent. Sci. 3 (2017) 163–175, https://doi.org/10.1021/acscentsci.6b00371.

[12] N.K. Rajendran, S.S.D. Kumar, N.N. Houreld, H. Abrahamse, A review on nanoparticle based treatment for wound healing, J. Drug Deliv. Sci. Technol. 44 (2018) 421–430, https://doi.org/10.1016/j.jddst.2018.01.009.

[13] D. Medina-Cruz, B. Saleh, A. Vernet-Crua, A. Ajo, A.K. Roy, T.J. Webster, Drug-delivery nanocarriers for skin wound-healing applications, in: Wound Healing, Tissue Repair, and Regeneration in Diabetes, Elsevier, 2020, pp. 439–488, https://doi.org/10.1016/b978-0-12-816413-6.00022-8.

[14] E.P. Figueiredo, J.M. Ribeiro, E.K. Nishio, S. Scandorieiro, A.F. Costa, V.F. Cardozo, A.-G. Oliveira, N. Durán, L.A. Panagio, R.K.T. Kobayashi, G. Nakazato, New approach for simvastatin as an antibacterial: synergistic effect with bio-synthesized silver nanoparticles against multidrug-resistant bacteria, Int. J. Nanomedicine 14 (2019) 7975–7985, https://doi.org/10.2147/IJN.S211756.

[15] A. Gupta, S. Mumtaz, C.H. Li, I. Hussain, V.M. Rotello, Combatting antibiotic-resistant bacteria using nanomaterials, Chem. Soc. Rev. 48 (2019) 415–427, https://doi.org/10.1039/c7cs00748e.

[16] A. Singh, P.K. Gautam, A. Verma, V. Singh, P.M. Shivapriya, S. Shivalkar, A.K. Sahoo, S.K. Samanta, Green synthesis of metallic nanoparticles as effective alternatives to treat antibiotics resistant bacterial infections: a review, Biotechnol. Rep. 25 (2020), https://doi.org/10.1016/j.btre.2020.e00427, e00427.

[17] M.M. Mihai, M.B. Dima, B. Dima, A.M. Holban, Nanomaterials for wound healing and infection control, Materials (Basel) 12 (2019), https://doi.org/10.3390/ma12132176.

[18] M.M. Mohamed, S.A. Fouad, H.A. Elshoky, G.M. Mohammed, T.A. Salaheldin, Antibacterial effect of gold nanoparticles against *Corynebacterium pseudotuberculosis*, Int. J. Vet. Sci. Med. 5 (2017) 23–29, https://doi.org/10.1016/j.ijvsm.2017.02.003.

[19] Y. Liu, F. Li, Z. Guo, Y. Xiao, Y. Zhang, X. Sun, T. Zhe, Y. Cao, L. Wang, Q. Lu, J. Wang, Silver nanoparticle-embedded hydrogel as a photothermal platform for combating bacterial infections, Chem. Eng. J. 382 (2020), https://doi.org/10.1016/j.cej.2019.122990, 122990.

[20] G. Fang, W. Li, X. Shen, J.M. Perez-Aguilar, Y. Chong, X. Gao, Z. Chai, C. Chen, C. Ge, R. Zhou, Differential Pd-nanocrystal facets demonstrate distinct antibacterial activity against gram-positive and gram-negative bacteria, Nat. Commun. 9 (2018) 1–9, https://doi.org/10.1038/s41467-017-02502-3.

[21] C. Vijilvani, M.R. Bindhu, F.C. Frincy, M.S. AlSalhi, S. Sabitha, K. Saravanakumar, S. Devanesan, M. Umadevi, M.J. Aljaafreh, M. Atif, Antimicrobial and catalytic activities of biosynthesized gold, silver and palladium nanoparticles from Solanum nigurum leaves, J. Photochem. Photobiol. B Biol. 202 (2020), https://doi.org/10.1016/j.jphotobiol.2019.111713, 111713.

[22] K. Niska, E. Zielinska, M.W. Radomski, I. Inkielewicz-Stepniak, Metal nanoparticles in dermatology and cosmetology: interactions with human skin cells, Chem. Biol. Interact. 295 (2018) 38–51, https://doi.org/10.1016/j.cbi.2017.06.018.

[23] H. Itohiya, Y. Matsushima, S. Shirakawa, S. Kajiyama, A. Yashima, T. Nagano, K. Gomi, Organic resolution function and effects of platinum nanoparticles on bacteria and organic matter, PLoS One 14 (2019), https://doi.org/10.1371/journal.pone.0222634, e0222634.

[24] Z. Qiao, Y. Yao, S. Song, M. Yin, J. Luo, Silver nanoparticles with pH induced surface charge switchable properties for antibacterial and antibiofilm applications, J. Mater. Chem. B 7 (2019) 830–840, https://doi.org/10.1039/c8tb02917b.

[25] V. Vijayakumar, S.K. Samal, S. Mohanty, S.K. Nayak, Recent advancements in biopolymer and metal nanoparticle-based materials in diabetic wound healing management, Int. J. Biol. Macromol. 122 (2019) 137–148, https://doi.org/10.1016/j.ijbiomac.2018.10.120.

[26] S. Shaikh, N. Nazam, S.M.D. Rizvi, K. Ahmad, M.H. Baig, E.J. Lee, I. Choi, Mechanistic insights into the antimicrobial actions of metallic nanoparticles and their implications for multidrug resistance, Int. J. Mol. Sci. 20 (2019), https://doi.org/10.3390/ijms20102468.

[27] S.K. Nethi, S. Das, C.R. Patra, S. Mukherjee, Recent advances in inorganic nanomaterials for wound-healing applications, Biomater. Sci. 7 (2019) 2652–2674, https://doi.org/10.1039/c9bm00423h.

[28] M.G. Arafa, R.F. El-Kased, M.M. Elmazar, Thermoresponsive gels containing gold nanoparticles as smart antibacterial and wound healing agents, Sci. Rep. 8 (2018) 1–16, https://doi.org/10.1038/s41598-018-31895-4.

[29] A. Chen, C. Ostrom, Palladium-based nanomaterials: synthesis and electrochemical applications, Chem. Rev. 115 (2015) 11999–12044, https://doi.org/10.1021/acs.chemrev.5b00324.

[30] T.T.V. Phan, T.C. Huynh, P. Manivasagan, S. Mondal, J. Oh, An up-to-date review on biomedical applications of palladium nanoparticles, Nanomaterials 10 (2020), https://doi.org/10.3390/nano10010066.

[31] S. Ullah, A. Ahmad, A. Khan, J. Zhang, M. Raza, A.u. Rahman, M. Tariq, U. Ali khan, S. Zada, Q. Yuan, Palladium nanoparticles synthesis, characterization using glucosamine as the reductant and stabilizing agent to explore their antibacterial & catalytic applications, Microb. Pathog. 125 (2018) 150–157, https://doi.org/10.1016/j.micpath.2018.09.020.

[32] C.P. Adams, K.A. Walker, S.O. Obare, K.M. Docherty, Size-dependent antimicrobial effects of novel palladium nanoparticles, PLoS One 9 (2014), https://doi.org/10.1371/journal.pone.0085981, e85981.

[33] B. Murugesan, N. Pandiyan, M. Arumugam, J. Sonamuthu, S. Samayanan, C. Yurong, Y. Juming, S. Mahalingam, Fabrication of palladium nanoparticles anchored polypyrrole functionalized reduced graphene oxide nanocomposite for antibiofilm associated orthopedic tissue engineering, Appl. Surf. Sci. 510 (2020) 145403, https://doi.org/10.1016/j.apsusc.2020.145403.

[34] M. Jeyaraj, S. Gurunathan, M. Qasim, M.H. Kang, J.H. Kim, A comprehensive review on the synthesis, characterization, and biomedical application of platinum nanoparticles, Nanomaterials 9 (2019), https://doi.org/10.3390/nano9121719.

[35] T. Deng, H. Zhao, M. Shi, Y. Qiu, S. Jiang, X. Yang, Y. Zhao, Y. Zhang, Photoactivated trifunctional platinum nanobiotics for precise synergism of multiple antibacterial modes, Small 15 (2019) 1902647, https://doi.org/10.1002/smll.201902647.

[36] D. Kharaghani, D. Dutta, P. Gitigard, Y. Tamada, A. Katagiri, D.N. Phan, M.D.P. Willcox, I.S. Kim, Development of antibacterial contact lenses containing metallic nanoparticles, Polym. Test. 79 (2019), https://doi.org/10.1016/j.polymertesting.2019.106034, 106034.

[37] S. Sasidharan, L. Pottail, Antimicrobial activity of metal and non-metallic nanoparticles from *Cyperus rotundus* root extract on infectious disease causing pathogens, J. Plant Biochem. Biotechnol. 29 (2020) 134–143, https://doi.org/10.1007/s13562-019-00523-1.

[38] A. Attar, M. Altikatoglu Yapaoz, Biosynthesis of palladium nanoparticles using *Diospyros kaki* leaf extract and determination of antibacterial efficacy, Prep. Biochem. Biotechnol. 48 (2018) 629–634, https://doi.org/10.1080/10826068.2018.1479862.

[39] M. Safdar, M. Ozaslan, R.A. Khailany, S. Latif, Y. Junejo, M. Saeed, M.S. Al-Attar, B.O. Kanabe, Synthesis, characterization and applications of a novel platinum-based nanoparticles: catalytic, antibacterial and cytotoxic studies, J. Inorg. Organomet. Polym. Mater. 30 (2019) 2430–2439, https://doi.org/10.1007/s10904-019-01387-7.

[40] S. Bharathan, N.S. Sundaramoorthy, H. Chandrasekaran, G. Rangappa, G.P. ArunKumar, S.B. Subramaniyan, A. Veerappan, S. Nagarajan, Sub lethal levels of platinum nanoparticle cures plasmid and in combination with carbapenem, curtails carbapenem resistant *Escherichia coli*, Sci. Rep. 9 (2019) 1–13, https://doi.org/10.1038/s41598-019-41489-3.

[41] S. Ali, A.S. Sharma, W. Ahmad, M. Zareef, M.M. Hassan, A. Viswadevarayalu, T. Jiao, H. Li, Q. Chen, Noble metals based bimetallic and trimetallic nanoparticles: controlled synthesis, antimicrobial and anticancer applications, Crit. Rev. Anal. Chem. 0 (2020) 1–28, https://doi.org/10.1080/10408347.2020.1743964.

[42] B.S. Sivamaruthi, V.S. Ramkumar, G. Archunan, C. Chaiyasut, N. Suganthy, Biogenic synthesis of silver palladium bimetallic nanoparticles from fruit extract of *Terminalia chebula* – in vitro evaluation of anticancer and antimicrobial activity, J. Drug Delivery Sci. Technol. 51 (2019) 139–151, https://doi.org/10.1016/j.jddst.2019.02.024.

[43] A. Al-Asfar, Z. Zaheer, E.S. Aazam, Eco-friendly green synthesis of Ag@Fe bimetallic nanoparticles: antioxidant, antimicrobial and photocatalytic degradation of bromothymol blue, J. Photochem. Photobiol. B Biol. 185 (2018) 143–152, https://doi.org/10.1016/j.jphotobiol.2018.05.028.

[44] D. Lomelí-Marroquín, D.M. Cruz, A. Nieto-Argüello, A.V. Crua, J. Chen, A. Torres-Castro, T.J. Webster, J.L. Cholula-Díaz, Starch-mediated synthesis of mono- and bimetallic silver/gold nanoparticles as antimicrobial and anticancer agents, Int. J. Nanomedicine 14 (2019) 2171–2190, https://doi.org/10.2147/IJN.S192757.

[45] Q. Li, F. Lu, H. Ye, K. Yu, B. Lu, R. Bao, Y. Xiao, F. Dai, G. Lan, Silver inlaid with gold nanoparticles: enhanced antibacterial ability coupled with the ability to visualize antibacterial efficacy, ACS Sustain. Chem. Eng. 6 (2018) 9813–9821, https://doi.org/10.1021/acssuschemeng.8b00931.

[46] Y. Chang, Y. Cheng, Y. Feng, K. Li, H. Jian, H. Zhang, Upshift of the d band center toward the Fermi level for promoting silver ion release, Bacteria inactivation, and

wound healing of alloy silver nanoparticles, ACS Appl. Mater. Interfaces 11 (2019) 12224–12231, https://doi.org/10.1021/acsami.8b21768.

[47] E. Bolea, J. Jiménez-Lamana, F. Laborda, I. Abad-Álvaro, C. Bladé, L. Arola, J.R. Castillo, Detection and characterization of silver nanoparticles and dissolved species of silver in culture medium and cells by AsFlFFF-UV-vis-ICPMS: application to nanotoxicity tests, Analyst 139 (2014) 914–922, https://doi.org/10.1039/c3an01443f.

[48] X. Dai, Q. Guo, Y. Zhao, P. Zhang, T. Zhang, X. Zhang, C. Li, Functional silver nanoparticle as a benign antimicrobial agent that eradicates antibiotic-resistant Bacteria and promotes wound healing, ACS Appl. Mater. Interfaces 8 (2016) 25798–25807, https://doi.org/10.1021/acsami.6b09267.

[49] D. Sharma, S. Kanchi, K. Bisetty, Biogenic synthesis of nanoparticles: a review, Arab. J. Chem. 12 (2019) 3576–3600, https://doi.org/10.1016/j.arabjc.2015.11.002.

[50] P. Manivasagan, J. Venkatesan, K. Senthilkumar, K. Sivakumar, S.K. Kim, Biosynthesis, antimicrobial and cytotoxic effect of silver nanoparticles using a novel Nocardiopsis sp. MBRC-1, Biomed. Res. Int. 2013 (2013), https://doi.org/10.1155/2013/287638.

[51] D. Paredes, C. Ortiz, R. Torres, Synthesis, characterization, and evaluation of antibacterial effect of Ag nanoparticles against *escherichia coli* O157:H7 and methicillin-resistant *staphylococcus aureus* (MRSA), Int. J. Nanomedicine 9 (2014) 1717–1729, https://doi.org/10.2147/IJN.S57156.

[52] R. Ferrando, J. Jellinek, R.L. Johnston, Nanoalloys: from theory to applications of alloy clusters and nanoparticles, Chem. Rev. 108 (2008) 845–910, https://doi.org/10.1021/cr040090g.

[53] A. Gour, N.K. Jain, Advances in green synthesis of nanoparticles, Artif. Cells Nanomed. Biotechnol. 47 (2019) 844–851, https://doi.org/10.1080/21691401.2019.1577878.

[54] K.L. Smitten, S.D. Fairbanks, C.C. Robertson, J. Bernardino De La Serna, S.J. Foster, J.A. Thomas, Ruthenium based antimicrobial theranostics-using nanoscopy to identify therapeutic targets and resistance mechanisms in: *Staphylococcus aureus*, Chem. Sci. 11 (2020) 70–79, https://doi.org/10.1039/c9sc04710g.

[55] Q. Xin, H. Shah, A. Nawaz, W. Xie, M.Z. Akram, A. Batool, L. Tian, S.U. Jan, R. Boddula, B. Guo, Q. Liu, J.R. Gong, Antibacterial carbon-based nanomaterials, Adv. Mater. 31 (2019) 1804838, https://doi.org/10.1002/adma.201804838.

[56] F.A.G. da Silva, C.M.S. de Araújo, J.J. Alcaraz-Espinoza, H.P. de Oliveira, Toward flexible and antibacterial piezoresistive porous devices for wound dressing and motion detectors, J. Polym. Sci. B Polym. Phys. 56 (2018) 1063–1072, https://doi.org/10.1002/polb.24626.

[57] T.S. John, P.K. Yadav, D. Kumar, S.K. Singh, S.H. Hasan, Highly fluorescent carbon dots from wheat bran as a novel drug delivery system for bacterial inhibition, Luminescence (2020), https://doi.org/10.1002/bio.3801.

[58] C. Li, R. Ye, J. Bouckaert, A. Zurutuza, D. Drider, T. Dumych, S. Paryzhak, V. Vovk, R.O. Bilyy, S. Melinte, M. Li, R. Boukherroub, S. Szunerits, Flexible nanoholey patches for antibiotic-free treatments of skin infections, ACS Appl. Mater. Interfaces 9 (2017) 36665–36674, https://doi.org/10.1021/acsami.7b12949.

[59] P. Paramanantham, V.T. Anju, M. Dyavaiah, B. Siddhardha, Applications of carbon-based nanomaterials for antimicrobial photodynamic therapy, Springer, Cham, 2019, pp. 237–259, https://doi.org/10.1007/978-3-030-16534-5_12.

[60] G. Hong, S. Diao, A.L. Antaris, H. Dai, Carbon nanomaterials for biological imaging and nanomedicinal therapy, Chem. Rev. 115 (2015) 10816–10906, https://doi.org/10.1021/acs.chemrev.5b00008.

[61] J. Duch, M. Golda-Cepa, A. Kotarba, Evaluating the effect of oxygen groups attached to the surface of graphenic sheets on bacteria adhesion: the role of the electronic factor, Appl. Surf. Sci. 463 (2019) 1134–1140, https://doi.org/10.1016/j.apsusc.2018.08.237.

[62] W. Pajerski, J. Duch, D. Ochonska, M. Golda-Cepa, M. Brzychczy-Wloch, A. Kotarba, Bacterial attachment to oxygen-functionalized graphenic surfaces, Mater. Sci. Eng. C 113 (2020), https://doi.org/10.1016/j.msec.2020.110972, 110972.

[63] S. Yick, A. Mai-Prochnow, I. Levchenko, J. Fang, M.K. Bull, M. Bradbury, A.B. Murphya, K. Ostrikov, The effects of plasma treatment on bacterial biofilm formation on vertically-aligned carbon nanotube arrays, RSC Adv. 5 (2015) 5142–5148, https://doi.org/10.1039/c4ra08187k.

[64] F. Zou, H. Zhou, D.Y. Jeong, J. Kwon, S.U. Eom, T.J. Park, S.W. Hong, J. Lee, Wrinkled surface-mediated antibacterial activity of graphene oxide nanosheets, ACS Appl. Mater. Interfaces 9 (2017) 1343–1351, https://doi.org/10.1021/acsami.6b15085.

[65] N. Lin, P. Berton, C. Moraes, R.D. Rogers, N. Tufenkji, Nanodarts, nanoblades, and nanospikes: mechano-bactericidal nanostructures and where to find them, Adv. Colloid Interf. Sci. 252 (2018) 55–68, https://doi.org/10.1016/j.cis.2017.12.007.

[66] H. Li, J. Huang, Y. Song, M. Zhang, H. Wang, F. Lu, H. Huang, Y. Liu, X. Dai, Z. Gu, Z. Yang, R. Zhou, Z. Kang, Degradable carbon dots with broad-spectrum antibacterial activity, ACS Appl. Mater. Interfaces 10 (2018) 26936–26946, https://doi.org/10.1021/acsami.8b08832.

[67] M. Grinholc, J. Nakonieczna, G. Fila, A. Taraszkiewicz, A. Kawiak, G. Szewczyk, T. Sarna, L. Lilge, K.P. Bielawski, Antimicrobial photodynamic therapy with fulleropyrrolidine: photoinactivation mechanism of *Staphylococcus aureus*, in vitro and in vivo studies, Appl. Microbiol. Biotechnol. 99 (2015) 4031–4043, https://doi.org/10.1007/s00253-015-6539-8.

[68] Y. Zhu, J. Xu, Y. Wang, C. Chen, H. Gu, Y. Chai, Y. Wang, Silver nanoparticles-decorated and mesoporous silica coated single-walled carbon nanotubes with an enhanced antibacterial activity for killing drug-resistant bacteria, Nano Res. 13 (2020) 389–400, https://doi.org/10.1007/s12274-020-2621-3.

[69] X. Zhao, B. Guo, H. Wu, Y. Liang, P.X. Ma, Injectable antibacterial conductive nanocomposite cryogels with rapid shape recovery for noncompressible hemorrhage and wound healing, Nat. Commun. 9 (2018) 2784, https://doi.org/10.1038/s41467-018-04998-9.

[70] Y. Zhou, R. Chen, T. He, K. Xu, D. Du, N. Zhao, X. Cheng, J. Yang, H. Shi, Y. Lin, Biomedical potential of ultrafine ag/AgCl nanoparticles coated on graphene with special reference to antimicrobial performances and burn wound healing, ACS Appl. Mater. Interfaces 8 (2016) 15067–15075, https://doi.org/10.1021/acsami.6b03021.

[71] W.Y. Pan, C.C. Huang, T.T. Lin, H.Y. Hu, W.C. Lin, M.J. Li, H.W. Sung, Synergistic antibacterial effects of localized heat and oxidative stress caused by hydroxyl radicals mediated by graphene/iron oxide-based nanocomposites, nanomedicine nanotechnology, Biol. Med. 12 (2016) 431–438, https://doi.org/10.1016/j.nano.2015.11.014.

[72] P. Tang, L. Han, P. Li, Z. Jia, K. Wang, H. Zhang, H. Tan, T. Guo, X. Lu, Mussel-inspired electroactive and antioxidative scaffolds with incorporation of polydopamine-reduced graphene oxide for enhancing skin wound healing, ACS Appl. Mater. Interfaces 11 (2019) 7703–7714, https://doi.org/10.1021/acsami.8b18931.

[73] I. Altinbasak, R. Jijie, A. Barras, B. Golba, R. Sanyal, J. Bouckaert, D. Drider, R. Bilyy, T. Dumych, S. Paryzhak, V. Vovk, R. Boukherroub, A. Sanyal, S. Szunerits, Reduced graphene-oxide-embedded polymeric nanofiber mats: an "on-demand" photothermally triggered antibiotic release platform, ACS Appl. Mater. Interfaces 10 (2018) 41098–41106, https://doi.org/10.1021/acsami.8b14784.

[74] M. Di Giulio, R. Zappacosta, S. Di Lodovico, E. Di Campli, G. Siani, A. Fontana, L. Cellini, Antimicrobial and antibiofilm efficacy of graphene oxide against chronic wound microorganisms, Antimicrob. Agents Chemother. 62 (2018), https://doi.org/10.1128/AAC.00547-18.

[75] S. Chen, H. Wang, Z. Jian, G. Fei, W. Qian, G. Luo, Z. Wang, H. Xia, Novel poly(vinyl alcohol)/chitosan/modified graphene oxide biocomposite for wound dressing application, Macromol. Biosci. 20 (2020) 1900385, https://doi.org/10.1002/mabi.201900385.
[76] N. Mahmoudi, N. Eslahi, A. Mehdipour, M. Mohammadi, M. Akbari, A. Samadikuchaksaraei, A. Simchi, Temporary skin grafts based on hybrid graphene oxide-natural biopolymer nanofibers as effective wound healing substitutes: pre-clinical and pathological studies in animal models, J. Mater. Sci. Mater. Med. 28 (2017) 1–13, https://doi.org/10.1007/s10856-017-5874-y.
[77] T. Mitra, P.J. Manna, S.T.K. Raja, A. Gnanamani, P.P. Kundu, Curcumin loaded nano graphene oxide reinforced fish scale collagen-a 3D scaffold biomaterial for wound healing applications, RSC Adv. 5 (2015) 98653–98665, https://doi.org/10.1039/c5ra15726a.
[78] C. Yang, Z. Yan, Y. Lian, J. Wang, K. Zhang, Graphene oxide coated shell-core structured chitosan/PLLA nanofibrous scaffolds for wound dressing, J. Biomater. Sci. Polym. Ed. 31 (2020) 622–641, https://doi.org/10.1080/09205063.2019.1706149.
[79] W. Han, Z. Wu, Y. Li, Y. Wang, Graphene family nanomaterials (GFNs)—promising materials for antimicrobial coating and film: a review, Chem. Eng. J. 358 (2019) 1022–1037, https://doi.org/10.1016/j.cej.2018.10.106.
[80] P.M. Costa, M. Bourgognon, J.T.W. Wang, K.T. Al-Jamal, Functionalized carbon nanotubes: from intracellular uptake and cell-related toxicity to systemic brain delivery, J. Control. Release 241 (2016) 200–219, https://doi.org/10.1016/j.jconrel.2016.09.033.
[81] S.-W. Kim, Y.K. Lee, S.-H. Kim, J.-Y. Park, D.U. Lee, J. Choi, J.H. Hong, S. Kim, D. Khang, Covalent, non-covalent, encapsulated nanodrug regulate the fate of intra- and extracellular trafficking: impact on Cancer and Normal cells, Sci. Rep. 7 (2017) 6454, https://doi.org/10.1038/s41598-017-06796-7.
[82] M. Mortimer, N. Devarajan, D. Li, P.A. Holden, Multiwall carbon nanotubes induce more pronounced transcriptomic responses in *Pseudomonas aeruginosa* PG201 than graphene, exfoliated boron nitride, or carbon black, ACS Nano 12 (2018) 2728–2740, https://doi.org/10.1021/acsnano.7b08977.
[83] K.H. Tan, S. Sattari, S. Beyranvand, A. Faghani, K. Ludwig, K. Schwibbert, C. Böttcher, R. Haag, M. Adeli, Thermoresponsive amphiphilic functionalization of thermally reduced graphene oxide to study graphene/bacteria hydrophobic interactions, Langmuir 35 (2019) 4736–4746, https://doi.org/10.1021/acs.langmuir.8b03660.
[84] H.E. Karahan, C. Wiraja, C. Xu, J. Wei, Y. Wang, L. Wang, F. Liu, Y. Chen, Graphene materials in antimicrobial nanomedicine: current status and future perspectives, Adv. Healthc. Mater. 7 (2018) 1701406, https://doi.org/10.1002/adhm.201701406.
[85] D.P. Linklater, M. De Volder, V.A. Baulin, M. Werner, S. Jessl, M. Golozar, L. Maggini, S. Rubanov, E. Hanssen, S. Juodkazis, E.P. Ivanova, High aspect ratio nanostructures kill bacteria via storage and release of mechanical energy, ACS Nano 12 (2018) 6657–6667, https://doi.org/10.1021/acsnano.8b01665.
[86] A. Benko, D. Medina-Cruz, J. Duch, T. Popiela, S. Wilk, M. Bińczak, M. Nocuń, E. Menaszek, L.D. Geoffrion, G. Guisbiers, A. Kotarba, T.J. Webster, Conductive all-carbon nanotube layers: results on attractive physicochemical, anti-bacterial, anticancer and biocompatibility properties, Mater. Sci. Eng. C Mater. Biol. Appl. 120 (2021) 111703, https://doi.org/10.1016/j.msec.2020.111703.
[87] A. Kassem, G.M. Ayoub, L. Malaeb, Antibacterial activity of chitosan nano-composites and carbon nanotubes: a review, Sci. Total Environ. 668 (2019) 566–576, https://doi.org/10.1016/j.scitotenv.2019.02.446.
[88] W.B. Liechty, D.R. Kryscio, B.V. Slaughter, N.A. Peppas, Polymers for drug delivery systems, Annu. Rev. Chem. Biomol. Eng. 1 (2010) 149–173, https://doi.org/10.1146/annurev-chembioeng-073009-100847.
[89] D.A. Resendiz-Lara, G.R. Whittell, E.M. Leitao, I. Manners, Catalytic synthesis, characterization, and properties of polyaminoborane homopolymers and random

copolymers, Macromolecules 52 (2019) 7052–7064, https://doi.org/10.1021/acs.macromol.9b01139.

[90] A.B. Cook, S. Perrier, Branched and dendritic polymer architectures: functional nanomaterials for therapeutic delivery, Adv. Funct. Mater. 30 (2020) 1901001, https://doi.org/10.1002/adfm.201901001.

[91] K. Aoki, N. Saito, Biodegradable polymers as drug delivery systems for bone regeneration, Pharmaceutics 12 (2020), https://doi.org/10.3390/pharmaceutics12020095.

[92] M. Parhizkar, S. Mahalingam, S. Homer-Vanniasinkam, M. Edirisinghe, Latest developments in innovative manufacturing to combine nanotechnology with healthcare, Nanomedicine 13 (2018) 5–8, https://doi.org/10.2217/nnm-2017-0283.

[93] B.L. Banik, P. Fattahi, J.L. Brown, Polymeric nanoparticles: the future of nanomedicine, WIREs Nanomed. Nanobiotechnol. 8 (2016) 271–299, https://doi.org/10.1002/wnan.1364.

[94] C.A. Marangon, V.C.A. Martins, M.H. Ling, C.C. Melo, A.M.G. Plepis, R.L. Meyer, M. Nitschke, Combination of rhamnolipid and chitosan in nanoparticles boosts their antimicrobial efficacy, ACS Appl. Mater. Interfaces 12 (2020) 5488–5499, https://doi.org/10.1021/acsami.9b19253.

[95] M.G. Arafa, H.A. Mousa, N.N. Afifi, Preparation of PLGA-chitosan based nanocarriers for enhancing antibacterial effect of ciprofloxacin in root canal infection, Drug Deliv. 27 (2020) 26–39, https://doi.org/10.1080/10717544.2019.1701140.

[96] P. Lueth, S.L. Haughney, A.M. Binnebose, A.S. Mullis, N. Peroutka-Bigus, B. Narasimhan, B.H. Bellaire, Nanotherapeutic provides dose sparing and improved antimicrobial activity against *Brucella melitensis* infections, J. Control. Release 294 (2019) 288–297, https://doi.org/10.1016/j.jconrel.2018.12.024.

[97] K. Perveen, F. Masood, A. Hameed, Preparation, characterization and evaluation of antibacterial properties of epirubicin loaded PHB and PHBV nanoparticles, Int. J. Biol. Macromol. 144 (2020) 259–266, https://doi.org/10.1016/j.ijbiomac.2019.12.049.

[98] M. Chen, S. Xie, J. Wei, X. Song, Z. Ding, X. Li, Antibacterial micelles with vancomycin-mediated targeting and pH/lipase-triggered release of antibiotics, ACS Appl. Mater. Interfaces 10 (2018) 36814–36823, https://doi.org/10.1021/acsami.8b16092.

[99] P. Walvekar, R. Gannimani, M. Salih, S. Makhathini, C. Mocktar, T. Govender, Self-assembled oleylamine grafted hyaluronic acid polymersomes for delivery of vancomycin against methicillin resistant *Staphylococcus aureus* (MRSA), Colloids Surf. B Biointerfaces 182 (2019) 110388, https://doi.org/10.1016/j.colsurfb.2019.110388.

[100] N. Boehnke, S. Correa, L. Hao, W. Wang, J.P. Straehla, S.N. Bhatia, P.T. Hammond, Theranostic layer-by-layer nanoparticles for simultaneous tumor detection and gene silencing, Angew. Chem. Ausg. 132 (2020) 2798–2805, https://doi.org/10.1002/ange.201911762.

[101] D. Lombardo, P. Calandra, L. Pasqua, S. Magazù, Self-assembly of organic nanomaterials and biomaterials: the bottom-up approach for functional nanostructures formation and advanced applications, Materials (Basel). 13 (2020) 1048, https://doi.org/10.3390/ma13051048.

[102] M. Osorio, E. Martinez, T. Naranjo, C. Castro, Recent advances in polymer nanomaterials for drug delivery of adjuvants in colorectal cancer treatment: a scientific-technological analysis and review, Molecules 25 (2020), https://doi.org/10.3390/molecules25102270.

[103] M. Morey, A. Pandit, Responsive triggering systems for delivery in chronic wound healing, Adv. Drug Deliv. Rev. 129 (2018) 169–193, https://doi.org/10.1016/j.addr.2018.02.008.

[104] A. Ali, S. Ahmed, A review on chitosan and its nanocomposites in drug delivery, Int. J. Biol. Macromol. 109 (2018) 273–286, https://doi.org/10.1016/j.ijbiomac.2017.12.078.

[105] Y. Liu, L. Shi, L. Su, H.C. Van der Mei, P.C. Jutte, Y. Ren, H.J. Busscher, Nanotechnology-based antimicrobials and delivery systems for biofilm-infection control, Chem. Soc. Rev. 48 (2019) 428–446, https://doi.org/10.1039/c7cs00807d.

[106] F.Y. Alqahtani, F.S. Aleanizy, E. El Tahir, B.T. Alquadeib, I.A. Alsarra, J.S. Alanazi, H.G. Abdelhady, Preparation, characterization, and antibacterial activity of diclofenac-loaded chitosan nanoparticles, Saudi Pharm. J. 27 (2019) 82–87, https://doi.org/10.1016/j.jsps.2018.08.001.

[107] J. Cao, Y. Zhao, Y. Liu, S. Tian, C. Zheng, C. Liu, Y. Zhai, Y. An, H.J. Busscher, L. Shi, Y. Liu, Phosphorylcholine-based polymer encapsulated chitosan nanoparticles enhance the penetration of antimicrobials in a staphylococcal biofilm, ACS Macro Lett. 8 (2019) 651–657, https://doi.org/10.1021/acsmacrolett.9b00142.

[108] M. Ashrafi, M. Bayat, P. Mortazavi, S.J. Hashemi, A. Meimandipour, Antimicrobial effect of chitosan–silver–copper nanocomposite on *Candida albicans*, J. Nanostruct. Chem. 10 (2020) 87–95, https://doi.org/10.1007/s40097-020-00331-3.

[109] T. Nalini, S.K. Basha, A.M. Mohamed Sadiq, V.S. Kumari, K. Kaviyarasu, Development and characterization of alginate/chitosan nanoparticulate system for hydrophobic drug encapsulation, J. Drug Deliv. Sci. Technol. 52 (2019) 65–72, https://doi.org/10.1016/j.jddst.2019.04.002.

[110] C.C. Pola, A.R.F. Moraes, E.A.A. Medeiros, R.F. Teófilo, N.F.F. Soares, C.L. Gomes, Development and optimization of pH-responsive PLGA-chitosan nanoparticles for triggered release of antimicrobials, Food Chem. 295 (2019) 671–679, https://doi.org/10.1016/j.foodchem.2019.05.165.

[111] V.C. Kalia, S. Ray, S.K.S. Patel, M. Singh, G.P. Singh, Applications of polyhydroxyalkanoates and their metabolites as drug carriers, in: Biotechnological Applications of Polyhydroxyalkanoates, Springer, Singapore, 2019, pp. 35–48, https://doi.org/10.1007/978-981-13-3759-8_3.

[112] F. Qi, J. Wu, H. Li, G. Ma, Recent research and development of PLGA/PLA microspheres/nanoparticles: a review in scientific and industrial aspects, Front. Chem. Sci. Eng. 13 (2019) 14–27, https://doi.org/10.1007/s11705-018-1729-4.

[113] M. Hill, R.N. Cunningham, R.M. Hathout, C. Johnston, J.G. Hardy, M.E. Migaud, Formulation of antimicrobial tobramycin loaded PLGA nanoparticles via complexation with AOT, J. Funct. Biomater. 10 (2019), https://doi.org/10.3390/jfb10020026.

[114] I.M. Chu, T.H. Liu, Y.R. Chen, Preparation and characterization of sustained release system based on polyanhydride microspheres with core/shell-like structures, J. Polym. Res. 26 (2019) 1–6, https://doi.org/10.1007/s10965-018-1657-5.

[115] S.M. Kelly, A. Mitra, S. Mathur, B. Narasimhan, Synthesis and characterization of rapidly degrading Polyanhydrides as vaccine adjuvants, ACS Biomater. Sci. Eng. 6 (2020) 265–276, https://doi.org/10.1021/acsbiomaterials.9b01427.

[116] L.Y. Guo, S.Z. Yan, X. Tao, Q. Yang, Q. Li, T.S. Wang, S.Q. Yu, S.L. Chen, Evaluation of hypocrellin A-loaded lipase sensitive polymer micelles for intervening methicillin-resistant *Staphylococcus aureus* antibiotic-resistant bacterial infection, Mater. Sci. Eng. C. 106 (2020), https://doi.org/10.1016/j.msec.2019.110230, 110230.

[117] W. Li, X. Fan, X. Wang, X. Shang, Q. Wang, J. Lin, Z. Hu, Z. Li, Stereocomplexed micelle formation through enantiomeric PLA-based Y-shaped copolymer for targeted drug delivery, Mater. Sci. Eng. C. 91 (2018) 688–695, https://doi.org/10.1016/j.msec.2018.06.006.

[118] Y. Su, L. Zhao, F. Meng, Z. Qiao, Y. Yao, J. Luo, Triclosan loaded polyurethane micelles with pH and lipase sensitive properties for antibacterial applications and treatment of biofilms, Mater. Sci. Eng. C. 93 (2018) 921–930, https://doi.org/10.1016/j.msec.2018.08.063.

[119] G. Tripodo, S. Perteghella, P. Grisoli, A. Trapani, M.L. Torre, D. Mandracchia, Drug delivery of rifampicin by natural micelles based on inulin: physicochemical properties, antibacterial activity and human macrophages uptake, Eur. J. Pharm. Biopharm. 136 (2019) 250–258, https://doi.org/10.1016/j.ejpb.2019.01.022.

[120] M.S. Kadayifci, D. Gokkaya, M. Topuzogullari, S.D. Isoglu, T. Atabey, T. Arasoglu, M.-M. Ozmen, Core-crosslinking as a pathway to develop inherently antibacterial polymeric micelles, J. Appl. Polym. Sci. 137 (2020) 48393, https://doi.org/10.1002/app.48393.
[121] A. Lanzilotto, M. Kyropoulou, E.C. Constable, C.E. Housecroft, W.P. Meier, C.G. Palivan, Porphyrin-polymer nanocompartments: singlet oxygen generation and antimicrobial activity, J. Biol. Inorg. Chem. 23 (2018) 109–122, https://doi.org/10.1007/s00775-017-1514-8.
[122] L. Palanikumar, S. Al-Hosani, M. Kalmouni, V.P. Nguyen, L. Ali, R. Pasricha, F.N. Barrera, M. Magzoub, pH-responsive high stability polymeric nanoparticles for targeted delivery of anticancer therapeutics, Commun. Biol. 3 (2020) 1–17, https://doi.org/10.1038/s42003-020-0817-4.
[123] Q. Zhang, G. Kuang, S. He, H. Lu, Y. Cheng, D. Zhou, Y. Huang, Photoactivatable prodrug-backboned polymeric nanoparticles for efficient light-controlled gene delivery and synergistic treatment of platinum-resistant ovarian Cancer, Nano Lett. 20 (2020) 3039–3049, https://doi.org/10.1021/acs.nanolett.9b04981.
[124] J. Choi, Y. Rui, J. Kim, N. Gorelick, D.R. Wilson, K. Kozielski, A. Mangraviti, E. Sankey, H. Brem, B. Tyler, J.J. Green, E.M. Jackson, Nonviral polymeric nanoparticles for gene therapy in pediatric CNS malignancies, nanomedicine nanotechnology, Biol. Med. 23 (2020) 102115, https://doi.org/10.1016/j.nano.2019.102115.
[125] M.L. de Souza, W.M. dos Santos, A.L.M.D. de Sousa, V. de Albuquerque Wanderley Sales, F.P. Nóbrega, M.V.G. de Oliveira, P.J.R. Neto, Lipid nanoparticles as a skin wound healing drug delivery system: discoveries and advances, Curr. Pharm. Des. 26 (2020), https://doi.org/10.2174/1381612826666200417144530.
[126] S. Nasirizadeh, B. Malaekeh-Nikouei, Solid lipid nanoparticles and nanostructured lipid carriers in oral cancer drug delivery, J. Drug Delivery Sci. Technol. 55 (2020), https://doi.org/10.1016/j.jddst.2019.101458, 101458.
[127] H.M. Abdelaziz, M.S. Freag, A.O. Elzoghby, Chapter 5—Solid lipid nanoparticle-based drug delivery for lung cancer, in: P. Kesharwani (Ed.), Nanotechnology-Based Targeted Drug Delivery Systems for Lung Cancer, Academic Press, 2019, pp. 95–121, https://doi.org/10.1016/B978-0-12-815720-6.00005-8.
[128] A. Mandal, R. Bisht, D. Pal, A.K. Mitra, Chapter 4—Diagnosis and drug delivery to the brain: Novel strategies, in: A.K. Mitra, K. Cholkar, A. Mandal (Eds.), Emerging Nanotechnologies for Diagnostics, Drug Delivery and Medical Devices, Elsevier, Boston, MA, 2017, pp. 59–83, https://doi.org/10.1016/B978-0-323-42978-8.00004-8.
[129] C. Carbone, V. Fuochi, A. Zielińska, T. Musumeci, E.B. Souto, A. Bonaccorso, C. Puglia, G. Petronio Petronio, P.M. Furneri, Dual-drugs delivery in solid lipid nanoparticles for the treatment of *Candida albicans* mycosis, Colloids Surf. B Biointerfaces 186 (2020), https://doi.org/10.1016/j.colsurfb.2019.110705, 110705.
[130] L. Kaushik, S. Srivastava, A. Panjeta, D. Chaudhari, R. Ghadi, K. Kuche, R. Malik, S. Preet, S. Jain, K. Raza, Exploration of docetaxel palmitate and its solid lipid nanoparticles as a novel option for alleviating the rising concern of multi-drug resistance, Int. J. Pharm. 578 (2020), https://doi.org/10.1016/j.ijpharm.2020.119088, 119088.
[131] C. Dumont, S. Bourgeois, H. Fessi, P.-Y. Dugas, V. Jannin, In-vitro evaluation of solid lipid nanoparticles: ability to encapsulate, release and ensure effective protection of peptides in the gastrointestinal tract, Int. J. Pharm. 565 (2019) 409–418, https://doi.org/10.1016/j.ijpharm.2019.05.037.
[132] E. Oner, M. Kotmakci, A.G. Kantarci, A promising approach to develop nanostructured lipid carriers from solid lipid nanoparticles: preparation, characterization, cytotoxicity and nucleic acid binding ability, Pharm. Dev. Technol. (2020) 1–13, https://doi.org/10.1080/10837450.2020.1759630.

[133] V.T. Arantes, A.A.G. Faraco, F.B. Ferreira, C.A. Oliveira, E. Martins-Santos, P. Cassini-Vieira, L.S. Barcelos, L.A.M. Ferreira, G.A.C. Goulart, Retinoic acid-loaded solid lipid nanoparticles surrounded by chitosan film support diabetic wound healing in in vivo study, Colloids Surf. B Biointerfaces 188 (2020) 110749, https://doi.org/10.1016/j.colsurfb.2019.110749.

[134] R.S. Kalhapure, S.J. Sonawane, D.R. Sikwal, M. Jadhav, S. Rambharose, C. Mocktar, T. Govender, Solid lipid nanoparticles of clotrimazole silver complex: an efficient nano antibacterial against *Staphylococcus aureus* and MRSA, Colloids Surf. B Biointerfaces 136 (2015) 651–658, https://doi.org/10.1016/j.colsurfb.2015.10.003.

[135] G. Gainza, M. Pastor, J.J. Aguirre, S. Villullas, J.L. Pedraz, R.M. Hernandez, M. Igartua, A novel strategy for the treatment of chronic wounds based on the topical administration of rhEGF-loaded lipid nanoparticles: in vitro bioactivity and in vivo effectiveness in healing-impaired db/db mice, J. Control. Release 185 (2014) 51–61, https://doi.org/10.1016/j.jconrel.2014.04.032.

[136] K. Khezri, M.R. Farahpour, S. Mounesi Rad, Efficacy of *Mentha pulegium* essential oil encapsulated into nanostructured lipid carriers as an in vitro antibacterial and infected wound healing agent, Colloids Surf. A Physicochem. Eng. Asp. 589 (2020), https://doi.org/10.1016/j.colsurfa.2020.124414, 124414.

[137] N. Osman, C.A. Omolo, R. Gannimani, A.Y. Waddad, S. Rambharose, C. Mocktar, S. Singh, R. Parboosing, T. Govender, Novel fatty acid-based pH-responsive nanostructured lipid carriers for enhancing antibacterial delivery, J. Drug Deliv. Sci. Technol. 53 (2019), https://doi.org/10.1016/j.jddst.2019.101125, 101125.

[138] M. Liu, W. Chen, X. Zhang, P. Su, F. Yue, S. Zeng, S. Du, Improved surface adhesion and wound healing effect of madecassoside liposomes modified by temperature-responsive PEG-PCL-PEG copolymers, Eur. J. Pharm. Sci. 151 (2020), https://doi.org/10.1016/j.ejps.2020.105373, 105373.

[139] J.U. Choi, S.W. Lee, R. Pangeni, Y. Byun, I.S. Yoon, J.W. Park, Preparation and in vivo evaluation of cationic elastic liposomes comprising highly skin-permeable growth factors combined with hyaluronic acid for enhanced diabetic wound-healing therapy, Acta Biomater. 57 (2017) 197–215, https://doi.org/10.1016/j.actbio.2017.04.034.

[140] A. Gauthier, A. Fisch, K. Seuwen, B. Baumgarten, H. Ruffner, A. Aebi, M. Rausch, F. Kiessling, M. Bartneck, R. Weiskirchen, F. Tacke, G. Storm, T. Lammers, M.G. Ludwig, Glucocorticoid-loaded liposomes induce a pro-resolution phenotype in human primary macrophages to support chronic wound healing, Biomaterials 178 (2018) 481–495, https://doi.org/10.1016/j.biomaterials.2018.04.006.

[141] V. Choudhary, H. Shivakumar, H. Ojha, Curcumin-loaded liposomes for wound healing: preparation, optimization, in-vivo skin permeation and bioevaluation, J. Drug Deliv. Sci. Technol. 49 (2019) 683–691, https://doi.org/10.1016/j.jddst.2018.12.008.

[142] Q. Xiang, J. Xiao, H. Zhang, X. Zhang, M. Lu, H. Zhang, Z. Su, W. Zhao, C. Lin, Y. Huang, X. Li, Preparation and characterisation of bFGF-encapsulated liposomes and evaluation of wound-healing activities in the rat, Burns 37 (2011) 886–895, https://doi.org/10.1016/j.burns.2011.01.018.

[143] L. Risaliti, A. Kehagia, E. Daoultzi, D. Lazari, M.C. Bergonzi, S. Vergkizi-Nikolakaki, D. Hadjipavlou-Litina, A.R. Bilia, Liposomes loaded with Salvia triloba and *Rosmarinus officinalis* essential oils: in vitro assessment of antioxidant, antiinflammatory and antibacterial activities, J. Drug Deliv. Sci. Technol. 51 (2019) 493–498, https://doi.org/10.1016/j.jddst.2019.03.034.

[144] Y. Zhang, C. Pu, W. Tang, S. Wang, Q. Sun, Gallic acid liposomes decorated with lactoferrin: characterization, in vitro digestion and antibacterial activity, Food Chem. 293 (2019) 315–322, https://doi.org/10.1016/j.foodchem.2019.04.116.

[145] A. Gharib, Z. Faezizadeh, M. Godarzee, In vitro and in vivo activities of ticarcillin-loaded nanoliposomes with different surface charges against *pseudomonas aeruginosa* (ATCC 29248), DARU, J. Pharm. Sci. 20 (2012) 41, https://doi.org/10.1186/2008-2231-20-41.

[146] B.E. Kim, E. Goleva, C.F. Hall, S.H. Park, U.H. Lee, A.M. Brauweiler, J.E. Streib, B.N. Richers, G. Kim, D.Y.M. Leung, Skin wound healing is accelerated by a lipid mixture representing major lipid components of *Chamaecyparis obtusa* plant extract, J. Invest. Dermatol. 138 (2018) 1176–1186, https://doi.org/10.1016/j.jid.2017.11.039.

[147] H.A. Gad, F.A.A. Abd El-Rahman, G.M. Hamdy, Chamomile oil loaded solid lipid nanoparticles: a naturally formulated remedy to enhance the wound healing, J. Drug Deliv. Sci. Technol. 50 (2019) 329–338, https://doi.org/10.1016/j.jddst.2019.01.008.

[148] S. Hallaj-Nezhadi, M. Hassan, Nanoliposome-based antibacterial drug delivery, Drug Deliv. 22 (2015) 581–589, https://doi.org/10.3109/10717544.2013.863409.

CHAPTER

3

Nanocomposite scaffolds and coatings for wound healing and infection control

Sebastian Wilk[a,*], *Agata Przekora*[b,*], *Paulina Kazimierczak*[b,*], *David Medina-Cruz*[c,*], *Lidia Martínez*[d], *Catherine O'Connell*[c], *Linh B. Truong*[c], *Katarzyna Reczyńska-Kolman*[a], *Hamed Barabadi*[e], *Jorge L. Cholula-Diaz*[f], *Elżbieta Pamuła*[a], *Yves Huttel*[d], *José Miguel García-Martín*[g], *Thomas J. Webster*[c], *and Aleksandra Benko*[a,c]

[a]Faculty of Materials Science and Ceramics, AGH University of Science and Technology, Krakow, Poland, [b]Independent Unit of Tissue Engineering and Regenerative Medicine, Medical University of Lublin, Lublin, Poland, [c]Department of Chemical Engineering, Northeastern University, Boston, MA, United States, [d]Instituto de Ciencia de Materiales de Madrid, ICMM-CSIC, Madrid, Spain, [e]Department of Pharmaceutical Biotechnology, School of Pharmacy, Shahid Beheshti University of Medical Sciences, Tehran, Iran, [f]School of Engineering and Sciences, Tecnologico de Monterrey, Monterrey, NL, Mexico, [g]Instituto de Micro y Nanotecnología, IMN-CNM, CSIC (CEI UAM+CSIC), Tres Cantos, Spain

[*]These authors contributed equally to this work.

https://doi.org/10.1016/B978-0-12-821637-8.00007-9
Copyright © 2023 Elsevier Inc. All rights reserved.

1 Introduction: Coatings and scaffolds—Another way nanotechnology can influence infection control

While the previous chapter outlined various advancements in nanostructures with optimistic antimicrobial efficacy, it is important to discuss methods of delivery that can maximize these promises. The incorporation of nanostructures with inherent biomedical benefits within a smart delivery matrix can exponentially expedite wound healing and deter microbial infections [1]. In order to design such matrices, an understanding of the skin and its healing processes are critical. During any damaging events, the skin undergoes specific stages of repair, including: (a) homeostasis—a formation of clot to control further damage, (b) inflammation—an innate response against potential infections, (c) proliferation—an initiation of new dermal layers by the mobilization of fibroblasts, and (d) remodeling—a maturation of the new layers of the skin [2]. Multiple cell types and processes are involved unproportionally in different stages of wound healing, and this complexity can lead to difficulties in generating any universal solutions for consistent and quick wound healing [3].

Such complexity and involvement also leads to various factors that can impair the healing processes. These factors can be generally divided into internal factors, where the risk comes from the inherent state of the patients, and external factors, where the risk comes from the surroundings. Internal factors can include age, sex, existing medical conditions (diabetes, cancers, heredity diseases, etc.), and lifestyle (alcohol, drug consumption, stress, and nutrition) [4–7]. These vary by patient, and so, any deterrence should be customized. On the other hand, external factors include oxygenation, moisture, and infections. A lack of oxygen at a wound site can delay and inhibit cellular metabolism, and prevent any of the critical wound healing stages from appropriate completion [8]. Meanwhile, a moist environment facilitates angiogenesis, prevents the buildup of dead tissues, and ensures hydration of cells [9]. Lastly, infections not only derail the healing process by driving inflammation responses, but they also pose a significant threat to both local and systemic tissues [10]. While inflammation is a part of the healing process, a lengthened and persisted state of inflammation can damage the ECM and prevent growth factors to initiate the formation of new fibroblast layers at the site [11].

Unlike internal risk factors, the prevention and mitigation of external risk factors are both possible and potentially applicable across different types of wounds, and in different patients. Therefore, dressings have been a popular approach to expedite wound healing, by maintaining optimal environmental conditions and protecting damaged sites from infections. However, current dressings, such as cotton and bandages, are inadequate in both of the mentioned tasks. Crucially, a new understanding of the different mechanisms in different stages enable engineering of improved scaffolds as dressings. Overall, a well-designed scaffold shows appropriate mechanical properties that are biomimetic to those of the ECM, with tensile strength and elasticity of particular importance [12–14]. Additionally, the porosity, wettability, and water uptake ability are also of significance, as these properties ensure proper attachment of cells, adequate oxygen exchange, and appropriate moisture of wound environments [15–17]. Lastly, in the case of incorporation of any therapeutic agents, biodegradability, release efficiency, and biocompatibility of the scaffolds must be considered. These properties can be controlled by designing the 3D structure, selecting specific materials, or alterating the scaffold synthesis pathways [18].

While the scaffolds act as layers of protection to expedite wound healing, they can benefit greatly from the introduction of antimicrobial agents such as the nanomaterials (NMs), as described in Chapter 2. This chapter will discuss different approaches in the design of scaffolds that incorporate engineered NMs. Special emphasis will be put on the therapeutic effect of the as-obtained composites, with respect to wound healing and infection control. While many different types of scaffolds can be designed, these can be categorized into three main groups, which will be the focus of this chapter. These groups are hydrogels, nanofibers, and films.

2 Hydrogels

Hydrogels are defined as hydrophilic materials with a macromolecular structure of a gel, that are highly absorbent (may increase their weight several times compared to the dry state), have a great swelling ability, and contain even 99% of biological fluids in a wet state. Hydrogels have a form of three-dimensional (3D), highly hydrated interconnected matrices that are created by crosslinking of natural or synthetic polymer chains [19,20]. The porous network of hydrogels is ideal for the immobilization of materials/enzymes/NPs and various bioactive agents by entrapment methods, and thus they have been given special attention in the field of biomaterial engineering for biomedical applications. Depending on the production method, two types of hydrogels may be distinguished: (1) chemical gels (thermosetting) created by covalent crosslinking (e.g., polymerization with the use of a crosslinker) and (2) physical gels (thermoplastic) produced by noncovalent interactions (e.g., Van der Waals forces, hydrogen bonding). Unlike chemical gels, physical hydrogels dissolve in an aqueous environment and melt when heated. The main properties of the hydrogels include swelling tendency, mechanical properties, and diffusion rates of the entrapped agents, all of which depend on the crosslinking density, macromolecular structures in the resultant gel, and presence of residues (e.g., monomers, initiators) [20]. To improve the biological properties of hydrogels designed for biomedical applications, they are often produced with embedded bioactive molecules (e.g., growth factors, cytokines, cell adhesive proteins, antimicrobial agents, etc.). A recent trend in biomaterial engineering involves the preparation of hydrogels by physical/chemical crosslinking and modifying them with nanoscaled features (nanoparticles, nanofibers, or nanolayers), leading to the formation of nanocomposite hydrogels with new properties (Fig. 1).

Chronic wounds are at high risk of persistent infections. Most of the pathogens present in the chronic wound microenvironment (such as *S. aureus* and *P. aeruginosa*) have the ability to form a bacterial biofilm, which can have 500 times higher resistance to antibiotics than the freely living bacteria (planktonic cells). Despite the presence of persistent infections at the chronic wound site, topical administration of antibiotics is not recommended for wound management unless a serious bacteria colonization is observed [21]. Topical delivery of antibiotics may lead to antibiotic resistance, impaired healing, or a delayed hypersensitivity reaction. Thus,

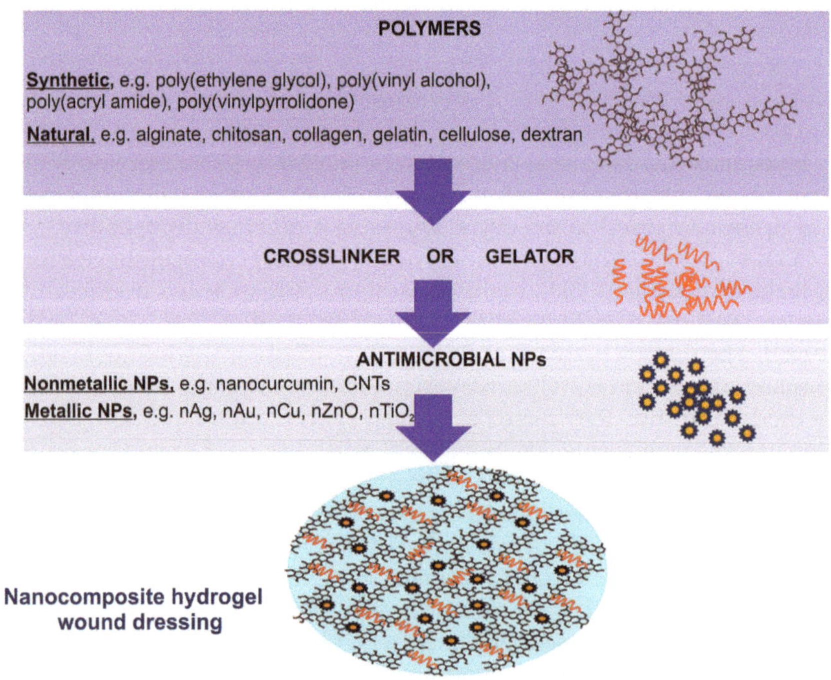

FIG. 1 Graphical representation of an antibacterial nanocomposite hydrogel for biomedical applications.

the development of antibiotic-free systems for the management of infected wounds, which have the antimicrobial properties and are capable of inhibiting bacterial biofilm formation, is of high importance. An ideal wound dressing for chronic wound treatment possesses the ability to absorb and remove excess exudate, while maintaining optimal moisture at the wound-dressing interface. It is also flexible and comfortable, providing "intimate contact" with the wound surface (without any "dead spaces") that facilitates the regeneration process and reduces the risk of infections. Also, optimal wound dressings should prevent secondary infections and allow dressing changes without disturbing the wound bed [21,22]. Hydrogel wound dressings fit all these criteria very well. Since the incorporation of antibiotics into the structure of the wound dressings is not recommended, nanocomposite hydrogel dressings made of a polymer matrix reinforced with antimicrobial nanoparticles have been recently developed for the management of infected wounds (Table 1).

As metallic NPs are known for their broad antimicrobial spectrum, mainly through a free radical formation mechanism, nanocomposite hydrogel dressings containing silver (AgNPs), gold (AuNPs), copper (CuNPs), zinc oxide (ZnONPs), titania (TiO_2NPs), or iron oxide (Fe_3O_4NP) nanoparticles have been recently developed to provide antimicrobial activity and improved wound healing. Among various metallic NPs, AgNPs has been the most extensively applied for the production of antimicrobial nanocomposite hydrogel wound dressings [33]. Dai et al. [23] used a cationic dendrimer as a template for the synthesis of stable AgNPs. The produced dendrimer-encapsulated AgNPs were subsequently used for the preparation of a nanocomposite hydrogel via Schiff base linkage with an oxidized dextran. The developed hydrogel showed pH-responsive release of a cationic polymer and silver ions that revealed a synergistic antibacterial activity against Gram-negative (*E. coli, P. aeruginosa*) and Gram-positive (*S. epidermidis, S. aureus*) bacteria. The nanocomposite also exhibited a superior in vivo antibacterial efficacy against *S. aureus* compared to a commercial silver hydrogel in an infected wound mouse model. Shin et al. [24] produced an antimicrobial nanocomposite by embedding AgNPs-immobilized nanocellulose fibers within the alginate gel that was crosslinked with calcium chloride ($CaCl_2$). The resultant hydrogel showed sustained release (for 7 days) of Ag^+ ions that inhibited *E. coli* growth without inducing a cytotoxic effect against eukaryotic cells.

Alternatively, Bartel et al. [25] developed antibacterial gel-shell nanoparticles for biomedical applications that were made of sulfonated polystyrene decorated with AgNPs which demonstrated activity against *E. coli, P. aeruginosa, S. aureus,* and *S. epidermidis*. Kar et al. [26] produced a nanocomposite wound dressing made of aminated guar gum, alginate, and gelatin that were mixed 1:1:1, and crosslinked with $CaCl_2$, which preceded freezing and lyophilization. In these studies, aminated guar gum

TABLE 1 Representative examples of nanocomposite hydrogels for the control of bacterial infections in wound healing applications.

System	Synthesis method	NP size	Antimicrobial activity	References
Oxidized dextran with dendrimer-encapsulated nAg	Schiff base linkage between oxidized dextran and cationic dendrimers encapsulated with nAg	3–4 nm	*EC* *PA* *SA* *SE*	[23]
Injectable alginate loaded with nAg-incorporated nanocellulose fibers	Loose crosslinking of alginate with $CaSO_4$, followed by complete crosslinking of the composite with $CaCl_2$	12–15 nm	*EC*	[24]
Nanobeads made of a polystyrene core surrounded by a hydrogel shell with embedded nAg	Sulfonation of polystyrene followed by the reduction of Ag^+ accumulated within the hydrogel shell with sodium borohydride	4–5 nm	*EC* *PA* *SA* *SE*	[25]
Aminated guar gum/sodium alginate/gelatin matrix with embedded nAg and epigallocatechin gallate	Crosslinking of the composite with $CaCl_2$, followed by lyophilization	6–12 nm	*EC* *PA* *SA* *BS*	[26]
Poly(gamma-glutamic acid) with embedded nAg	Crosslinking with ethylene glycol diglycidyl ether	35–55 nm	Not tested	[27]
Injectable dual-network silica-based scaffold made of polyethylene glycol diacrylate and dynamic network formed between sodium alginate and Cu-doped bioactive glass NPs	UV light crosslinking using I2959 as a photoinitiator and additional crosslinking between NPs and sodium alginate	Not provided	*EC* *SA*	[28]
Terpolymer made of 2-acrylamide-2-methylpropane sulfonic acid, acrylamide, acrylonitrile, and acrylic acid monomers with embedded nFe_3O_4	Free radical copolymerization using benzoyl peroxide as an initiator and ethylene glycol dimethacrylate as a crosslinker	15–20 nm	*EC* *BS*	[29]

Poly(*N*-vinyl pyrrolidone) hydrogel with embedded $nTiO_2$	Sol-gel method using tetraethoxysilane as a crosslinking agent and lactic acid as a catalyst	10–15nm	*SA*	[19]
Thiolated chitosan-alginate with embedded ZnO	Initial precipitation of thiolated chitosan with NaOH, followed by lyophilization of the entire composite mixture	50–70nm	*SA*	[30]
Injectable glycidyl methacrylate functionalized quaternized chitosan cryogel with embedded CNT	Cryopolymerization of modified chitosan with the diacrylate functionalized (poly(ethylene glycol)-*b*-poly(propylene glycol)-*b*-poly (ethylene glycol))/CNT dispersion liquid using APS/TEMED as redox initiator system	10–20nm	*EC* *SA* *PA*	[31]
Injectable chitosan-*g*-pluronic hydrogel with embedded nanocurcumin	Copolymerization between activated pluronic and chitosan	20–50nm	*EC* *SA* *PA* *ST*	[32]

BS, Bacillus subtilis; EC, Escherichia coli; MRSA, methicillin-resistant *Staphylococcus aureus; PA, Pseudomonas aeruginosa; SA, Staphylococcus aureus; SE, Staphylococcus epidermidis.*

acted as a reductant and stabilizer for Ag^+ ions to obtain stable and size-controlled AgNPs. Moreover, antioxidant epigallocatechin gallate was embedded within the nanocomposite matrix to achieve antiinflammatory effects. The obtained nanocomposite hydrogel showed antimicrobial activity (against *E. coli, P. aeruginosa, S. aureus,* and *B. subtilis*), noncytotoxicity against eukaryotic cells, and accelerated healing of subcutaneous wounds in a rat model. The dressing promoted better wound contraction, collagen deposition, and augmented vascularization of the wound region by modulating the release of inflammatory cytokines and growth factors. Wang et al. [27] fabricated a poly(gamma-glutamic acid) hydrogel embedded with AgNPs, and crosslinked with ethylene glycol diglycidyl ether. The nanocomposite exhibited a sustained release of Ag^+ ions. Although the studies did not present antibacterial tests, these materials have the potential of providing an effective antibacterial protection. Moreover, the nanocomposite hydrogel revealed the ability to accelerate wound healing and promote reepithelialization in a mouse model.

Apart from AgNPs, other metallic nanoparticles are widely used for the production of antimicrobial nanocomposite hydrogels. Farag et al. produced an antimicrobial terpolymer nanocomposite hydrogel using Fe_3O_4NPs and four monomers: 2-acrylamido-2-methylpropane sulfonic acid, acrylamide, acrylonitrile, and acrylic acid. The hydrogel was crosslinked by a free radical copolymerization method using benzoyl peroxide as an initiator and ethylene glycol dimethacrylate as a crosslinker. These nanocomposite hydrogels demonstrated activity against *E. coli, B. subtilis,* and *C. albicans* [29]. Timaeva et al. developed a poly(*N*-vinyl pyrrolidone) hydrogel wound dressing containing TiO_2NPs that exhibited resistance to bacteria colonization and antibacterial properties against *S. aureus* [19]. Meanwhile, Arshad et al. fabricated a porous nanocomposite wound dressing made of thiolated chitosan, alginate, and ZnONPs, through lyophilization. The resultant wound dressing revealed antibacterial activity against *S. aureus,* low cytotoxicity against human cervical cancer cells (HeLa cell line) in vitro, and an accelerated wound healing in a mouse model [30].

Antibiotic-free nanocomposite hydrogel systems with strong antimicrobial activity can be prepared by reinforcing polymers with metallic NPs. Nonmetallic NPs have also been reported to be promising antimicrobial nanofillers of a hydrogel matrix. Li et al. used bioactive glass nanoparticles containing Cu^{2+} to prepare a nanocomposite hydrogel system [28]. Sodium alginate was crosslinked with Cu-doped bioactive glass NPs to form an antimicrobial dynamic network, whereas a photo-crosslinked polyethylene glycol diacrylate created the second network of the scaffold. The hydrogel showed a sustained release of Cu^{2+} ions, providing antibacterial activity against *E. coli* and *S. aureus.* Moreover, the hydrogel significantly enhanced the proliferation and angiogenic ability of endothelial progenitor cells in vitro and efficiently restored blood vessels through augmented HIF-1α/VEGF expression in vivo. Meanwhile, Dang et al. incorporated nanocurcumin within the matrix of a chitosan-*g*-pluronic copolymer to obtain an injectable hydrogel with antibacterial activity against *E. coli, S. aureus, P. aeruginosa,* and *S. typhimurium* [32]. The nanocomposite hydrogel was nontoxic against human dermal fibroblasts in vitro and promoted collagen deposition and wound healing of burn wounds in a mouse model.

In summary, hydrogel-based materials are typically modified with various NP types to provide materials with new functionalities. Dependent on the type of nanofiller, these may be divided into several groups, such as carbon-, inorganic-, or metallic-based materials. Each type of nanofiller is characterized by different properties that determine biomedical applications of the resultant nanocomposite. For instance, carbon-based nanofillers are usually used to produce nanocomposites with an intended use in electrically conductive tissues, but they also have the potential to provide sufficient mechanical reinforcement and intrinsic bacteriostatic properties for the treatment of wounds and hemorrhages in vital organs. Meanwhile, inorganic-based (e.g., bioglass) nanocomposite hydrogels are characterized by good mechanical properties and are widely applied as bone scaffolds, but the employment of various network modifiers can also provide antibacterial effects. Still, antimicrobial nanocomposite hydrogels are most frequently prepared by the incorporation of metallic nanofillers within the polymeric matrix [20].

The antimicrobial activity of this class of materials may be regulated by the content and size of NPs within the polymer matrix and is known to be highly dependent on the release profile of metal ions from the nanocomposite. Sustained release is believed to provide the best therapeutic effects. Initial burst release may be harmful to the surrounding tissues, hindering the healing process, but may be desired in case of invasive infection. The release profile and biological properties of the hydrogel nanocomposites may be modified by using different fabrication techniques, obtaining an appropriate crosslinking density. For instance, hydrogel wound dressings for exudate wound treatment should be produced by lyophilization to obtain highly absorbent nanocomposites with the gradual release of ions, protecting against secondary infections. However, chronic wounds with critical bacteria colonization should be treated with a topical nanocomposite hydrogel because lyophilized dressings may not provide sufficient antimicrobial protection [33]. Nevertheless, despite recent advances in the field of biomaterials engineering, there are still difficulties in the preparation of nanocomposite hydrogels with controlled antimicrobial activity and also in providing long-term noncytotoxicity toward eukaryotic cells. Thus, this area of research is still very challenging but an important direction in the field of regenerative medicine.

3 Nanofibers

In the biomedical field of wound healing, nanofibers (NFs) present themselves as a relatively new class of materials that can offer many specific and distinctive advantages that are often unavailable for other types of nanomaterials. Nanofibers are materials with diameters ranging from a few to several hundred nanometers. The very small diameter of the cross-section, which is at least hundreds of times smaller than the total fiber length, is also a specific nanofiber feature [34]. NFs possess various extraordinary properties like high surface area-to-volume ratio, easy chemical functionalization, and high mechanical properties [35]. These properties make them very interesting scaffold candidates for many different biomedical applications, including tissue engineering, membrane technologies, drug delivery systems, wound healing, and cell culturing [36,37]. NFs might be obtained from various materials, both synthetic and natural, creating a possibility to ensure many diverse properties that can be selected for specific purposes [38]. There are many different methods to produce nanofibers such as electrospinning, drawing, template synthesis, self-assembly, or phase separation. Among these methods, electrospinning seems to be the most popular and superior technique, due to versatility, cost-efficiency, and simplicity, which enables one to obtain ultrafine fibers with controlled porosity levels, pore sizes, and fiber diameters [36,39].

The extraordinary properties of nanofibers can provide various advantages in wound dressing and antimicrobial applications. For example, 3D scaffold created from a biocompatible nanofibrous mesh possesses the ability to mimic both the morphology and functionality of the native extracellular matrix (ECM), which is responsible for cell migration, cell proliferation, cellular response, and neo tissue genesis in the body environment. From a chemical point of view, the natural ECM is composed of various proteins and sugar-protein complexes, mostly collagen, elastin, and fibrillin. All of these proteins have a fibrous morphology, with fiber bundle diameters ranging from tens to hundreds of nanometers [40–42]. Hence, nanofibrous scaffolds possessing similar architecture and morphology are able to support cell adhesion and ensure guided cellular growth, proliferation, and behavior during the regeneration process in the wound [41,43]. The high surface area-to-volume ratio additionally provides a large area for numerous scaffold-tissue interactions in a wound environment that can promote cell adhesion even further, activating cell signaling and attracting fibroblasts for the secretion of ECM components [44,45]. High porosity, which can be easily obtained via electrospinning, enables not only efficient gas exchange, water permeability, and interchange of nutrients and other important substances like, e.g., growth factors, but also provides fluid loss control. Those properties are extremely important for assuring the right homeostasis at the wound site, to enhance angiogenesis and prevent dehydration of the cells, which are crucial in the wound healing process [39,44,45].

Nanoscale porosity and the high surface area of nanofibers are features that provide other important benefits. Interconnected pores of small diameters protect the wound against microbial infiltration and bacterial infection. They can also be used as a drug delivery system, with a high drug loading capacity, and release profile easily controlled by pore size and biodegradation rate of the fibers [45]. High surface area favors incorporation of various antibiotics, drugs, and antimicrobial agents as well as provides a suitable matrix for specifically targeted chemical functionalization, which might be able to steer the cellular reaction. Numerous substances might be incorporated into nanofibrous dressings to treat different wound types, for example, burn or diabetic wounds. This is achieved by either enhancing the healing process, reducing the inflammation response, or causing antimicrobial and bactericidal effects [38,39,46].

Nanofibers represent a class of materials that can provide numerous benefits in wound dressing and inflammation control applications. This section aims to introduce some of the recent research developments in the use of nanofibers for the wound healing purposes, with examples gathered in Table 2.

One of the most common strategies in manufacturing infection controlling scaffolds is to incorporate various types of antibiotics into the biocompatible nanofibrous matrix (Fig. 2A and B). Bakhsheshi-Rad et al. created an electrospun chitosan-alginate (1:1 ratio) nanofibrous scaffold, containing various amounts (0–10 wt%) of gentamicin (Gn), and crosslinked by glutaraldehyde vapors [47]. The developed materials revealed antimicrobial activity against Gram-positive (*S. aureus*) and Gram-negative (*E. coli*) bacteria, which increased with antibiotic concentration. Samples loaded with lower amounts of Gn also promoted L929 fibroblast cell attachment and proliferation in vitro. In vivo tests, carried out on mouse models, showed that the material incorporated with 3 wt% of Gn provided an excellent burn wound healing capability and promoted skin reepithelialization. Ajmal et al. produced ciprofloxacin, hydrochloride, and quercetin-loaded poly(ε-caprolactone)-gelatin nanofibers via the electrospinning technique [45]. The use of two different antibiotics in one material can counteract certain antibiotic-resistant bacteria. It was found that gelatin enhanced the hydrophilicity and degradation rate of the fibers, which resulted in a faster drug release, and improved the cell viability on the scaffold. Antibacterial assessment against *S. aureus* revealed significantly reduced bacteria growth in vitro. Moreover, an open-wound healing experiment, carried out on Wistar rats, revealed that within 16 days the

TABLE 2 Representative examples of nanofibers used for the control of bacterial infections in wound healing applications.

System	Synthesis method	NF diameter	Antimicrobial activity	References
Gentamicin-loaded chitosan-alginate nanofibers	Electrospinning (solvents: acetic acid and deionized water), then crosslinking by glutaraldehyde vapors	301–398nm	*SA* *EC*	[47]
Ciprofloxacin-, hydrochloride-, and quercetin-loaded PCL-gelatin nanofibers	Electrospinning (solvents: 1,1,1,3,3,3-hexafluoro-isopropanol (HFIP), glacial acetic acid, and dimethyl sulfoxide)	524–928nm	*SA*	[45]
Silver sulfadiazine-loaded zein nanofibers	Electrospinning (solvent: aqueous solution of ethanol and water (75:25 by weight)) with glutaraldehyde as a crosslinker, then crosslinking by HCl foaming	226–236nm	*EC*	[48]
Silver sulfadiazine-loaded cellulose acetate nanofibers	Electrospinning (solvent: DMF and acetone solution (6:4 ratio)), then electrospun mats were decentralized by NaOH	286–292nm	*BS* *EC*	[49]
ZnO NP-loaded chitosan-PVA nanofibers	Electrospinning (solvents: acetic acid, deionized water)	279–892nm	*EC* *PA* *BS* *SA*	[50]
ZnAg NP-loaded PU nanofibers	Electrospinning (solvent: DMF)	181–221nm	*EC* *SA* *BS*	[51]
ZnO NP-loaded PEO-chitosan nanofibers and ciprofloxacin-loaded PEO-chitosan nanofibers	Electrospinning (solvent: aqueous solution of acetic acid and water (50:50 ratio))	142–277nm	*SA* *EC*	[46]

Honey-loaded alginate-PVA nanofibers	Electrospinning (solvent: distilled water), then crosslinking in glutaraldehyde vapors	379–528 nm	*SA* *EC*	[52]
Two-layer asymmetric membrane: Top layer—silk fibroin and PCL nanofibers Bottom layer—thymol-loaded silk fibroin and hyaluronic acid nanofibers	Electrospinning of the top layer (solvents: formic acid and trifluoroethanol (2:1)), then bottom layer electrospinning directly on the top of the previous layer (solvents: formic acid and water (2:1), thymol solvent: PEO), after that: crosslinking in ethanol vapor (75%)	Top layer: 425–806 nm Bottom layer: 306–519 nm	*SA* *PA*	[53]
Superhydrophobic CNF nanocomposite coatings	Spray coating of CNFs/PTFE or CNFs/PDMS mixture on different substrate surfaces, thermal treatment afterwards	100 nm	*EC*	[54]

BS, Bacillus subtilis; EC, Escherichia coli; PA, Pseudomonas aeruginosa; SA, Staphylococcus aureus.

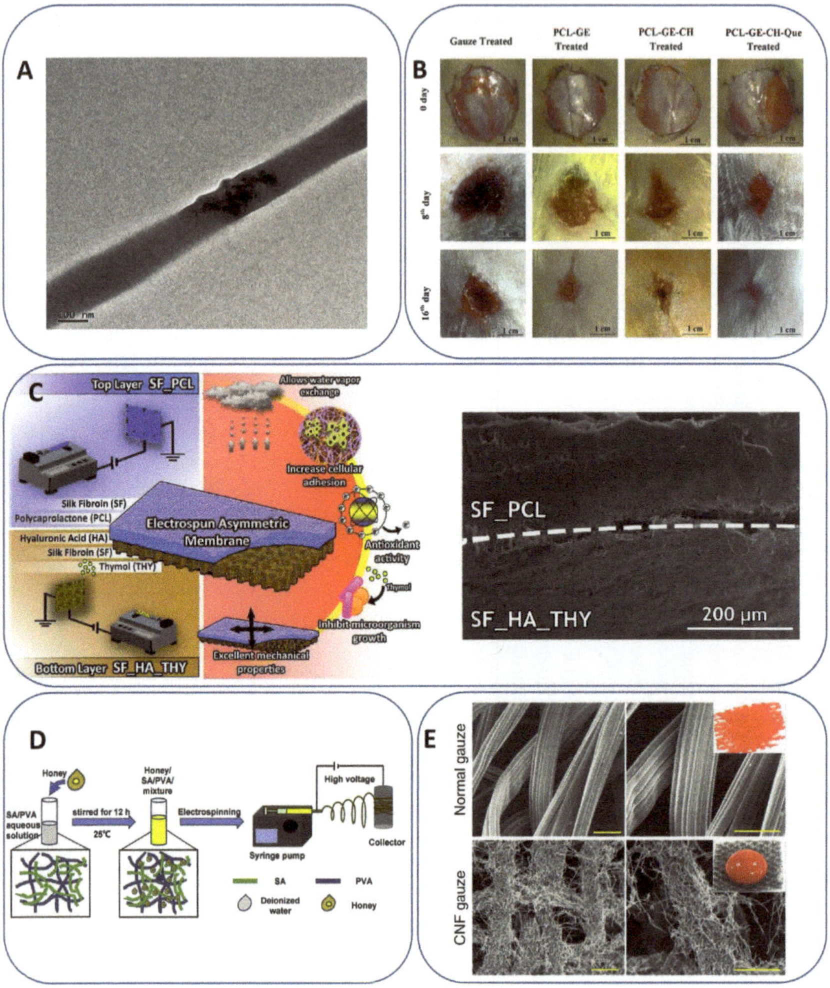

FIG. 2 Examples of different approaches used in fabricating multifunctional, fibrous wound dressings. (A) TEM image of the drug-loaded (ciprofloxacin) chitosan nanofiber for the pH-controllable drug delivery system; (B) in vivo superior performance of the 2-drug (ciprofloxacin and quercetin) loaded composite nanofibers (PCL-gelatin) used for effective treatment of full-thickness wounds in rats [45]; (C) graphical representation and SEM image of a two-layered material, designed to mimic the epidermis (top—dense sf_pcl) and dermis (bottom—fibrous sf_ha, loaded with an herbal drug), aimed to provide optimal conditions for tissue regeneration combined with an antibacterial effect [53]; (D) graphical representation of honey-loaded nanofibers, designed to obtain a highly biocompatible and antibacterial wound dressing [52]; (E) SEM images of a standard gauze, which absorbs blood, and the superhydrophobic CNF-coated gauze, which repels blood. Such materials may be used to effectively inhibit bacterial adhesion and reduce blood loss to control severe hemorrhages [54]. *(Panel A) Reproduced from S. Abid, T. Hussain, A. Nazir, A. Zahir, S. Ramakrishna, M. Hameed, N. Khenoussi, Enhanced antibacterial activity of PEO-chitosan nanofibers with potential application in burn infection management, Int. J. Biol. Macromol. 135 (2019) 1222–1236, https://doi.org/10.1016/j.ijbiomac.2019.06.022 with permission from Elsevier (2019).*

(Continued)

developed material ensured 100% full-thickness wound closure and enhanced healing and inflammation control.

Ullah et al. [48] prepared electrospun zein fibers scaffold incorporated with silver sulfadiazine (AgSD)—an antibiotic commonly used in burn wound treatment. Due to its high biocompatibility and biodegradability, zein is being used in many different applications, including tissue engineering. The study found that the material with a 0.6% concentration of AgSD in zein fibers showed strong antimicrobial activity against *E. coli* and *Bacillus*, Gram-negative and Gram-positive bacteria, respectively. An investigation on drug release showed that the release of Ag gradually increased with the concentration of AgSD in the zein nanofibers. Silver sulfadiazine was also used by Khan et al. [49]. They manufactured AgSD-loaded cellulose acetate nanofibers via electrospinning. In vitro antimicrobial tests showed that the materials possessed better bactericidal properties at lower AgSD concentrations. Furthermore, Gram-negative *E. coli* bacteria were less resistant and had reduced viability in contact with AgSD-zein scaffold than the Gram-positive *B. subtilis*. Such differences could be caused by differences in cell membrane morphology and chemistry.

Along with antibiotics, various types of metallic nanoparticles (MNPs) are also being incorporated into the nanofibrous matrix to obtain antimicrobial properties. Nanofibrous chitosan-poly(vinyl alcohol) mats incorporated with 40-nm zinc oxide MNPs were developed by Ahmed et al. via an electrospinning technique [50]. The antimicrobial tests against *E. coli, P. aeruginosa, B. subtilis,* and *S. aureus* bacterial strains revealed significantly better antibacterial potential of a chitosan/PVA/ZnO nanocomposite when compared to a pure chitosan/PVA material. Scaffolds loaded with MNPs also showed enhanced collagen formation. Accelerated healing in the presence of material were confirmed by in vivo tests on diabetic

FIG. 2, cont'd *(Panel B) Reprodiced from G. Ajmal, G.V. Bonde, P. Mittal, G. Khan, V.K. Pandey, B.V. Bakade, B. Mishra, Biomimetic PCL-gelatin based nanofibers loaded with ciprofloxacin hydrochloride and quercetin: a potential antibacterial and anti-oxidant dressing material for accelerated healing of a full thickness wound, Int. J. Pharm. 567 (2019) 118480, https://doi.org/10.1016/j.ijpharm.2019.118480, with permission from Elsevier (2019). (Panel C) Reproduced from S.P. Miguel, D. Simões, A.F. Moreira, R.S. Sequeira, I.J. Correia, Production and characterization of electrospun silk fibroin based asymmetric membranes for wound dressing applications, Int. J. Biol. Macromol. 121 (2019) 524–535, https://doi.org/10.1016/j.ijbiomac.2018.10.041 with permission from Elsevier (2019). (Panel D) Reproduced from Y. Tang, X. Lan, C. Liang, Z. Zhong, R. Xie, Y. Zhou, X. Miao, H. Wang, W. Wang, Honey loaded alginate/PVA nanofibrous membrane as potential bioactive wound dressing, Carbohydr. Polym. 219 (2019) 113–120, https://doi.org/10.1016/j.carbpol.2019.05.004, with permission from Elsevier (2019). (Panel E) Reproduced from Z. Li, A. Milionis, Y. Zheng, M. Yee, L. Codispoti, F. Tan, D. Poulikakos, C.H. Yap, Superhydrophobic hemostatic nanofiber composites for fast clotting and minimal adhesion, Nat. Commun. 10 (2019) 5562, https://doi.org/10.1038/s41467-019-13512-8 under a Creative Commons Attribution 4.0.*

rabbit models, by exhibiting increased diabetic skin wound enclosures during 12 days long trials. Jatoi manufactured polyurethane (PU) nanofibers filled with ZnAg nanoparticles [51]. Bimetallic NPs, with an average size of 11.3 nm, were created from ZnO nanoparticles (by a two-step synthesis with the use of dopamine hydrochloride (Dopa) and silver nitrate ($AgNO_3$) solutions), added into dissolved PU, and electrospun to obtain PU-ZnAg nanofibrous mats. Antimicrobial tests, carried out on *E. coli, S. aureus,* and *B. subtilis* bacterial strains, showed extraordinary bactericidal and bacteriostatic properties for the developed material, which were significantly better than the PU-ZnO nanofibrous reference scaffold. Abid et al. developed poly(ethylene oxide)-chitosan (PEO-CS) based nanofibers, loaded with either antibiotic (ciprofloxacin) or ZnO nanoparticles, in order to compare their antimicrobial activities. It was found that the antibiotic-loaded mats exhibited better bactericidal effects against *S. aureus* and *E. coli* bacteria than the MNPs-loaded ones. Moreover, in vitro tests revealed that the nanofibers with ciprofloxacin had low cytotoxicity and moderately high cell viability when tested on HaCaT (human keratinocytes) and HDF (human dermal fibroblast) cell lines.

Substances harvested from natural sources, like honey, plant-based essential oils, or herbal drugs, are often widely used in modern clinical wound dressing systems, due to their high biocompatibility, green synthesis and handling methods (Fig. 2C and D). Honey is known for its antiinflammatory, antioxidant, and bactericidal properties, attributable to a high content of antioxidant phenols. For this reason, Tang et al. developed alginate-poly(vinyl alcohol) nanofibrous mats, loaded with various concentrations of acacia honey (0%–20% v/v), and fabricated by electrospinning [52]. In vitro tests revealed that the antioxidant and bactericidal properties of the material increased with increasing honey concentration. The material proved to be effective against both Gram-negative and Gram-positive bacterial strains (*E. coli* and *S. aureus,* respectively) but exhibited better activity against the latter. Miguel et al. manufactured a two-layered, asymmetric, nanofibrous membrane [53]. The top layer, which was supposed to act like a human epidermis, was electrospun from silk fibroin (SF) and poly(ε-caprolactone). The bottom layer was supposed to act as the human dermis and was created from SF and hyaluronic acid (HA) nanofibers, electrospun directly on top of the previous one, and loaded with an herbal drug—thymol (THY). Hence, a skin-like scaffold was obtained, with the top layer acting as a physical barrier, and the bottom one having antimicrobial properties and the ability to promote cell adhesion and proliferation, essential for wound healing. The resultant material possessed an optimal porosity and pore size, allowing for free gas exchange and successful bacteria inhibition in the top layer, and effective cell adhesion and proliferation in the bottom one. The dermis-like layer also possessed enhanced antimicrobial activity against *S. aureus* and *P. aeruginosa* bacterial strains.

A different approach was presented by Li et al. [54]. By using the spray-coating technique, they created superhydrophobic carbon nanofiber (CNF) layers on various surfaces (including woven cotton gauze) to obtain antibacterial and easy to peel wound dressings which could prevent blood loss and inhibit bacterial adhesion (Fig. 2E). Tests showed that the obtained material greatly enhanced fibrin fiber formation and accelerated the blood coagulation process, in comparison to a commonly used gauze wound dressing. Antimicrobial tests, carried out with *E. coli* strain, showed exceptionally low bacteria adhesion to the material surface. In vivo tests were performed on rats with the back-bleeding model. Excellent blood-repelling and coagulation-promoting properties were revealed, and the material was summarized as safe for hemostatic use.

In recent years, the interest in nanofibers for wound healing applications has been growing. This is because of their morphology, extraordinary physicochemical properties, and high biomimicry toward the native ECM [35,41,43]. Electrospinning is the most popular synthesis method, exhibiting numerous benefits, such as high versatility, cost-effectiveness, and easy application [39]. Nanofibrous scaffolds are obtained from either natural or synthetic polymers, or their blends. Choosing the right polymer is an extremely important consideration because it determines not only the final material properties, but also the substances which must be used during the synthesis process, and which can affect the final product's biocompatibility [44]. Many solvents, like HFIP or acetic acid, crosslinkers, like glutaraldehyde or hydroxy-1-(4-(hydroxyethoxy) phenyl)-2 methyl-1-propanone, are toxic and their usage in the process poses a risk of detrimental residues in the final product. Thus, the polymers, which can be dissolved in biocompatible solvents and crosslinked with nontoxic methods, should be highly favored when used in wound healing applications. Antibiotics, metallic nanoparticles, and substances from natural sources are commonly added for antimicrobial and bactericidal purposes. Any type of those additives can be introduced into the nanofibrous material to enhance its wound healing and infection control abilities for the treatment of specific wound types. The use of nanofibers incorporated with specific bioactive molecules and drugs creates the possibility of delivering cargo at a certain place, with stable and controllable release kinetics [38]. Due to the high surface area-to-volume ratio, the loading capacity of the nanofibers can be significantly higher than that of the microfibers [43]. These features reduce the risks of pathogens developing immunity, and of unwanted side effects at the wound site. Moreover, introduction of more than one bioactive substance allows for the development of new strategies toward drug-resistant microorganisms.

More advanced approaches incorporate different strategies for manufacturing an ideal wound dressing material. Double-layered asymmetric skin-like scaffolds [53], or superhydrophobic CNF coating on gauze [54], are some examples of new generation scaffolds with high potential in healing specific, deep, or complex wounds. Nevertheless, there are still many difficulties in developing perfectly biocompatible materials with antimicrobial abilities for both long- and short-term use. Gram-negative and Gram-positive bacteria types very often show different resistance when in contact with developed systems [52]. More in vitro and in vivo trials are also required to fully understand the complicated antioxidant, bactericidal, and wound healing properties of the manufactured materials. Still, the use of nanofibers in the treatment of hard-to-heal infected wounds seems to be a very promising avenue, offering many advantages over other alternatives.

4 Films

Due to their specific physical properties, films are some of the main choices among all accessible wound dressings. Film dressings are semipermeable sheets that are permeable to gas and impermeable to liquid and extraneous bacteria. Moreover, the transparency of the films allows inspection of the wound without dressing removal. The important feature of the films is also their flexibility, permitting easy adjustment to the patient's body [55,56]. Film dressings are generally used as primary or secondary dressings [56], and they can be used for the management of minor lacerations and abrasions, superficial burns, low exuding wounds, or postoperative sutured wounds [55].

Synthetic polymers, such as polylactic acid (PLA), polyethylene glycol (PEG), poly(vinyl alcohol) (PVA), and poly(*N*-vinylpyrrolidone) (PVP) [57–59] or natural polymers, such as chitosan, gellan gum, cellulose, and agar [60–68], are commonly used to fabricate film dressings. In recent years, there has been a growing interest in dressing films based on polymers containing nanoparticles (NPs), which boost films' efficiency in the wound healing process by enhancing their biocompatibility, antibacterial ability, and mechanical properties (Fig. 3). As presented in Table 3, several reports exist on the incorporation of inorganic NPs (silver (AgNPs), copper (CuNPs), titanium dioxide (TiO_2NPs), graphene oxide (GO)) and organic NPs (nanostarch, nanochitosan) into polymeric films, that could find applications as wound dressings for preventing the growth of pathogenic microorganisms. For instance, poly(vinyl alcohol)/GO-AgNPs nanocomposites (PVA/GO-AgNPs) with different GO-AgNPs content (0.5–5 wt%) have been developed [58]. The PVA/GO-AgNPs

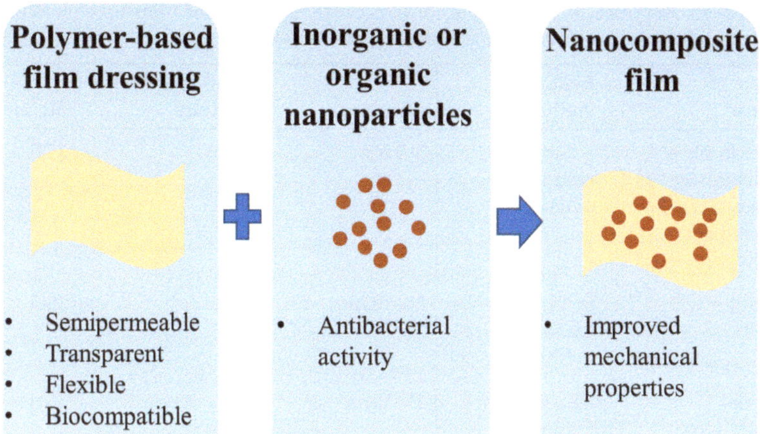

FIG. 3 Schematic illustration showing the main concept of fabrication of nanocomposite films for potential applications as antibacterial wound dressings.

TABLE 3 Representative examples of nanoparticle-embedded films for the control of bacterial infections in wound healing applications.

System	Synthesis method	NP size	Antimicrobial activity	References
PLA/PEG film with AgNPs	Solvent volatilization method	26.2–38.2 nm	*SA* *PA*	[57]
PVA film with AgNPs-coated graphene oxide	Solvent casting method	2.3–3.9 nm	*SA* *EC*	[58]
A quaternized-chitosan film with AgNPs	Solution casting method	93.4–103.8 nm	*SA* *EC* *PA*	[60]
Chitosan/sericin/AgNPs film with moxifloxacin	Solvent casting method	18.39–96.93 nm	*SA* *MRSA* *SE* *PA* *AB*	[61]
Bacterial nanocellulose film with AgNPs	Solution casting method	20–50 nm	*SA* *SE* *PA*	[65]

Continued

TABLE 3 Representative examples of nanoparticle-embedded films for the control of bacterial infections in wound healing applications—cont'd

System	Synthesis method	NP size	Antimicrobial activity	References
Chitosan films with chitosan-capped CuONPs and chitosan-capped CuNPs	Solution casting method	5–9 nm	*EC*	[62]
Agar film with CuNPs	Solution casting method	50–100 nm	*EC*	[68]
Bacterial nanocellulose film with magnetic NPs (Fe_3O_4NPs)	Solution casting method	15–30 nm	*SA* *SE* *PA*	[66]
Gellan gum film with TiO_2NPs	Evaporative casting method	20–26 nm	*SA* *EC*	[63]
Polyurethane film with nanochitosan and TiO_2NPs	Phase inversion technique	Chitosan-NPs (141.2 nm) TiO_2-NPs (<25 nm)	*SA* *PA*	[64]
Graphene oxide/ isophorone diisocyanate with Ag-coated ZnONPs blended with hydroxypropyl cellulose	Solution casting method	Lack of data	*SA* *EC*	[67]
Chitosan/PVP/ stearic acid film with nanostarch	Salt leaching method	6–35 nm	*SA* *BS* *EC* *PA*	[59]

AB, Acinetobacter baumannii; EC, Escherichia coli; MRSA, methicillin-resistant *Staphylococcus aureus; PA, Pseudomonas aeruginosa; SA, Staphylococcus aureus; SE, Staphylococcus epidermidtis.*

nanocomposites displayed an ability to suppress the growth of *S. aureus* and *E. coli* with the strongest antibacterial activity detected for the film with the highest GO-AgNPs content (5 wt%). Rahimi et al. developed quaternized chitosan (with imidazolium functional groups) films loaded with different amounts of AgNPs (0.125, 0.25, and 0.5 wt%) and demonstrated that the addition of AgNPs to the chitosan film enhanced its antibacterial activity against tested microorganisms (*S. aureus*, *E. coli*, *P. aeruginosa*, and *C. albicans*), when compared to the intrinsic antimicrobial activity of pure quaternized chitosan film [60]. Furthermore, the developed AgNPs-containing nanocomposite films displayed a nontoxic character to human skin fibroblasts, the most favorable one being that of the lowest AgNPs content (0.125 wt%).

Shah et al. produced chitosan/sericin/AgNPs nanocomposite films incorporating moxifloxacin, revealing antibacterial activity against *S. aureus*, methicillin-resistant *Staphylococcus aureus* (MRSA), *S. epidermidis*, *P. aeruginosa*, and *A. baumannii* [61]. The developed nanocomposites also revealed remarkable burn wound healing activity in a rat model. In another study, Jayaramudu et al. developed two different kinds of chitosan-based films, incorporated with chitosan-capped CuONPs and chitosan-capped CuNPs, using a solution casting method [62]. The chitosan-capped CuONPs and CuNPs were synthesized by mixing copper sulfate with chitosan as a capping material, through chemical reduction using nucleation agents, such as sodium hydroxide and ascorbic acid. The developed nanocomposite films displayed high antibacterial activity against *E. coli* and *Bacillus*, with stronger effects observed for the chitosan-capped CuONPs-chitosan nanocomposite films. In a different approach, Ismail et al. produced a gellan gum film containing 1 wt% TiO_2NPs (GG + TiO_2NPs) via an evaporative casting technique [63]. Their results revealed that the introduction of TiO_2NPs into a gellan gum film inhibited the growth of *S. aureus* and *E. coli*. More importantly, the authors found that the inhibition zones were comparable to a penicillin control sample. An in vitro study using mouse fibroblasts proved the cytocompatibility of the nanocomposite film, while an in vivo wound healing study, using Sprague Dawley rats, displayed that the GG + TiO_2NPs film accelerated wound healing with no scar formation. Wang et al. fabricated nanocomposite films composed of a hydroxypropyl cellulose matrix and GO/isophorone diisocyanate with Ag-coated ZnONPs, using a solution blending method [67]. Antibacterial activity showed that the developed nanocomposite films inactivated *S. aureus* and *E. coli*, even at the lowest loading of Ag-coated ZnONPs (0.5 wt%). Additionally, the authors reported accelerated wound healing in a mouse model. In a different study, Tang and coworkers loaded CuNPs into an agar film [68]. The developed agar-CuNPs nanocomposite film exhibited excellent toxicity against *E. coli*. Nevertheless, there is a lack of information about the cytotoxicity of such materials against healthy cells.

The use of nanoscale polymer particles for medical applications is a novel trend in wound dressing engineering [64]. For instance, Poonguzhali and coworkers developed asymmetric (with a stearic acid coating on one side of the film) chitosan/PVP films reinforced with various amounts of nanostarch, fabricated by a salt leaching method [59]. Their results revealed that a stearic acid coating on the film (the top layer of the nanocomposite film) provided a highly smooth hydrophobic surface, which was characterized by bacterial antiadhesion properties, which is important for wound dressings as it could prevent wound infection from an external contamination. Meanwhile, the uncoated side of the films (the bottom layer of the nanocomposite film) had a hydrophilic characteristic, possessed noncytotoxicity, and superior antibacterial properties, which was due to the nanostarch content being able to efficiently inhibit bacterial growth on the wound surface. In vivo wound healing analysis showed that the developed chitosan/PVP/nanostarch films promoted wound healing in a rat model. In other work, Jafari et al. reported the incorporation of chitosan NPs (obtained by ionic gelation method) and TiO_2NPs into polyurethane membranes [64]. The obtained polyurethane nanocomposites displayed antibacterial activity against *S. aureus* and no cytotoxicity to human fibroblast cells. Furthermore, the addition of chitosan NPs caused an augmentation of hydrophilicity and biocompatibility of the nanocomposite polyurethane membrane.

Recently, eco-friendly nanocomposite films with antibacterial activity and excellent healing properties were developed as potential wound dressings. For example, Bardania et al. produced a polylactic acid/polyethylene glycol (PLA/PEG) film loaded with AgNPs, which was green synthesized using a *Teucrium polium* extract as a reducing agent [57]. The PLA/PEG/AgNPs film exhibited efficient free radical scavenging capacity and antibacterial activity against *S. aureus* and *P. aeruginosa.* Moreover, the concentration of AgNPs equal to 62.5 μg/mL showed an antibacterial activity with no significant cytotoxicity to a macrophage-like cell line. Moniri et al. reported the in situ green synthesis of AgNPs using a *Citrullus colocynthis* fruit extract in bacterial nanocellulose (BNC) that was obtained from the fermentation of Gram-negative *Gluconacetobacter xylinus* [65]. The resultant BNC/AgNPs nanocomposite films containing 8.0 wt% AgNPs significantly reduced *S. aureus, S. epidermidis,* and *P. aeruginosa* colonization in vitro, while having low cytotoxicity against human dermal fibroblasts in vitro. In another study, Moniri et al. fabricated BNC/magnetic nanoparticles (BNC/Fe_3O_4NPs) nanocomposite films, using an *Aloe vera* extract for the green synthesis of Fe_3O_4NPs [66]. The BNC/Fe_3O_4NPs nanocomposite film containing 8.0 wt% Fe_3O_4NPs showed strong antibacterial activity against *S. aureus, S. epidermidis,* and *P. aeruginosa,* and a lack of cytotoxicity.

Polymer-based nanocomposite films reinforced with inorganic and organic NPs have gained significant interest in the engineering of wound dressings to prevent and control the growth of pathogenic microorganisms. The incorporation of NPs into a polymeric matrix improves the antibacterial efficiency of the nanocomposite films without the use of antibiotics. Nevertheless, in most studies, the antibacterial activity of the developed nanocomposite films was confirmed only by in vitro assays. Thus, the antibacterial efficiency of the nanocomposite dressings in a complex in vivo environment is not evident. Furthermore, it is well known that the antibacterial efficiency of NPs in the nanocomposites depends on the type and amount of NPs, their release rate and possible interactions with other components, the nanocomposite fabrication method, and, finally, the bacterial strain used for analysis. Although the nanocomposite films developed so far have been reported to possess strong antibacterial activity, their applications in medical practice are partly limited by the cytotoxicity of NPs at high concentrations. In turn, nanocomposite films with lower NP content show cytocompatibility but insufficient antibacterial protection. Thus, it is a huge challenge to develop ideal nanocomposite films which would be characterized by both appropriate physical properties (like semipermeability, flexibility, and transparency) and desired biological properties (like high biocompatibility with strong antibacterial activity).

5 Nanocoatings

Nowadays, the large adaptive capacity of bacteria to colonize wounds and the surfaces of biomaterials is a problem of the utmost importance. Due to the increasing occurrence of antibiotic-resistant bacterial strains, the use of nanostructures has appeared as a current strategy to fight the problem of bacterial contamination of wounds and medical implants. A large variety of nanostructures can be produced nowadays in laboratories, using several nanofabrication techniques (such as electronic lithography, ion beam lithography, etc.). However, in order to be used in clinical applications, nanostructures must be manufactured on a large scale, using methods that offer high performance at a low cost. Among the chemical methods, besides the well-known electrochemical routes, green synthesis is gaining attention because the nanostructures are produced in an environmentally friendly manner [69,70]. Among the physical techniques, glancing angle deposition with magnetron sputtering (GLAD-MS) is able to produce nanostructured systems based on nanocolumns, which are cost effective, safe, and environmentally friendly [71–73].

The development of infections occurs in two steps: first, the attachment of bacteria onto the surface, and second, biofilm maturation, which can reach an irreversible stage. So, the best way to prevent infection is to fight bacterial adhesion and interfere in the early stages of biofilm formation. Such a fight can be done following two routes [74,75]. On one hand, using nonfouling surfaces that dramatically decrease the number of bacteria and inhibit their growth. On the other hand, using nanostructures that exhibit biocide behavior, i.e., induce bacterial killing. These two routes are combined in the so-called synergic coatings, or, in other words, nanostructured surfaces that not only prevent the adhesion of bacteria, but also play an active role in killing those few that still attach.

In the initial process of attachment, several properties are involved, such as chemical composition, roughness, and hydrophilicity/hydrophobicity of the surface. Several surface modification studies have been inspired by the wings of the cicada [76] and the dragonfly [77], which are natural nanotextured surfaces onto which bacteria cannot adhere and proliferate. For instance, in a study by Izquierdo-Barba et al., a nanostructured coating of titanium nanocolumns mimicked the highly hydrophobic and antibacterial behavior of the cicada wings [78]. In recent work, Jenkins et al. studied the multifactorial antibacterial effect of a nanotopography closely mimicking the nanoprotrusions found on dragonfly wings, which consisted of titania nanopillars fabricated by anodization [79].

In recent years, the antibacterial action of several NPs have been reported, especially silver [69,80], copper [81], and zinc [82]. The main mechanisms involved in their biocide effects are the enhanced production of reactive oxygen species (ROS) and metallic ion release [83]. ROS are oxygen-derived species (superoxide, hydrogen peroxide, etc.) that are naturally generated in the process of cell metabolism. MNPs are able to induce the overproduction of such species in their vicinity, which leads to several mechanisms: membrane disruption, oxidative stress and, in some cases, even a DNA damage. As a result, bacterial cell death is achieved. On the other hand, the MNPs are usually composed of a pure metallic core and an outer oxide shell. The metal ion of the shell can also be released and cause irreparable cell damage that triggers bacterial cell death. In such coatings, obtaining a perfect balance between bacteria-killing efficiency and low toxicity toward healthy cells is a challenge of the utmost importance.

Regarding nanostructured synergic systems, different routes have been explored. For instance, surfaces possessing nanocontainers that release antimicrobial drugs is the approach followed by Auñón and coworkers [84], who prepared a nanostructured Ti-based alloy, loaded with gentamicin and vancomycin. Another possibility is the preparation of coatings impregnated with antimicrobial nanoparticles that cause contact killing. This is the case of the nanocolumnar titanium surfaces with tellurium nanorods on top, developed by Medina et al. [81]. All of these approaches can be used as novel nanocoatings on traditional medical devices.

6 Conclusion and future directions

Open abrasions or wounds are extremely susceptible to bacterial infection during wound healing. There are four steps to the wound healing process: hemostasis, inflammation, proliferation, and maturation. During the first two steps, blood clotting is activated and platelets are aggregated at the wound site. Next, neutrophils enter the wound to destroy bacterial infections. However, the immune system can become overwhelmed, thus treatments need to be administered. Administering antibiotics could result in bacteria developing antibiotic resistance, and so it is imperative to develop new methods for facilitating wound healing. Recently, nanomaterials have been suggested as alternatives in the treatment of bacterial infection while also supporting for the wound healing process. These mainly include MNPs, carbon nanomaterials, polymeric NPs, and lipidic NPs, used either as coatings or as matrix modifiers of scaffolds.

NM-modified scaffolds for wound healing applications can be divided by their chemical structure/morphology into three main categories: hydrogels, fibers, and films. In hydrogels, a crosslinking with nanoscale structures can be employed. Still, controlled antimicrobial activity and long-term noncytotoxicity toward healthy cells can be challenging. Nanofibers have also been developed for treating the chronic wounds. These NMs have a high surface area-to-volume ratio, can be easily functionalized, and have high mechanical properties, characteristics which significantly reduce the risk of pathogens developing immunity to various treatments. On the other hand, nanofilms possess important characteristics, such as flexibility and transparency, which allow them to be molded to the patient's body while ensuring feasible inspection of the wound. However, these films are limited by their cytotoxicity due to large concentrations of NPs necessary to achieve ideal antimicrobial protections.

In conclusion, there are various new methods for treating and preventing microbial infection that seem promising, but require further evaluation before they could be commercially applied. Furthermore, their production technologies need to be developed to a higher extent, so that good quality and satisfactory repeatability can be reached. Efforts should also be made to reduce the use of toxic solvents and crosslinkers. This review discussed several promising solutions from recent years, while indicating the necessity for further investigation.

Acknowledgments

S.W. would like to acknowledge financial support from the National Centre for Research and Development under grant no. LIDER/7/0020/L-11/19/NCBR/2020. A.P. and P.K. would like to acknowledge the National Science Centre (NCN) in Poland for support (grant no. UMO-2018/31/B/ST8/00945). K. R-K. and E.P. would like to acknowledge financial support from the National Science Center, Poland, under grants no. 2018/29/N/ST5/01543 and 2019/35/B/ST5/01103, the Program "Excellence Initiative–Research University" for the AGH University of Science and Technology, Kraków, Poland, and the subsidy of the Ministry of Education and Science for the AGH University of Science and Technology in Kraków (Project No 16.16.160.557). A. B. would like to acknowledge financial support from the National Science Center, Poland, under grant no. UMO-2017/24/C/ST8/00400.

References

[1] H. Naderi, M.M. Matin, A.R. Bahrami, Review paper: Critical issues in tissue engineering: biomaterials, cell sources, angiogenesis, and drug delivery systems, J. Biomater. Appl. 26 (2011) 383–417, https://doi.org/10.1177/0885328211408946.
[2] H. Sinno, S. Prakash, Complements and the wound healing cascade: an updated review, Plast. Surg. Int. (2013) 146764, https://doi.org/10.1155/2013/146764.
[3] T. Velnar, T. Bailey, V. Smrkolj, The wound healing process: an overview of the cellular and molecular mechanisms, J. Int. Med. Res. 37 (2009) 1528–1542, https://doi.org/10.1177/147323000903700531.
[4] K.A. Radek, E.J. Kovacs, R.L. Gallo, L.A. DiPietro, Acute ethanol exposure disrupts VEGF receptor cell signaling in endothelial cells, Am. J. Physiol. Heart Circ. Physiol. 295 (2008), https://doi.org/10.1152/AJPHEART.00699.2007.
[5] H. Brem, M. Tomic-Canic, Cellular and molecular basis of wound healing in diabetes, J. Clin. Invest. 117 (2007) 1219–1222, https://doi.org/10.1172/JCI32169.
[6] A. Gosain, L.A. DiPietro, Aging and wound healing, World J. Surg. 28 (2004) 321–326, https://doi.org/10.1007/S00268-003-7397-6.
[7] S. Guo, L.A. DiPietro, Factors affecting wound healing, J. Dent. Res. 89 (2010) 219, https://doi.org/10.1177/0022034509359125.
[8] A. Bishop, Role of oxygen in wound healing, J. Wound Care 17 (2008) 399–402, https://doi.org/10.12968/JOWC.2008.17.9.30937.
[9] C.K. Field, M.D. Kerstein, Overview of wound healing in a moist environment, Am. J. Surg. 167 (1994) S2–S6, https://doi.org/10.1016/0002-9610(94)90002-7.
[10] R. Edwards, K.G. Harding, Bacteria and wound healing, Curr. Opin. Infect. Dis. 17 (2004) 91–96, https://doi.org/10.1097/00001432-200404000-00004.
[11] N.B. Menke, K.R. Ward, T.M. Witten, D.G. Bonchev, R.F. Diegelmann, Impaired wound healing, Clin. Dermatol. 25 (2007) 19–25, https://doi.org/10.1016/J.CLINDERMATOL.2006.12.005.
[12] L. Lancerotto, D.P. Orgill, Mechanoregulation of angiogenesis in wound healing, Adv. Wound Care 3 (2014) 626–634, https://doi.org/10.1089/wound.2013.0491.
[13] S. Luebberding, N. Krueger, M. Kerscher, Mechanical properties of human skin in vivo: a comparative evaluation in 300 men and women, Skin Res. Technol. 20 (2014) 127–135, https://doi.org/10.1111/SRT.12094.
[14] H. Zahouani, C. Pailler-Mattei, B. Sohm, R. Vargiolu, V. Cenizo, R. Debret, Characterization of the mechanical properties of a dermal equivalent compared with human skin in vivo by indentation and static friction tests, Skin Res. Technol. 15 (2009) 68–76, https://doi.org/10.1111/J.1600-0846.2008.00329.X.
[15] S. Guo, X. Zhu, M. Li, L. Shi, J.L.T. Ong, D. Jańczewski, K.G. Neoh, Parallel control over surface charge and wettability using polyelectrolyte architecture: effect on protein adsorption and cell adhesion, ACS Appl. Mater. Interfaces 8 (2016) 30552–30563, https://doi.org/10.1021/ACSAMI.6B09481.
[16] R. Xu, H. Xia, W. He, Z. Li, J. Zhao, B. Liu, Y. Wang, Q. Lei, Y. Kong, Y. Bai, Z. Yao, R. Yan, H. Li, R. Zhan, S. Yang, G. Luo, J. Wu, Controlled water vapor transmission rate promotes wound-healing via wound re-epithelialization and contraction enhancement, Sci. Rep. 6 (1) (2016) 1–12, https://doi.org/10.1038/srep24596.
[17] T.M. Freyman, I.V. Yannas, L.J. Gibson, Cellular materials as porous scaffolds for tissue engineering, Prog. Mater. Sci. 46 (2001) 273–282, https://doi.org/10.1016/S0079-6425(00)00018-9.
[18] M.J. Whitaker, R.A. Quirk, S.M. Howdle, K.M. Shakesheff, Growth factor release from tissue engineering scaffolds, J. Pharm. Pharmacol. 53 (2001) 1427–1437, https://doi.org/10.1211/0022357011777963.
[19] O. Timaeva, I. Pashkin, S. Mulakov, G. Kuzmicheva, P. Konarev, R. Terekhova, N. Sadovskaya, O. Czakkel, S. Prevost, Synthesis and physico-chemical properties of

poly(N-vinyl pyrrolidone)-based hydrogels with titania nanoparticles, J. Mater. Sci. 55 (2020) 3005–3021, https://doi.org/10.1007/s10853-019-04230-z.

[20] S. Rafieian, H. Mirzadeh, H. Mahdavi, M.E. Masoumi, A review on nanocomposite hydrogels and their biomedical applications, IEEE J. Sel. Top. Quantum Electron. 26 (2019) 154–174, https://doi.org/10.1515/secm-2017-0161.

[21] L. Atkin, Chronic wounds: the challenges of appropriate management, Br. J. Community Nurs. 24 (2019) S26–S32, https://doi.org/10.12968/bjcn.2019.24.Sup9.S26.

[22] V. Vivcharenko, M. Wojcik, A. Przekora, Cellular response to vitamin C-enriched chitosan/agarose film with potential application as artificial skin substitute for chronic wound treatment, Cells 9 (2020) 1185, https://doi.org/10.3390/cells9051185.

[23] T. Dai, C. Wang, Y. Wang, W. Xu, J. Hu, Y. Cheng, A nanocomposite hydrogel with potent and broad-spectrum antibacterial activity, ACS Appl. Mater. Interfaces 10 (2018) 15163–15173, https://doi.org/10.1021/acsami.8b02527.

[24] J.U. Shin, J. Gwon, S.Y. Lee, H.S. Yoo, Silver-incorporated nanocellulose fibers for antibacterial hydrogels, ACS Omega 3 (2018) 16150–16157, https://doi.org/10.1021/acsomega.8b02180.

[25] M. Bartel, K. Markowska, M. Strawski, K. Wolska, M. Mazur, Silver-decorated gel-shell nanobeads: physicochemical characterization and evaluation of antibacterial properties, Beilstein J. Nanotechnol. 11 (2020) 620–630, https://doi.org/10.3762/bjnano.11.49.

[26] A.K. Kar, A. Singh, N. Dhiman, M.P. Purohit, P. Jagdale, M. Kamthan, D. Singh, M. Kumar, D. Ghosh, S. Patnaik, Polymer-assisted in situ synthesis of silver nanoparticles with epigallocatechin gallate (EGCG) impregnated wound patch potentiate controlled inflammatory responses for brisk wound healing, Int. J. Nanomedicine 14 (2019) 9837–9854, https://doi.org/10.2147/IJN.S228462.

[27] Y. Wang, C. Dou, G. He, L. Ban, L. Huang, Z. Li, J. Gong, J. Zhang, P. Yu, Biomedical potential of ultrafine Ag nanoparticles coated on poly (gamma-glutamic acid) hydrogel with special reference to wound healing, Nanomaterials 8 (2018) 324, https://doi.org/10.3390/nano8050324.

[28] Y. Li, T. Xu, Z. Tu, W. Dai, Y. Xue, C. Tang, W. Gao, C. Mao, B. Lei, C. Lin, Bioactive antibacterial silica-based nanocomposites hydrogel scaffolds with high angiogenesis for promoting diabetic wound healing and skin repair, Theranostics 10 (2020) 4929–4943, https://doi.org/10.7150/thno.41839.

[29] R.K. Farag, A. Labena, S.H. Fakhry, G. Safwat, A. Diab, A.M. Atta, Antimicrobial activity of hybrids terpolymers based on magnetite hydrogel nanocomposites, Materials (Basel) 12 (2019) 3604, https://doi.org/10.3390/ma12213604.

[30] R. Arshad, M.F. Sohail, H.S. Sarwar, H. Saeed, I. Ali, S. Akhtar, S.Z. Hussain, I. Afzal, S. Jahan, Anees-Ur-Rehman, G. Shahnaz, ZnO-NPs embedded biodegradable thiolated bandage for postoperative surgical site infection: in vitro and in vivo evaluation, PLoS One 14 (2019) 1–18, https://doi.org/10.1371/journal.pone.0217079.

[31] X. Zhao, B. Guo, H. Wu, Y. Liang, P.X. Ma, Injectable antibacterial conductive nanocomposite cryogels with rapid shape recovery for noncompressible hemorrhage and wound healing, Nat. Commun. 9 (2018) 2784, https://doi.org/10.1038/s41467-018-04998-9.

[32] L.H. Dang, T.H. Nguyen, H.L.B. Tran, V.N. Doan, N.Q. Tran, Injectable nanocurcumin-formulated chitosan-g-pluronic hydrogel exhibiting a great potential for burn treatment, J. Healthc. Eng. 2018 (2018) 5754890, https://doi.org/10.1155/2018/5754890.

[33] A. Mohandas, S. Deepthi, R. Biswas, R. Jayakumar, Chitosan based metallic nanocomposite scaffolds as antimicrobial wound dressings, Bioact. Mater. 3 (2018) 267–277, https://doi.org/10.1016/j.bioactmat.2017.11.003.

[34] Y. Zhang, C.T. Lim, S. Ramakrishna, Z.-M. Huang, Recent development of polymer nanofibers for biomedical and biotechnological applications, J. Mater. Sci. Mater. Med. 16 (2005) 933–946, https://doi.org/10.1007/s10856-005-4428-x.

[35] J. Du, D. Liu, S. Chen, D. Wan, H. Pu, A novel method for fabricating continuous polymer nanofibers, Polymer 102 (2016) 209–213, https://doi.org/10.1016/j.polymer.2016.09.018.
[36] K. Kalantari, A.M. Afifi, H. Jahangirian, T.J. Webster, Biomedical applications of chitosan electrospun nanofibers as a green polymer—review, Carbohydr. Polym. (2019), https://doi.org/10.1016/j.carbpol.2018.12.011.
[37] S. Fujita, Y. Wakuda, M. Matsumura, S. Suye, Geometrically customizable alginate hydrogel nanofibers for cell culture platforms, J. Mater. Chem. B 7 (2019) 6556–6563, https://doi.org/10.1039/C9TB01353A.
[38] R.S. Ambekar, B. Kandasubramanian, Advancements in nanofibers for wound dressing: a review, Eur. Polym. J. 117 (2019) 304–336, https://doi.org/10.1016/j.eurpolymj.2019.05.020.
[39] S. Homaeigohar, A.R. Boccaccini, Antibacterial biohybrid nanofibers for wound dressings, Acta Biomater. 107 (2020) 25–49, https://doi.org/10.1016/j.actbio.2020.02.022.
[40] X. Wang, B. Ding, B. Li, Biomimetic electrospun nanofibrous structures for tissue engineering, Mater. Today 16 (2013) 229–241, https://doi.org/10.1016/j.mattod.2013.06.005.
[41] P.X. Ma, Biomimetic materials for tissue engineering, Adv. Drug Deliv. Rev. 60 (2008) 184–198, https://doi.org/10.1016/j.addr.2007.08.041.
[42] E. Stodolak-Zych, Materiały dla inżynierii tkankowej i medycyny regeneracyjnej, in: S. Błażewicz, J. Marciniak (Eds.), Inżynieria Biomed. Pod. i Zastos. Tom 4 Biomateriały, 2016th ed., Akademicka Oficyna Wydawnicza EXIT, Warszawa, 2016, pp. 733–785.
[43] M. Stevens, J. George, Exploring and engineering the cell-surface interface, Science 310 (2005) 1135–1138, https://doi.org/10.1126/science.1106587.
[44] S.P. Miguel, D.R. Figueira, D. Simões, M.P. Ribeiro, P. Coutinho, P. Ferreira, I.J. Correia, Electrospun polymeric nanofibres as wound dressings: a review, Colloids Surf. B Biointerfaces 169 (2018) 60–71, https://doi.org/10.1016/j.colsurfb.2018.05.011.
[45] G. Ajmal, G.V. Bonde, P. Mittal, G. Khan, V.K. Pandey, B.V. Bakade, B. Mishra, Biomimetic PCL-gelatin based nanofibers loaded with ciprofloxacin hydrochloride and quercetin: a potential antibacterial and anti-oxidant dressing material for accelerated healing of a full thickness wound, Int. J. Pharm. 567 (2019) 118480, https://doi.org/10.1016/j.ijpharm.2019.118480.
[46] S. Abid, T. Hussain, A. Nazir, A. Zahir, S. Ramakrishna, M. Hameed, N. Khenoussi, Enhanced antibacterial activity of PEO-chitosan nanofibers with potential application in burn infection management, Int. J. Biol. Macromol. 135 (2019) 1222–1236, https://doi.org/10.1016/j.ijbiomac.2019.06.022.
[47] H.R. Bakhsheshi-Rad, Z. Hadisi, A.F. Ismail, M. Aziz, M. Akbari, F. Berto, X.B. Chen, In vitro and in vivo evaluation of chitosan-alginate/gentamicin wound dressing nanofibrous with high antibacterial performance, Polym. Test. 82 (2020) 106298, https://doi.org/10.1016/j.polymertesting.2019.106298.
[48] S. Ullah, M. Hashmi, M.Q. Khan, D. Kharaghani, Y. Saito, T. Yamamoto, I.S. Kim, Silver sulfadiazine loaded zein nanofiber mats as a novel wound dressing, RSC Adv. 9 (2019) 268–277.
[49] M.Q. Khan, D. Kharaghani, Sanaullah, A. Shahzad, Y. Saito, T. Yamamoto, H. Ogasawara, I.S. Kim, Fabrication of antibacterial electrospun cellulose acetate/silver-sulfadiazine nanofibers composites for wound dressings applications, Polym. Test. 74 (2019) 39–44, https://doi.org/10.1016/j.polymertesting.2018.12.015.
[50] R. Ahmed, M. Tariq, I. Ali, R. Asghar, P. Noorunnisa Khanam, R. Augustine, A. Hasan, Novel electrospun chitosan/polyvinyl alcohol/zinc oxide nanofibrous mats with antibacterial and antioxidant properties for diabetic wound healing, Int. J. Biol. Macromol. 120 (2018) 385–393, https://doi.org/10.1016/j.ijbiomac.2018.08.057.

[51] A.W. Jatoi, Polyurethane nanofibers incorporated with ZnAg composite nanoparticles for antibacterial wound dressing applications, Compos. Commun. 19 (2020) 103–107, https://doi.org/10.1016/j.coco.2020.03.004.
[52] Y. Tang, X. Lan, C. Liang, Z. Zhong, R. Xie, Y. Zhou, X. Miao, H. Wang, W. Wang, Honey loaded alginate/PVA nanofibrous membrane as potential bioactive wound dressing, Carbohydr. Polym. 219 (2019) 113–120, https://doi.org/10.1016/j.carbpol.2019.05.004.
[53] S.P. Miguel, D. Simões, A.F. Moreira, R.S. Sequeira, I.J. Correia, Production and characterization of electrospun silk fibroin based asymmetric membranes for wound dressing applications, Int. J. Biol. Macromol. 121 (2019) 524–535, https://doi.org/10.1016/j.ijbiomac.2018.10.041.
[54] Z. Li, A. Milionis, Y. Zheng, M. Yee, L. Codispoti, F. Tan, D. Poulikakos, C.H. Yap, Superhydrophobic hemostatic nanofiber composites for fast clotting and minimal adhesion, Nat. Commun. 10 (2019) 5562, https://doi.org/10.1038/s41467-019-13512-8.
[55] C.D. Weller, V. Team, G. Sussman, First-line interactive wound dressing update: a comprehensive review of the evidence, Front. Pharmacol. 11 (2020) 1–13, https://doi.org/10.3389/fphar.2020.00155.
[56] K. Vowden, P. Vowden, Wound dressings: principles and practice, Surgery (Oxford) 35 (2017) 489–494, https://doi.org/10.1016/j.mpsur.2017.06.005.
[57] H. Bardania, R. Mahmoudi, H. Bagheri, Z. Salehpour, M.H. Fouani, B. Darabian, S.S. Khoramrooz, A. Mousavizadeh, M. Kowsari, S.E. Moosavifard, G. Christiansen, D. Javeshghani, M. Alipour, M. Akrami, Facile preparation of a novel biogenic silver-loaded Nanofilm with intrinsic anti-bacterial and oxidant scavenging activities for wound healing, Sci. Rep. 10 (2020) 1–14, https://doi.org/10.1038/s41598-020-63032-5.
[58] M. Cobos, I. De-La-Pinta, G. Quindós, M.J. Fernández, M.D. Fernández, Synthesis, physical, mechanical and antibacterial properties of nanocomposites based on poly(vinyl alcohol)/graphene oxide–silver nanoparticles, Polymers (Basel) 12 (2020), https://doi.org/10.3390/polym12030723.
[59] R. Poonguzhali, S. Khaleel Basha, V. Sugantha Kumari, Fabrication of asymmetric nanostarch reinforced Chitosan/PVP membrane and its evaluation as an antibacterial patch for in vivo wound healing application, Int. J. Biol. Macromol. 114 (2018) 204–213, https://doi.org/10.1016/j.ijbiomac.2018.03.092.
[60] M. Rahimi, R. Ahmadi, H. Samadi Kafil, V. Shafiei-Irannejad, A novel bioactive quaternized chitosan and its silver-containing nanocomposites as a potent antimicrobial wound dressing: structural and biological properties, Mater. Sci. Eng. C 101 (2019) 360–369, https://doi.org/10.1016/j.msec.2019.03.092.
[61] A. Shah, M. Ali Buabeid, E.S.A. Arafa, I. Hussain, L. Li, G. Murtaza, The wound healing and antibacterial potential of triple-component nanocomposite (chitosan-silver-sericin) films loaded with moxifloxacin, Int. J. Pharm. (2019), https://doi.org/10.1016/j.ijpharm.2019.04.046.
[62] T. Jayaramudu, K. Varaprasad, R.D. Pyarasani, K.K. Reddy, K.D. Kumar, A. Akbari-Fakhrabadi, R.V. Mangalaraja, J. Amalraj, Chitosan capped copper oxide/copper nanoparticles encapsulated microbial resistant nanocomposite films, Int. J. Biol. Macromol. 128 (2019) 499–508, https://doi.org/10.1016/j.ijbiomac.2019.01.145.
[63] N.A. Ismail, K.A.M. Amin, F.A.A. Majid, M.H. Razali, Gellan gum incorporating titanium dioxide nanoparticles biofilm as wound dressing: physicochemical, mechanical, antibacterial properties and wound healing studies, Mater. Sci. Eng. C 103 (2019) 109770, https://doi.org/10.1016/j.msec.2019.109770.
[64] A. Jafari, S. Hassanajili, M.B. Karimi, A. Emami, F. Ghaffari, N. Azarpira, Effect of organic/inorganic nanoparticles on performance of polyurethane nanocomposites for potential wound dressing applications, J. Mech. Behav. Biomed. Mater. 88 (2018) 395–405, https://doi.org/10.1016/j.jmbbm.2018.09.001.

[65] M. Moniri, A.B. Moghaddam, S. Azizi, R.A. Rahim, S.W. Zuhainis, M. Navaderi, R. Mohamad, In vitro molecular study of wound healing using biosynthesized bacteria nanocellulose/silver nanocomposite assisted by bioinformatics databases, Int. J. Nanomedicine 13 (2018) 5097–5112, https://doi.org/10.2147/IJN.S164573.
[66] M. Moniri, A. Boroumand Moghaddam, S. Azizi, R. Abdul Rahim, W. Zuhainis Saad, M. Navaderi, P. Arulselvan, R. Mohamad, Molecular study of wound healing after using biosynthesized BNC/Fe_3O_4 nanocomposites assisted with a bioinformatics approach, Int. J. Nanomedicine 13 (2018) 2955–2971, https://doi.org/10.2147/IJN.S159637.
[67] Y. Wang, L. Shi, H. Wu, Q. Li, W. Hu, Z. Zhang, L. Huang, J. Zhang, D. Chen, S. Deng, S. Tan, Z. Jiang, Graphene oxide-IPDI-Ag/ZnO@hydroxypropyl cellulose nanocomposite films for biological wound-dressing applications, ACS Omega 4 (2019) 15373–15381, https://doi.org/10.1021/acsomega.9b01291.
[68] L. Tang, L. Zhu, F. Tang, C. Yao, J. Wang, L. Li, Mild synthesis of copper nanoparticles with enhanced oxidative stability and their application in antibacterial films, Langmuir 34 (2018) 14570–14576, https://doi.org/10.1021/acs.langmuir.8b02470.
[69] G. Chinnasamy, S. Chandrasekharan, S. Bhatnagar, Biosynthesis of silver nanoparticles from *Melia azedarach*: enhancement of antibacterial, wound healing, antidiabetic and antioxidant activities, Int. J. Nanomedicine 14 (2019) 9823–9836, https://doi.org/10.2147/IJN.S231340.
[70] D. Medina Cruz, W. Tien-Street, B. Zhang, X. Huang, A. Vernet Crua, A. Nieto-Argüello, J.L. Cholula-Díaz, L. Martínez, Y. Huttel, M.U. González, J.M. García-Martín, T.J. Webster, Citric juice-mediated synthesis of tellurium nanoparticles with antimicrobial and anticancer properties, Green Chem. 21 (2019) 1982–1998, https://doi.org/10.1039/c9gc00131j.
[71] R. Alvarez, S. Muñoz-Piña, M.U. González, I. Izquierdo-Barba, I. Fernández-Martínez, V. Rico, D. Arcos, A. García-Valenzuela, A. Palmero, M. Vallet-Regi, A.R. González-Elipe, J.M. García-Martín, Antibacterial nanostructured Ti coatings by magnetron sputtering: from laboratory scales to industrial reactors, Nanomaterials 9 (2019), https://doi.org/10.3390/nano9091217.
[72] N. Ziegler, C. Sengstock, V. Mai, T.A. Schildhauer, M. Köller, A. Ludwig, Glancing-angle deposition of nanostructures on an implant material surface, Nanomaterials 9 (2019), https://doi.org/10.3390/nano9010060.
[73] K.A. Prosolov, M.A. Khimich, J.V. Rau, D.V. Lychagin, Y.P. Sharkeev, Influence of oblique angle deposition on Cu-substituted hydroxyapatite nano-roughness and morphology, Surf. Coat. Technol. 394 (2020) 125883, https://doi.org/10.1016/j.surfcoat.2020.125883.
[74] M. Vallet-Regí, B. González, I. Izquierdo-Barba, Nanomaterials as promising alternative in the infection treatment, Int. J. Mol. Sci. 20 (2019), https://doi.org/10.3390/ijms20153806.
[75] J. Sjollema, S.A.J. Zaat, V. Fontaine, M. Ramstedt, R. Luginbuehl, K. Thevissen, J. Li, H.C. van der Mei, H.J. Busscher, In vitro methods for the evaluation of antimicrobial surface designs, Acta Biomater. 70 (2018) 12–24, https://doi.org/10.1016/j.actbio.2018.02.001.
[76] E.P. Ivanova, J. Hasan, H.K. Webb, V.K. Truong, G.S. Watson, J.A. Watson, V.A. Baulin, S. Pogodin, J.Y. Wang, M.J. Tobin, C. Löbbe, R.J. Crawford, Natural bactericidal surfaces: mechanical rupture of *pseudomonas aeruginosa* cells by cicada wings, Small 8 (2012) 2489–2494, https://doi.org/10.1002/smll.201200528.
[77] C.D. Bandara, G. Ballerin, M. Leppänen, T. Tesfamichael, K.(.K.). Ostrikov, C.B. Whitchurch, Resolving bio-nano interactions of E. coli bacteria-dragonfly wing interface with helium ion and 3D-structured illumination microscopy to understand bacterial death on nanotopography, ACS Biomater Sci. Eng. (2020), https://doi.org/10.1021/acsbiomaterials.9b01973.

[78] I. Izquierdo-Barba, J.M. García-Martín, R. Álvarez, A. Palmero, J. Esteban, C. Pérez-Jorge, D. Arcos, M. Vallet-Regí, Nanocolumnar coatings with selective behavior towards osteoblast and *Staphylococcus aureus* proliferation, Acta Biomater. 15 (2015) 20–28, https://doi.org/10.1016/j.actbio.2014.12.023.
[79] J. Jenkins, J. Mantell, C. Neal, A. Gholinia, P. Verkade, A.H. Nobbs, B. Su, Antibacterial effects of nanopillar surfaces are mediated by cell impedance, penetration and induction of oxidative stress, Nat. Commun. 11 (2020) 1–14, https://doi.org/10.1038/s41467-020-15471-x.
[80] H. Xu, F. Qu, H. Xu, W. Lai, Y.A. Wang, Z.P. Aguilar, H. Wei, Role of reactive oxygen species in the antibacterial mechanism of silver nanoparticles on *Escherichia coli* O157: H7, Biometals 25 (2012) 45–53, https://doi.org/10.1007/s10534-011-9482-x.
[81] D. Medina-Cruz, M.U. González, W. Tien-Street, M. Fernández-Castro, A. Vernet-Crua, I. Fernández-Martínez, L. Martínez, Y. Huttel, T.J. Webster, J.M. García-Martín, Synergic antibacterial coatings combining titanium nanocolumns and tellurium nanorods, Nanomed. Nanotechnol. Biol. Med. 17 (2019) 36–46, https://doi.org/10.1016/j.nano.2018.12.009.
[82] W. Salem, D.R. Leitner, F.G. Zingl, G. Schratter, R. Prassl, W. Goessler, J. Reidl, S. Schild, Antibacterial activity of silver and zinc nanoparticles against *Vibrio cholerae* and enterotoxic *Escherichia coli*, Int. J. Med. Microbiol. 305 (2015) 85–95, https://doi.org/10.1016/j.ijmm.2014.11.005.
[83] L. Wang, C. Hu, L. Shao, The antimicrobial activity of nanoparticles: present situation and prospects for the future, Int. J. Nanomedicine 12 (2017) 1227–1249, https://doi.org/10.2147/IJN.S121956.
[84] Á. Auñón, J. Esteban, A.L. Doadrio, M. Boiza-Sánchez, A. Mediero, D. Eguibar-Blázquez, J. Cordero-Ampuero, A. Conde, M.Á. Arenas, J.J. de-Damborenea, J.J. Aguilera-Correa, *Staphylococcus aureus* prosthetic joint infection is prevented by a fluorine- and phosphorus-doped nanostructured Ti–6Al–4V alloy loaded with gentamicin and vancomycin, J. Orthop. Res. 38 (2020) 588–597, https://doi.org/10.1002/jor.24496.

CHAPTER

4

Antiviral potential of nanomaterials: The fight against viruses

Walid Azab

Institute of Virology, Center for Infection Medicine, Freie Universitaet Berlin, Berlin, Germany

1 Introduction

Infectious viral diseases represent the greatest threat to humankind. The recent several decades have witnessed the emergence of new viral infections that had a great impact on lifestyle globally. Respiratory viral infections, viral hemorrhagic fever infections, arthropod-borne viral infections, and bat-borne viral infections represent the major categories of emerging viral infections around the world. Several examples include the pandemic severe acute respiratory syndrome SARS-CoV-1 and SARS-CoV-2, influenza viruses, dengue, chikungunya, Lassa virus, Zika virus, Hantavirus, Nipah viral disease, and Ebola virus [1–3]. The great diversity of viruses poses a prominent barrier in designing and developing antiviral compounds that fight against viral infection across different virus families. Understanding the morphology of viruses, the nature of virus–cell interaction, virus pathogenesis and replication will help to develop effective and broad-spectrum antiviral agents that can target different steps of the replication cycle. Although antiviral drugs are currently available, the continuous emergence of resistant strains threatens their regular use [4]. Consequently, there is a need for the development of new antiviral agents that keep up with the newly emerged viruses.

The recent advancement in nanotechnology could provide the tools needed to meet challenges associated with viral infections and combat the newly emerging ones. Nanotechnology can significantly improve

Antimicrobial Activity of Nanoparticles
https://doi.org/10.1016/B978-0-12-821637-8.00005-5

Copyright © 2023 Elsevier Inc. All rights reserved.

the design of antiviral therapeutics, modern vaccines, and drug delivery systems. In general, nanoparticles possess unique physical, chemical, and biological features, which can be ascribed to their nanoscale size and high surface-to-volume ratio [5]. Several platforms of nanoarchitectures have been investigated so far. These include metal nanoparticles such as gold, silver, copper, zinc, diamond, and platinum; organic nanoparticles such as nanogels, liposomes, and polymeric nanoparticles; and carbon-based nanoparticles such as graphene, carbon dots, and fullerenes [6–8]. Due their high surface-to-volume ratio, nanoparticles can be decorated with functional groups, polymers, and biomolecules that can be tailored based on the purpose of application [7]. The antiviral mechanisms of most nanoparticles are likely through blocking virus interaction with cell surface. Furthermore, nanoparticles can lead to the generation of reactive oxygen species (ROS) that induce oxidation of viral proteins or degradation of the viral genome and further stimulate the production of interferon (IFN)-stimulating genes (ISGs) and proinflammatory cytokines [9,10].

This chapter will provide an overview of the application of nanotechnology to develop antiviral therapeutics. The chapter will focus on the major and important nanomaterial platforms—and their mechanism of actions—that are investigated against various types of virus infections.

2 Metallic nanoparticles

Metal nanoparticles (NPs) have long been known and used as antimicrobial agents even before the discovery of antibiotics [11]. Currently, the progress in employing metal NPs as antimicrobial, particularly antiviral, agents has advanced rapidly owing to the fast growing technology of nanotherapeutics, the ability of metals to exert a multitarget effect on viruses, the minimal impact on the subsequent development of pathogen resistance, the broad-spectrum activity against pathogens, the adjustable physical and chemical properties, and the low cytotoxicity effect. Several metal NPs have been investigated for their role as antiviral agents, including silver, gold, copper, zinc, magnesium, titanium, diamond, selenium, tungsten, palladium, platinum, and iron. However, silver and gold nanoparticles are the ones that have received particular attention with intensive investigations [10,12].

The efficacy and activity of metal NPs as antiviral agents depend on several factors that include the size, shape, metal ions, zeta potential, and the downstream modifications. Particle size (ranging from 1 to 100 nm) is considered the most important factor influencing antiviral activities, cellular uptake, and accumulation capabilities in tissues. It was clear that the biological activities (i.e., antimicrobial and anticancer) of NPs significantly increase with decreasing the size of particles [7,13].

Nanoparticles ranging from 30 to 50 nm in size can better interact with cell surface receptors and subsequently up taken by the cell through receptor-mediated endocytosis [14]. Besides, the reduced particle size can be administered through intravenous route without any blockade of the blood vessels [15]. In addition to particle size, shape, composition, and surface charge have also an influence on biological activities, cellular uptake, and accumulation. For example, rod-shaped and positively charged NPs tend to accumulate more than disc, spherical, cylinder, and cubed-shaped particles or neutral and negatively charged NPs [10,16]. However, negatively charged NPs have shown more potent antiviral effect than the others [17].

Several mechanisms of action for metal NPs were identified either inside or outside the host cells. These include (i) interacting with virus particles with subsequent inhibition of virus attachment and entry into the cells, (ii) competing with viruses for host cell binding sites, (iii) inactivating (destroying) virus particles before entering the cells, (iv) interacting with viral genome and inhibition of replication, (v) generating reactive oxygen species (ROS) and subsequent capsid protein oxidation, (vi) generating hydroxyl radicals and degradation of viral proteins, and (vii) interacting with and inhibiting cellular factors needed for virus replication [12,13,17,18].

2.1 Gold NPs

Gold nanoparticles (GNPs) are broadly studied as antiviral agents and nanocarriers due to their excellent conductivity, biocompatibility, flexibility of surface modification, and affordable preparation methods [19]. Furthermore, GNPs have unique physical and chemical properties due to the gold core, which is inert and nontoxic, photophysical properties that can facilitate efficient drug release, and adaptable functionalization via thiol linkages [20,21]. GNPs can be prepared with different diameters ranging from 1 to 150 nm based on the application [7,22]. In a way to overcome the possible cytotoxicity of GNPs, Lysenko and coworkers developed different types of GNPs covered with silicon dioxide (SiO2) shell which is known of its high biocompatibility and biodistribution and complexed GNPs on these SiO2 nanoparticles—as a carrier—against human adenoviruses (HAdV-5). It was shown that both types of nanoparticles demonstrated good antiviral activities in cell culture. The complex GNPs on SiO2 nanoparticles had the strongest antiviral effect with no accumulation in living cells or cytotoxicity [23,24].

Middle East respiratory syndrome coronavirus (MERS-CoV) is among other coronaviruses that causes a severe acute respiratory syndrome-like illness with high mortality rates. The virus spike protein (S1 and S2) is essential for receptor binding, fusion with cell membrane, and subsequent entry. The S2 protein consists of three major domains, heptad repeat 1 (HR1), heptad repeat 2 (HR2), and fusion peptides (FP) that carry out virus–cell fusion process through the formation of a 6-helix bundle (6-HB), which brings the MERS-CoV envelope and the host cell membrane together to promote their fusion. Huang and coworkers have benefited from such mechanism and developed a series of heptad repeat 1 (HR1) peptide inhibitors to inhibit HR1/HR2-mediated membrane fusion. Particularly, the peptide pregnancy-induced hypertension (PIH) exhibited a potent inhibitory activity, which was even magnified by forming a PIH–gold nanorod complex (PIH–GNPs). The synthesized nanoparticles (with average size 5 nm) have induced a potent inhibition of S2-mediated cell fusion in vitro. PIH–GNPs have displayed and enhanced metabolic stability and biocompatibility both in vitro and in vivo and effectively prevented MERS-CoV infection [25]. Different GNPs coated with ligands ω-terminated with sugars bearing multiple sulfonate groups were synthesized. Variable chain length as well as the sulfonate group numbers presented by one ligand and different sugars as head group presenting the sulfonate groups in different spatial orientations to the virus were investigated in order to understand the mechanism of GNPs–virus interaction. Through virus inhibition assays, the data showed that a multisulfonated complex ligand with a glucose head group was moderately efficient in inhibiting dengue virus serotype 2 (DENV-2) in hepatocytes. This compound is considered as a promising antiviral drug as it induced a virucidal, rather than a reversible virustatic, effect. To understand the mechanism behind virus inhibition, molecular docking simulations on the ligand-binding pocket of the DENV-2 envelope protein were performed and suggested the presence of key interactions between the sulfonate substituents of the head groups and specific residues of the viral envelope protein [26]. Antidengue virus small interfering RNAs (anti-DENV-siRNAs) that target specific viral genes are considered a promising therapeutic agent against DENV infection; however, they are prone to degradation by serum nucleases and rapid renal excretion due to their small size and anionic character. To enhance siRNA delivery and stability, complexed anti-DENV siRNAs with biocompatible GNPs were synthesized to act as a delivery vehicle both in vivo and in vitro. The synthesized cationic GNPs–siRNA complexes were noncytotoxic and could enter the cells and significantly reduce DENV-2 replication and subsequent virus release under both pre- and postinfection conditions. In addition, GNPs were able to protect siRNA against RNase enzymatic degradation and the data showed that RNase-treated GNPs–siRNA complexes could still inhibit DENV-2 replication [27]. A hybrid combination of different sized gold nanoparticles (GNPs; 20, 40, and 80 nm) and domain III of the envelope glycoprotein of DENV-2 (EDIII) as a dengue subunit vaccine was synthesized. The efficacy of the EDIII-functionalized GNPs (GNP–EDIII) to induce anti-DENV neutralizing antibody in BALB/c mice was evaluated. The data showed that GNP–EDIII induced a high level of neutralizing antibodies that have mediated a serotype-specific neutralization of DENV-2. The titer of produced antibodies was dependent on both the size and the concentration of GNP–EDIII, which give it the flexibility to be modulated and fine-tuned through adjusting these parameters. The GNP–EDIII subunit vaccine has stimulated T-cell responses through the proliferation of IFN-γ and IL-4-producing T-cells, leading to the generation of anti-EDIII neutralizing antibody. More importantly, while GNPs were used as both EDIII vehicle and adjuvant to modulate the immune response, the data indicated a low accumulation of GNPs in major organs, which might suggest low potential side effects of GNPs [28].

Infection with herpesviruses, particularly herpes simplex type 1 and 2 (HSV-1 and HSV-2), is considered a major problem in human, but also animal, health. Several nanomaterials, such as copper, zinc, gold, and silver, have proven to be effective against herpesviruses [5,12,29,30]. Heparan sulfate (HS) is an important cell receptor that facilitates the entry of herpesviruses and several other viruses. The use of nanoparticles that mimic HS to inhibit virus infection has grabbed the attention of several research groups. GNPs capped with mercaptoethane sulfonate (GNPs–MES) were generated to mimic HS and have been used as effective inhibitors of HSV-1 infection. The developed GNPs–MES were noncytotoxic and were able to significantly inhibit HSV-1 infection of cell culture. Mechanistic studies revealed that GNPs–MES interfered with viral attachment, entry, and cell-to-cell spread, thereby blocking subsequent cycles of viral infection in vitro. However, the presence of GNPs–MES inside the cells did not interfere with virus replication or the overall infection rate. The design of these nanoparticles has ensured the orientation of functional groups anchored on GNPs, which is crucial to generate polyvalent interactions with virus particles at high specificity, strength, and efficiency [31]. In a way to "upgrade" the GNPs–MES, the short linkers in MES–GNPs were replaced with long and flexible ones (mercapto-undecane sulfonic acid; MUS) to achieve strong multivalent binding. A series of GNPs coated with MUS-containing ligands were generated to mimic heparan sulfate proteoglycans (HSPG) and compared to GNPs–MES. All GNPs–MUS have showed a strong and significant in vitro inhibition of many HSPG-dependent viruses, either enveloped (HSV, respiratory syncytial virus (RSV), lentivirus, and dengue virus) or naked (human papilloma virus; HPV) in vitro. In addition, a strong inhibitory ex vivo effect on human cervicovaginal histocultures infected by HSV-2 was seen as well as a strong in vivo effect in mice infected with RSV was recorded. It was interesting to show that the effect of the MES-NPs was temporary "virustatic" and was lost upon dilution, while all MUS-coated NPs showed a clear irreversible "virucidal" and no cytotoxic effect. It was postulated that the seen irreversible effect with MUS ligands compared to MES ligands can be explained as the MUS ligand has a long hydrophobic backbone terminating with a sulfonic acid, allowing its terminal group to move with some freedom. This makes GNPs coated with MUS ideal for the binding of multiple sulfonic acids to the HSPG-interacting motifs on envelope viral proteins. To unravel the mechanism of action, virucidal assays, electron microscopy images, and molecular dynamics simulations were conducted using HPV as a model. The data demonstrated that a selective multivalent binding has developed between the negative sulfonate groups of MUS ligand and the positive HSPG-binding lysine residues of L1 capsid protein complexes from the HPV-16 capsid. This multivalent binding can induce large stresses and local distortions of the L1 complexes that can eventually lead to global virus deformations [32]. GNPs conjugated with gallic acid (GNPs–GA) were produced after ultrasound-induced reduction of gallic acid, which leads to highly monodispersed nanoparticles. The antiviral efficacy of GNPs–GA against HSV infections in Vero cells showed that the nanoparticles were capable of inhibiting both HSV-1 and HSV-2 in a dose-dependent manner. The mechanism is speculated to be through the prevention of virus attachment and penetration into the cells. The cytotoxicity in Vero cells was determined to be significantly lower than acyclovir [33]. Not only herpes simplex viruses, but also GNPs were used to chemically conjugate a monoclonal antibody raised against human cytomegalo herpesvirus (CMV)-surface glycoprotein B (GNP-gB). Three different types of GNPs with different surface properties: spherical (GNPsp-gB), star (GNPst-gB), and popcorn (GNPop-gB) were synthesized. All three types of conjugated GNPs showed inhibitory effect against CMV infection in vitro; GNPs-gB could block multiple stages of virus life cycle, including viral entry, replication, and spread in cell culture without inducing cytotoxicity. Based on the transmission electron microscopy (TEM), high concentrations of GNPop-gB completely coated virus particles or agglomerate virus particles in clusters and blocked virus entry into cells. Furthermore, cells treated with GNPs-gB gain resistance to CMV infection. In addition, CMV-infected cells when bound to GNPs-gB could be selectively lysed after exposing them to specific wavelength of laser (nanophotothermolysis) [34]. The effect of surface modification of gold nanoclusters (GNCs) on the replication of pseudorabies virus (PRV) was also investigated. Two types of GNCs with different surface modification, histidine-stabilized (His–GNCs) and mercaptoethane sulfonate (MES–GNCs), were developed. His–GNCs could strongly inhibit the replication of PRV, when compared to MES–GNCs, through blocking the viral replication process rather than the steps of attachment, entry, or release. It has been shown that His–GNCs were mainly localized in the nucleus, where virus replication takes place, while MES–GNCs were strictly distributed in the cytoplasm [35].

Although many (around 30) approved antiretroviral drugs exists, HIV remains a big challenge and researchers are still looking for new therapeutic agents. Mercaptobenzoic acid-modified GNPs (2 nm in diameter) were developed as a nanoscale platform to construct multivalent therapeutics [36]. The mercaptobenzoic acid-coated GNPs were conjugated to SDC-1721, a derivative of the CCR5- (the principal entry coreceptor for most commonly transmitted strains of HIV-1) antagonist TAK-779. Conjugation of SDC-1721 to gold nanoparticles (GNPs–SDC-1721) at an average ratio of 12:1 resulted in a significant inhibition of HIV infection when compared to SDC-1721- or GNP-only treatments. The multivalent display of SDC-1721 on the surface of GNPs was a requirement for the antiviral activity. It was speculated that HIV inhibition was at the stage of viral entry, most probably due to inhibition of virus fusion to cell membrane [37]. GNPs stabilized with polyethylene glycol were proved to act as antiviral agent against HIV-1 in vitro. In a cell-based fusion assay, to mimic HIV-gp120-CD4-mediated fusion process, HeLa-CD4-LTR-B-gal cell line was used to determine the fusion process because of gp120-CD4 interaction through a B-gal reporter assay. Interestingly, stabilized GNPs were efficiently able to block the fusion process in a dose-dependent manner. Different HIV isolates, M-tropic, T-tropic, dual tropic and resistant isolates, were inhibited by the GNPs [38]. Anionic carbosilane dendrons decorated with sulfonate functions and one thiol moiety at the focal point were used to synthesize water-soluble GNPs. As a result, a series of dendronized anionic GNPs were synthesized and tested against HIV. By using luciferase-based assay, dendrons and GNPs were tested for their cytotoxicity and activity against R5-HIV-1 and X4-HIV-1 isolates in vitro. While proved to be nontoxic, only dendronized anionic GNPs could inhibit both isolates of HIV more effectively than free dendrons. The presence of extra sulfur atoms in the dendrons and the coordination of the $S(CH_2)_3SO_3$ chain may hinder the aggregation of GNPs so that making them effective to inhibit HIV [39]. GNPs are an interesting chemical tool to be used as smart and efficient drug delivery systems. Carbohydrate-coated GNPs were loaded with the nucleoside reverse transcriptase (abacavir and lamivudine; these drugs compete with endogenous nucleoside substrates acting as chain terminators in the DNA polymerization reaction catalyzed by HIV-1 reverse transcriptase) anti-HIV prodrug inhibitors. The prodrugs were converted to the corresponding thiol-ending ester derivatives and then conjugated to glucose-coated GNPs (around 3 nm in diameter) by means of "thiol-for-thiol" ligand place exchange reactions. The antiviral activity was tested by evaluating the replication of NL4–3 HIV in TZM-bl infected cells (derived HeLa cell line that expresses high levels of CD4 and the coreceptors CXCR4 and CCR5) using luciferase reporter gene assay. The drug-loaded GNPs showed a nucleoside analog reverse transcriptase inhibition (NRTi) activity similar to the free drugs, which suggested that the drug was delivered appropriately into the cells and competed with the natural substrate of reverse transcriptase inhibiting virus replication [40].

Giving its superior activity as a delivery carrier of various biopharmaceuticals, GNPs were used for the clinical treatment of hepatitis C virus (HCV) infection. A target-specific long-acting delivery of IFNα was successfully developed using the complex of GNPs and hyaluronic acid (GNPs-HA–IFNα). This complex showed a highly enhanced stability in human serum or even in mice liver when compared to polyethylene glycol (PEG)-conjugated INFα, the traditional treatment of hepatitis C infection. Accordingly, GNPs-HA–IFNα complex significantly enhanced the innate immune responses to viral infection in the liver through enhanced expression of 2′,5′-oligoadenylate synthetase 1 (OAS1) [41]. RNA interference is a gene regulatory mechanism derived by the cell and mediated by the RNA-induced silencing complex (RISC), which is an endonuclease-containing multiprotein complex that incorporates one strand of a small interfering RNA (siRNA) or microRNA to recognize, capture, and finally cleave a complementary messenger RNA (mRNA). Inspired by the structure and function of the RISC machinery, an artificial nanoparticle complex (called nanozyme), consisting of 13-nm GNPs, sequence nonspecific endoribonucleases, and single-stranded DNA oligonucleotides, was synthesized to mimic the function of the cellular RISC machinery system for inducing target (HCV) RNA cleavage. GNPs acted as the backbone of this nanozyme, which provided a large surface area to hold both endoribonucleases and DNA oligonucleotides at close proximity. Interestingly, the designed nanozyme could actively cleave HCV RNA in a sequence-specific manner. Other favorable characteristics of this nanozyme included less susceptibility to degradation by proteinases, effectively up taken by cultured hepatoma cells, no cytotoxicity to cell culture and a xenotransplantation mouse model, and no triggering of detectable cellular interferon responses. More than 99% decrease in HCV RNA levels in mice treated with the nanozyme was observed [42].

Highly stable GNPs (2 and 14 nm in diameter) were functionalized with sialic acid-terminated glycerol dendron. These nanoparticles have shown high binding affinity to influenza virus hemagglutinin (HA) in a size-dependent manner; only the 14-nm GNPs were able to interact with influenza viral particles, inhibit virus hemagglutination of RBCs, and further inhibit virus infection. It is worthy to mention that the synthesized GNPs were nontoxic to the cells and support the concept that multivalent interactions are strongly dependent on the size of the interacting particle surfaces [43]. Recently, porous GNPs fabricated via a surfactant-free emulsion method have shown to exhibit strong affinity for disulfide bonds due to gold–thiol interactions and provide extensive surface area for such interactions. Influenza virus HA, which is responsible for virus entry, is known to have 6 disulfide bonds that can easily bind with the surface of GNPs. The use of porous GNPs has shown a remarkable inhibitory effect on various influenza virus strains (H1N1, H3N2, and H9N2) in vitro. Furthermore, intracellular viral RNA copies as determined by real-time RT-PCR confirmed that porous GNPs successfully inhibited viral infection. Using TEM analysis, the mechanism of action is hypothesized to be through the inhibition of virus–membrane fusion and blocking viral entry process due to conformational deformation of HA protein [44].

Gold/copper sulfide (Au/CuS) core/shell nanoparticles were synthesized and have shown several advantages over the conventional metal nanoparticles due to the continuous release of copper ions into solution without reducing agents, low cost, and ready availability. Besides their known antimicrobial activity against the bacteria *Bacillus anthracis* [45], they were tested against human Norovirus (HuNoV) represented by the GI.1 VLPs as a model. Depending on the treatment time, Au/CuS nanoparticles could induce changes in the VLP particle morphology, size, and integrity, resulting in the inhibition of virus infectivity. Direct contact and physical damage to virus particles are common mechanisms known for many different NPs, but strongly suggested to be applied for Au/CuS nanoparticles [46]. Finally, green chemistry synthesis of GNPs using garlic extract (Allium sativa) as a reducing agent (GNPs-As) was evaluated against Measles virus (MeV). GNPs-As could actively inhibit MeV replication in Vero cells by binding directly to virus particles and blocking virus–cell interaction [47].

2.2 Silver NPs

Silver nanoparticles (SNPs) are the most effective among other metallic NPs against pathogenic microorganisms due to the inherent inhibitory and bactericidal potential of silver, their good conductivity, catalytic properties, and chemical stability [17]. SNPs (average size of 11 nm) have been shown to inhibit human adenovirus (HAdV-3) infection of HeLa cells in a dose-dependent manner. Treatment of HAdV-3 with SNPs could result in direct destruction of virus particles as seen by TEM [48]. A combination of SNPs and graphene oxide (GO–SNPs) was developed and tested against enveloped and naked viruses. The developed nanoparticles showed a moderate inhibitory effect on the enveloped feline coronavirus (FCoV) and the naked infectious bursal disease virus (IBDV) [48]. On the other hand, silver nanoclusters with silica (Ag_2S-NCs; less than 200 nm in diameter) were proved to possess highly efficient antiviral activity against another corona virus of pigs, porcine epidemic diarrhea virus (PEDV). The synthesized nanoclusters significantly inhibited virus infection in cell culture. Mechanistic studies indicated that Ag_2S-NCs could inhibit viral

RNA synthesis within the cells and further inhibited virus budding. In addition, Ag_2S-NCs were able to positively regulate the expression of IFN-stimulating genes (ISGs) and proinflammatory cytokines, which might block PEDV infection [49]. Interestingly, application of silver nanocluster silica composite sputtered coating to the facial FFP3 masks ultimately reduced the SARS-CoV-2 titer on cell culture [50]. This application opens the door for further investigation on the possible application of nanoparticles in combating emerging viruses. Further analysis on the possible effect on human health still needs to be conducted.

SNPs with aqueous leave and fruit extracts from *Ricinus communis* or *Lampranthus coccineus* showed a significant inhibitory effect on hepatitis A virus and Coxsackievirus, possibly by acting as a virus–cell fusion inhibitor or by interacting with the viral 3c protease [51,52]. SNPs capped with medicinal plant leaf extracts of *Carica papaya* or *Andrographis paniculata* were prepared and tested against Chikungunya virus (CHIKV) in vitro. The synthesized SNPs showed a significant inhibitory effect against CHIKV in Vero cells [10,53]. SNPs capped with different plant extracts from Ambrosia arborescens, paddy straw, Rhazya stricta, Daemia extensa, *Solanum mammosum* L. pilaquinga, and Hymenodictyon orixense were proved to have larvicidal activity against the chikungunya virus vector (mosquitos) [10,54–56]. Similar to the chikungunya virus, several studies were also conducted to control dengue virus or its insect (mosquito) vector. Biosynthesized silver NPs using algae (*Centroceras clavulatum*), plant (Bruguiera cylindrical), and seed (*Moringa oleifera*) extracts showed inhibitory effect against dengue virus in vitro [57–60]. In mosquitocidal assays, SNPs with *C. clavulatum* extract have showed acute toxic effect against larvae and pupae of the primary dengue vector *Aedes aegypti*. Furthermore, *C. clavulatum*-synthesized SNPs inhibited dengue virus (serotype-2; DEN-2) replication in Vero cells. SNPs showed no cytotoxicity on cells but reduced DEN-2 infection more than 80%. It was speculated that SNPs could inhibit virus internalization inside the cells [57].

In a similar approach to the GNPs, SNPs were capped with mercaptoethane sulfonate (SNPs–MES) to target HSV-1 entry and infection and to compete for its binding to cellular HS through their sulfonate end groups. SNPs–MES had no cytotoxic effects on host cells and efficiently inhibited HSV-1 infection in cell culture. The multivalent nature and spatially directed MES on the surface of SNPs have ensured effective polyvalent interaction with virus particles [61]. SNPs (30–40 nm in size) were also seen to significantly inhibit HSV-2 in vitro. The nanoparticles were nontoxic and at a concentration of 100 μg/mL were able to completely inhibit virus replication through a mechanism that speculated to be via inhibiting early stages of virus replication, most probably virus entry [62]. Tannic acid-modified SNPs (13 nm, 33 nm, and 46 nm of size) were synthesized and could reduce HSV-2 infectivity both in vitro and in vivo. The antiviral activity was size dependent and required direct interaction between SNPs and the virus before infection, in order to block virus attachment and subsequent entry. In mice, all tested tannic acid-modified SNPs could reduce both HSV-1 infection and inflammatory reaction when used at the time of infection or for a postinfection treatment. Smaller-sized SNPS were more efficient in induction of cytokines and chemokines production that act as antiviral factors [63]. In another approach, SNPs were biologically synthesized using different fungal species (*Alternaria* species, *F. oxysporum*, *Curvularia* species, *C. indicum*, and *Phoma* species) and tested against different viruses: HSV-1, HSV-2, and human parainfluenza virus type 3 (hPIV-3). Some of the fungal strains used have shown considerable toxicity; however, those produced from *F. oxysporum*, *Curvularia* species, and *C. indicum* have shown low cytotoxicity levels and promising antiviral activity. In virus inhibition assay, SNPs produced in *F. oxysporum*, *Curvularia* species, and *C. indicum* could reduce viral infectivity in a dose- and size-dependent as well as zeta potential manner. It was clear that SNPs produced by *F. oxysporum* and *Curvularia* species, having a size of 4–13 nm and 5–23 nm, respectively, had better antiviral activity (up to 90% inhibition) against HSV-1 and hPIV-3 viruses, probably by blocking virus–cell interaction [64]. Recently, SNPs showed higher cytotoxicity against the human oncogenic γ-herpesviruses, Kaposi's sarcoma-associated herpesvirus (KSHV)- and Epstein–Barr Virus (EBV)-latently infected cells through the reactivation of viral lytic replication. The mechanism relied on the induction of ROS generation and autophagy. Moreover, SNPs could block KSHV primary infection in vitro by directly destroying virion particles after direct interaction with the virions, which result in morphological changes in virus particles; they became swollen and flattened [65]. The inhibitory effect of SNPs on bovine herpesvirus type 1 (BHV-1) infection was also investigated. At a concentration of 24 μg/mL, SNPs were noncytotoxic for the MDBK cell culture and could inhibit BHV-1 infection when incubated with the virus before infecting the cells [66].

SNPs have been shown to interact with HIV-1 in a size-dependent manner; exclusively in the size range of 1–10 nm. SNPs were synthesized with three markedly different surface chemistries: foamy carbon, poly(N-vinyl-2-pyrrolidone) (PVP), and bovine serum albumin (BSA). High angle annular dark field (HAADF) scanning transmission electron microscopy was employed to study the interaction between the different SNPs and the virus. The data determined that the three different SNPs established regular spatial interaction with the viral envelope, most probably to the sulfur-bearing residues of the gp120 glycoprotein knobs. This might explain the size-dependent interaction manner and the inability of larger nanoparticles to bind the virus. All three nanoparticle preparations could reduce HIV infectivity in cell culture in a dose-dependent manner with moderate cell toxicity [67]. SNPs (30–50 nm diameter) were tested against a panel of HIV-1 isolates using a luciferase-based assay. The tested SNPs inhibited several HIV-1 strains, including T-tropic, M-tropic, dual-tropic, and resistant isolates at noncytotoxic concentrations. The mechanism underlying their HIV inhibitory activity suggested that SNPs exerted anti-HIV activity at an early stage of viral replication, most likely as an inhibitor of viral entry. SNPs could bind to gp120 in a manner that prevents CD4-dependent virion binding, fusion, and subsequently replication [68].

Monodisperse SNPs with mean diameters of approximately 10nm and 50nm were prepared from silver nitrate in HEPES buffer and tested against hepatitis B virus (HBV) in vitro. Both SNPs were able to significantly inhibit HBV infection and to reduce the formation of extracellular DNA. However, SNPs had little effect on the amount of covalently closed circular HBV DNA, but could inhibit the synthesis of intracellular viral RNA. In addition, TEM analysis showed that small-sized SNPs (10nm) could directly interact with the HBV viral particles that also added to the inhibitory effect [69].

Silver nanoparticles have demonstrated efficient inhibitory activities against H1N1 influenza A virus. It has been shown that SNPs (10nm in diameter) could rapidly and significantly block the binding between H1N1 influenza virus and chicken RBCs and thus inhibit virus hemagglutination of erythrocytes. SNPs could also reduce H1N1 influenza A virus-induced apoptosis of MDCK cells, which might indicate inhibition of virus infection [70]. Not only H1N1, but also SNPs have shown significant inhibitory effect on H3N2 influenza viruses. SNPs were determined to inhibit virus hemagglutination of RBCs, virus-induced apoptosis, and virus infection of MDCK cells. The mechanism of action is postulated to be due to virus–SNPs direct interaction, which resulted in destruction of the enveloped virus in a time-dependent manner (30–120min). In addition, intranasal application of SNPs in experimental mice significantly reduced the mortality rate after challenge with the H3N2 influenza virus due to reduced virus titers in lung tissues [71]. As a delivery system, SNPs decorated with the antiinfluenza drugs oseltamivir (OTV) or zanamivir (ZNV) were synthesized. The complex has shown a remarkable inhibition of H1N1 infection of MDCK with low cytotoxicity. The mechanism of action is speculated to be through blocking virus attachment to cell surface receptors, inhibition of apoptosis, DNA fragmentation, chromatin condensation, and caspase-3 activity, and inhibition of ROS accumulation in the cells (ROS-mediated p53 and AKT signaling pathways) [72,73]. The peptide "FluPep," a well-established inhibitor of influenza virus, was also conjugated with SNPs or GNPs (ca. 10nm in diameter) as delivery vehicles to enhance the antiviral activity. The functionalized SNPs- and GNPs-FluPep showed enhanced antiinfluenza viral activity compared to the free peptides in vitro [74].

The efficacy of SNPs and silver nitrate ($AgNO_3$) was evaluated to determine their effects on different caliciviruses, feline calicivirus (FCV), and murine norovirus (MNV). Exposure of the viruses to these nanoparticles significantly decreased FCV and MNV infectivity in a size- and dose-dependent manner. The small-sized (10 nm) SNPs effectively inactivated FCV in vitro when compared to the 75- and 110-nm-sized SNPs [75,76]. Decoration of micrometer-sized magnetic hybrid colloid (MHC) with variously sized SNPs (SNP–MHCs) resulted in a significant inactivation of different viruses, including bacteriophage ϕX174, MNV, and HAdV-2 [77]. The activity of SNPs against several other viruses such as poliovirus [78], respiratory syncytial virus [79], Rift Valley fever virus [80], arenaviruses [81], and monkey poxvirus [82] was also evaluated.

2.3 Other metal NPs

The "contact killing" mechanism was suggested to be initiated by the dissolved copper ions released from copper surfaces by the culture medium and causing damages of cell membranes as well as virus particles [83]. Several studies have shown that copper surfaces were able to inactivate and destroy virus particles when they come in contact, such as MNV and influenza viruses [84–86]. Impregnation of copper oxide in respiratory protective facemasks conferred biocidal activities against human and avian influenza A virus without altering the physical barrier properties of the masks. Similarly, it was demonstrated that copper oxide was effective against HIV-1 when passed through filters containing copper oxide powder or copper oxide-impregnated fibers [87,88]. Recently, copper–graphene (Cu–Gr) nanocomposite has been shown to impose a substantial reduction in viral infectivity (represented by influenza virus) through reduced viral entry, gene expression, and subsequent production of progeny virions. Material and biological characterization of the Cu–Gr nanocomposite suggested a unique metal oxide-embedded graphene sheet architecture can inactivate virus particles within 30 min after contact and subsequently block virus entry into host cells [89]. In biological systems, copper oxide nanoparticles were able to significantly inhibit the infectivity of HCV at a noncytotoxic concentration due to blocking virus attachment and entry into the cells [90]. The effect of copper oxide nanoparticles (CuO-NPs) on HSV-1 infection was also studied. Exposure of HSV-1 to CuO-NPs at the highest noncytotoxic concentration resulted in a significant and stark inhibition of the virus in vitro [29]. Treatment with zinc oxide has displayed antiviral activity against chikungunya virus by decreasing viral RNA transcription in the cytoplasm [91]. On the other hand, copper oxide NPs synthesized using leaf exacts from *Tridax procumbens* displayed larvicidal activity against the mosquito vector of chikungunya virus [92].

Tungsten carbide NPs have been described to significantly inhibit the infectivity of adenovirus type 5 in vitro. In addition, these NPs displayed significant antiviral activity against several other viruses such as modified vaccinia virus Ankara (MVA), poliovirus, and norovirus [45]. Anatase titanium dioxide NPs could also significantly reduce cell viability of MNV-infected macrophages in vitro [93].

Selenium NPs were found to inhibit enteroviruses. In addition, selenium NPs conjugated with the antiviral drug, oseltamivir, inhibited enterovirus infection by reducing the production of ROS in human astrocytoma cells. Other cellular pathways were also reported to be inhibited by selenium NPs, such as Jun amino-terminal kinase (JNK) signaling pathway and p38 kinase that lead to the reduction of viral protein synthesis and production [94,95]. Iron oxide NPs or selenium NPs either alone or conjugated with the broad-spectrum antiviral drug, ribavirin (RBV), could also inhibit influenza virus infection of MDCK cells [96,97].

Functionalized nanodiamonds (ca. 160 nm in diameter) with sulfate groups were found to be strong inhibitors of herpesviruses in vitro. These polyvalent hybrid architectures had high potential to bind virus particles and inhibit not only herpesvirus entry, but also, possibly, other known heparan sulfate-dependent viruses and even bacteria [98].

3 Carbon nanomaterials

Carbon-based architectures, especially graphene and its derivatives, have attracted attention as potential antiviral agents. In general, carbon nanomaterials have showed low cytotoxic effects and presented specific and potent antiviral activities. Carbon nanomaterials are considered relatively new in the nonmedical field; however, they have attracted intense research activities in the last years. There are several carbon nanomaterial platforms available to date, including graphene-based systems, carbon dots, carbon nanohorns, and fullerene and its derivatives. Graphene is a honeycomb-like lattice consisting of a flat single layer of carbon atoms arranged in a two-dimensional (2-D) structure. Due to its high flexibility, it can exist in different forms: 0-D fullerenes, 1-D nanotubes, or 3-D graphite [99–101]. Production of large-scale 2-D thermally reduced graphene oxide (TRGO) is characterized by its high flexibility, which can be further enhanced by the insertion of different ligands that can compete with virus infection [98,102–104]. In general, graphene has been shown to have superior properties, including (i) large surface area (around 2700 m^2/gram), (ii) utmost thinness, (iii) high flexibility, (iv) superb thermal and electrical conductivity, (v) superb light-to-heat conversion ability, (vi) ability to be tailored to different sizes, and (vii) easily modified to carry different functionalized groups such as magnetic nanoparticles, oxide, sialic, sulfonate or sulfate groups [100,105–108].

3.1 Graphene

Several studies have demonstrated the efficacy of graphene and its derivatives as potent antiviral agents. It has been sown that graphene oxide (GO) could inhibit enterovirus 71 (EV71) and influenza H9N2 virus in a temperature-dependent fashion. This inhibition was implied to a strong protein adsorption capacities of GO, which could influence the formation of chemical bonds between virus particles and graphene sheets. The study suggested that the unique physicochemical interactions between GO and viruses could primarily ascribe to hydrogen binding, electrostatic interactions, and even redox reaction, which resulted in virus (whether naked or enveloped) destruction [108]. It was further shown that reduced graphene oxide has showed a potent antiviral effect, as the naturally occurring sharp edges of the GO together with its negative charge can inactivate both RNA coronaviruses, represented by PEDV, and DNA herpesviruses, represented by PRV [109]. The inhibition activity was speculated to occur before virus entry into the cells due to physical disruption of viral envelope after direct (electrostatic) interaction with the sharp edges of the GO layers. In another study, GO and partially reduced sulfonated GO (rGO – SO3) have been shown to inhibit HSV-1 through a competitive inhibition mechanism. Both architectures (GO and rGO–SO3) bear multiple negative-charged groups (carboxyl and sulfonate) that mimic cell surface HS, the initial attachment receptors for HSV-1 [107]. Sulfonated magnetic nanoparticles functionalized with reduced graphene oxide (SMRGO) were shown to capture and photothermally inhibit HSV-1. Upon irradiation of the SMRGO with near-infrared light (NIR), SMRGO demonstrated superior photothermal antiviral activity that might be attributed to the increased capture efficiency due to electrostatic interaction with virus particles and the excellent photothermal properties of graphene that leads to virus inactivation after NIR irradiation [106]. The use of GO as photocatalysts was previously described. GO sheets that are covalently linked with aptamers were synthesized and characterized. GO nanosheets are quite engaging in photocatalysis due to their potential for trapping solar energy. Aptamers, on the other hand, specifically bound to targets (e.g. viruses), increased the stability of GO sheets, and promoted photocatalytic activities of GO. This GO–aptamer complex effectively and successfully photocatalyzed bacteriophage MS2 and damaged capsid proteins and DNA of viruses by electron and energy transfer in visible light irradiation [110]. In a similar approach, graphene–tungsten oxide composite thin films with sheet-like surface morphology were fabricated and applied in photoinactivation of viruses under visible light irradiation. The generated complex (GO–tungsten) resulted in a nearly complete destruction of viral (bacteriophage MS2) proteins and RNA efflux after light irradiation at room temperature [111].

Modifying graphene sheets with different chemical groups to mimic the extracellular cell matrix and cellular receptors has a great potential to modulate cellular responses and to bind and neutralize viruses. Dichlorotriazine groups were conjugated to the surface of graphene by a one-pot [2+1] nitrene cycloaddition reaction followed by conjugation of hyperbranched polyglycerol sulfate to produce highly negatively surface charged graphene nanomaterials, which mimic cell surface heparan sulfate, with a very good dispersibility in aqueous environment. These functionalized nanosheets revealed strong interactions with vesicular stomatitis virus (SVS), a rhabdovirus that uses heparan sulfate as entry receptors [112]. Similar strategy has further been developed to create a negatively charged HS-mimetic extracellular (2-D) matrix through functionalization of TRGO sheets with either linear or dendritic polyglycerol sulfates. The developed 2-D architectures could inhibit the entry of DNA-enveloped viruses of the family *Herpesviridae* (PRV) and *Asfarviridae* (African swine fever virus; ASFV) [113]. In another study, the effect of the size and degree of sulfation of TRGO sheets functionalized with multivalent dendritic polyglycerol sulfates (dPGS) on virus entry was investigated. Highly sulfated and smaller dPGS-TRGO sheets showed the strongest effect on virus entry and consequently infection. The correlation between the degree of sulfation and the inhibitory effect was explained by the stronger polyvalent effect resulting from high density of negatively charged sulfate groups conjugated to TRGO that enabled strong interactions with herpesvirus particles. On the other hand, the observed strong inhibition effect for the small-sized derivatives was attributed to the required bending energy for the functionalized TRGO sheet. This is consistent with the theory of Reynwar and coworkers who discussed that multiple binding is the essential aspect for remodeling large geometries such as cellular membranes [114]. TRGO sheets of the same size as virus particles are assumed to have a huge impact on the bending behavior, because areas without interactions could adversely affect the required bending energy. This explains how the smaller TRGO-functionalized sheets exhibit higher inhibition efficiencies than the larger ones because of the lower ratio of noninteracting to interacting areas [98]. Further functionalization of graphene derivatives with defined polyglycerol sulfate and fatty amine has been developed. These new derivatives kept the activity of electrostatic interactions with herpesvirus particles through polyglycerol sulfate groups while alkyl chains induced a high antiviral activity by secondary hydrophobic interactions. The length of alkyl chains had impact on the functionality of graphene sheets. The longer the alkyl chains

(C12–C18) the more potent antiviral activity and the more cytotoxic. The optimum synergistic effect between electrostatic and hydrophobic interactions was seen with graphene sheets functionalized with sulfate groups and C6–C9 long alkyl chains, which showed low cytotoxicity and high inhibition of herpesviruses [115].

In another approach, a β-cyclodextrin (CD) functionalized GO composite loaded with curcumin—curcumin is a natural polyphenol isolated from the rhizomes of the perennial herb *Curcuma longa* and has antioxidant, antifungal, antibacterial, antiinflammatory, and antiviral activities [116]—has been developed. The multifunctional curcumin-loaded GO was proved to be highly efficient in virus (RSV) inhibition with a great biocompatibility to the host cells. It was speculated that this graphene composite could prevent RSV from infecting the host cells by directly inactivating the virus and/or inhibiting virus attachment to cell surface receptors. The introduction of sulfonate groups to the functionalized GO surface can mimic cell surface structure and inhibit RSV infection through a competitive inhibition mechanism [117]. Furthermore, loading GO with hypericin—hypericin is a naturally occurring chromophore found in some species of the genus *Hypericum*, especially *Hypericum perforatum* L. (St. John's wort), and has broad pharmacological spectrum including antiviral activities [118]—has shown an effective capability to inhibit viral (novel duck reovirus) replication, which may be attributed to direct virus inactivation or inhibition of virus attachment [119].

3.2 Carbon dots

Carbon dots (C-dots) represent another large family of carbon-based nanostructures that include different types, including amorphous carbon nanoparticles, partially graphitized core-shell carbon nanoparticles, amorphous fluorescent polymeric nanoparticles, and graphene quantum dots (GQDs). There are several potential applications for the C-dots ranging from photocatalysis to bioimaging; however, C-dots have also shown an active role as antiviral nanoagents [120,121]. Surface-functionalized C-dots with boronic acid or amine showed a significant efficacy to interfere with the entry of HSV-1. It was suggested that C-dots are specifically acting on the early stages of virus entry through an interaction with the virus and probably the cells at the same time, which can interfere with the early step of interaction of HSV-1 with the cell resulting in an inhibition of the infection process [122]. Not only herpesviruses, but also C-dot nanoparticles have been shown to prevent HIV-1 infection via interaction with gp120 and subsequently eliminate virus–cell interaction. C-dots were associated with boronic acid-containing molecules, which were noncytotoxic, and enhanced the inhibitory activity by suppressing syncytium formation [123]. Modifying C-dots with PEG-diamine and ascorbic acid has shown encouraging results as antiviral agents. The generated C-dots were reported to significantly inhibit the replication of PRV and porcine reproductive and respiratory syndrome virus (PRRSV). The inhibitory effect was rationalized to be due to the activation of type I interferon (IFN-α) and the production of ISGs, which in turn inhibited virus replication [124]. Synthesis of highly biocompatible C-dots from glycyrrhizic acid—active ingredient of a Chinese herbal medicine—(Gly-C-dots) has shown a strong antiviral activity against PRRSV, PRV, and PEDV. Gly-C-dots could inhibit PRRSV entry and replication, stimulate antiviral innate immune responses, and inhibit the accumulation of intracellular ROS caused by PRRSV infection. Proteomics analysis demonstrated that Gly-C-dots can stimulate cellular regulation of some host restriction factors, including DDX53 and NOS3, which are directly related to PRRSV proliferation [125].

C-dots with different surface passivation molecules, 2,2′-(ethylenedioxy) bis(ethylamine) (EDA)-CDots and 3-ethoxypropylamine (EPA)-CDots, were synthesized and evaluated against human norovirus virus-like particles (VLPs). The results indicated both EDA- and EPA-C-dots were highly effective to inhibit VLPs' bindings to histo-blood group antigens (HBGA) receptors on human cells, with EDA-C-dots achieving 100% inhibition and EPA C-dots achieving 85%–99% inhibition. The positively charged EDA-C-dots exhibited higher inhibitory effect than the noncharged EPA-C-dots, suggesting the surface charge status of C-dots played a role in the electrostatic interactions between C-dots and the negatively charged VLPs. Both types of C-dots also exhibited inhibitory effect on the binding of VLPs to their respective antibodies; however, less effective than binding to HBGA receptors. Despite the electrostatic interaction between C-dots and VLPs, VLPs remained intact with no degradation being observed on VLPs' capsid proteins [126]. In another study, curcumin was applied to prepare uniform and stable cationic C-dots. The generated cationic C-dots were found to inhibit PEDV replication due to induced changes in the structure of envelope proteins, thereby inhibiting viral entry. Furthermore, the C-dots could also suppress the synthesis of negative-strand RNA of the virus, virus budding, and the accumulation of ROS by PEDV. In addition, the C-dots were shown to suppress viral replication by stimulating the production of ISGs and proinflammatory cytokines [127]. The antiviral activity of different C-dots for the treatment of human coronavirus HCoV-229E infections was also investigated. The C-dots were functionalized with hydrothermal carbonization of ethylenediamine/citric acid as carbon precursors and postmodified with boronic acid ligands. These nanostructures showed a concentration-dependent inactivation of HCoV-229E through blocking virus entry due to, possibly, interaction of the C-dots' functional groups with the virus spike protein, but also due to inhibition of viral replication process after entry [128]. Benzoxazine monomer-derived carbon dots (BZM-C-dots) demonstrated their inhibitory effect against the infection of the life-threatening flaviviruses (Japanese encephalitis, Zika, and dengue viruses) and the nonenveloped viruses (porcine parvovirus and adenovirus-associated virus). BZM-C-dots could directly bind to virus particles, and eventually block the initial virus–cell interaction [129].

Synthesis of C-dots through a simple one-step dry-heating treatment of curcumin (Cur-C-dots) has been shown to be an effective antiviral agent against enterovirus 71 (EV71). The surface properties of the synthesized Cur-C-dots, as well as their antiviral activity, were highly dependent on the heating temperature during the synthesis process. The one-step heating of curcumin at 180 °C preserved many of the polymeric moieties of curcumin on the surfaces of Cur-C-dots, resulting in superior antiviral characteristics. In vitro analysis (cytotoxicity and hemolysis assays) revealed high biocompatibility, superior solubility, and prominent antiviral activity of the produced Cur-C-dots against EV71. It was speculated that Cur-C-dots could inhibit EV71 mainly by binding to the virus to prevent viral attachment and entry to the host cell. Furthermore, Cur-C-dots could inhibit the viral replication process and prevent the EV71-induced termination of host translation. In vivo analysis showed that intraperitoneal administration of Cur-C-dots significantly decreased mortality rates and provided protection against virus-induced hind limb paralysis in newborn mice challenged with a lethal dose of EV71 [130].

The antiviral activity of C-dots alone or conjugated with reverse transcriptase inhibitors (RTI)—namely, CHI499 and CDF119; belong to the class of nonnucleoside reverse transcriptase inhibitors (NNRTI)—against HIV was investigated. Graphene-based nanomaterials were synthesized by using water-soluble C-dots synthesized from multiwalled carbon nanotubes through prolonged acidic oxidation and exfoliation. The RTI-conjugated compound C-dots-CHI499 has proven to have improved antiviral activities and act as a good potential candidate for HIV treatment through a mechanism similar to that exerted by NNRTI drugs. The conjugation of these poorly water-soluble drugs with C-dots was proved to afford high water dispersible systems, thus opening new opportunities in the use of other poorly water-soluble anti-HIV drugs endowed of critical cytotoxic side effects [131].

3.3 Fullerene

Fullerene and its derivatives were the first and most abundant class of carbon-based nanostructures to be discovered and used in academic and industrial research. It was clear that fullerene had enormous potential biological applications in different fields, including nanomedicine. However, one major problem has confronted the practical applications of fullerenes due to its difficult processibility and insolubility or only sparingly solubility in most solvents and its tendency to aggregate very easily [132]. Functionalization of fullerenes with different organic groups improved their solubility and different chemical and physical properties that allowed their applications in several fields. The first water-soluble fullerene derivative has been shown to be active against HIV in cell culture. This derivative could interact with and inhibit the in vitro activity of HIV-1 protease (HIV-1 PR) [133]. Since that time, research in the field of fullerenes-HIV inhibition has flourished and several studies were published describing the inhibitory effect of fullerene and its derivatives on HIV replication. The studied fullerene derivatives were reported to be noncytotoxic and, therefore, they are excellent candidates as anti-HIV-1 agents that worked through different mechanisms that include (i) protease inhibition through the interaction between fullerene and the hydrophobic pocket of HIV proteases, (ii) maturation inhibition through the strong interaction with the immature viral capsid, and (iii) reverse transcriptase inhibition [6,134,135].

Water-soluble tridecafullerenes decorated with 120 peripheral carbohydrate subunits (superballs) were synthesized from hexakis adducts of fullerene in one step by using copper-catalyzed azide–alkyne cycloaddition click chemistry. The product maintained a globular shape and allowed control over the size and multivalency. Infection assays showed that these superballs are potent inhibitors of cell infection by an artificial Ebola virus particle [136]. In another study, preparation of globular glycofullerenes with higher valency by a convergent strategy combining trimannosylated glycodendrons with a Bingel–Hirsch hexakis-adduct fullerene to obtain glycodendrofullerenes with 36 mannoses and two different spacers has shown the potential biological activity of this product as antiviral agent. This product could inhibit a DC-SIGN-dependent cell infection by pseudotyped Ebola virus particles [137]. Similar fullerene derivatives were also shown to inhibit both Zika and Dengue viruses through a similar mechanism [138]. The inhibitory activity of these compounds is most probably due to a more efficient interaction with cellular DC-SIGN receptors that could inhibit virus binding to cell surface. It is therefore important to combine an adequate scaffold to achieve the multivalency (globular fullerene) with the right ligand accessibility and flexibility.

Preparation and synthesis of highly water-soluble fullerene (C70) derivatives showed a pronounced antiviral activity against HIV and influenza A virus (subtypes H1N1 and H3N2). The compounds inhibited syncytia formation between persistently HIV-1-infected HUT-78/HIV-1 and uninfected CD4+-SupT1 cells. The compounds could interact with the envelope gp120 protein, which is essential for virus entry into the cells after binding to specific cell surface receptors responsible for virus entry (such as CD4, CCR5, or CXCR4). Binding to gp120 seems to be the key mechanism that inhibited HIV infection. Furthermore, the compounds suppressed different strains of influenza A virus subtypes H1N1 and H3N2 and it was also found to be active against different strains of influenza B virus [139].

Functionalization of fullerene C60 with poly(N-vinylpyrrolidone) (C60/PVP) is another water-soluble derivative that could inhibit influenza A and B viruses. These derivatives showed low cytotoxicity and did not change the metabolic and morphological properties of cells. It was speculated that these compounds interfere with the virus particle assembly [140]. Proline-type fullerene derivatives were shown to have the potential to inhibit HCV NS5B polymerase and HCV NS3/4A protease enzymes indicating that these fullerene derivatives might be dual inhibitors against NS5B and NS3/4A and could be novel lead compounds for the treatment of HCV infections [141]. Carboxylic fullerene C60 derivatives were also shown to be of low cytotoxicity and high antiviral activity against HSV and human cytomegalovirus [142]. Water-insoluble photosensitizer buckminsterfullerene (C60) was used to inactivate enveloped viruses, such as Semliki Forest virus (SFV, Togaviridae) and VSV (Rhabdoviridae) [143].

3.4 Carbon nanohorns

Single-Walled Carbon Nanohorns (SWCNHs) are other carbon conical horn-shaped nanostructures constructed from sp2 graphene sheets. Individual conical CNH structures are typically 2–5 nm in diameter and 40–50 nm in length. One important feature of SWCNHs is that they have extensive surface area and multitudes of horn interstices, which enable large amounts of adsorbed molecules and provides suitable sites to incorporate and protect drugs. Furthermore, SWCNHs have not shown any measurable toxicity in animals. SWCNHs have been used mainly as drug delivery systems for cancer therapy through delivery of chemotherapeutics, photothermal and photodynamic therapy, gene therapy, and immunosensing agents [144]. In addition, photo-induced nanomaterial containing carbon nanohorns (CNH) could also eliminate viruses (T7 bacteriophage) in vitro [145].

4 Conclusions

The field of antiviral therapeutics is now witnessing a remarkable development at all levels; starting from the use of different platforms to functionalization with different active molecules. Nanoparticles have shown significant and potent antiviral activities and can play a critical role in controlling current and emerging viruses. Various types of nanoarchitectures have been studied and evaluated for their prophylactic and therapeutic activity against many pathogens, including viruses. However, several challenges, including proper modifications and cytotoxicity, still exist and need to be bridled before achieving the full potential of nanoparticles to be successfully upgraded to the clinical trials.

References

[1] B.S. Graham, N.J. Sullivan, Emerging viral diseases from a vaccinology perspective: preparing for the next pandemic, Nat. Immunol. 19 (1) (2018) 20–28. Epub 2017/12/05 https://doi.org/10.1038/s41590-017-0007-9. 29199281. PMCPMC7097586.

[2] D.M. Morens, G.K. Folkers, A.S. Fauci, Emerging infections: a perpetual challenge, Lancet Infect. Dis. 8 (11) (2008) 710–719. Epub 2008/11/11 https://doi.org/10.1016/S1473-3099(08)70256-1. 18992407. PMCPMC2599922.

[3] A. Pekosz, G.E. Glass, Emerging viral diseases, Md. Med. 9 (1) (2008) 11. 3–6 18491811. PMCPMC2496997.

[4] E. De Clercq, Antiviral drugs in current clinical use, J. Clin. Virol. 30 (2) (2004) 115–133.

[5] L. Chen, J.G. Liang, An overview of functional nanoparticles as novel emerging antiviral therapeutic agents, Mat. Sci. Eng. C-Mater. 112 (2020), https://doi.org/10.1016/j.msec.2020.110924. PubMed PMID: WOS:000532843000001.

[6] P. Innocenzi, L. Stagi, Carbon-based antiviral nanomaterials: graphene, C-dots, and fullerenes. A perspective, Chem. Sci. 11 (26) (2020) 6606–6622, https://doi.org/10.1039/d0sc02658a. PubMed PMID: WOS:000552450400001.

[7] L. Singh, H.G. Kruger, G.E.M. Maguire, T. Govender, R. Parboosing, The role of nanotechnology in the treatment of viral infections, Ther. Adv. Infect. Dis. 4 (4) (2017) 105–131. Epub 2017/07/28 https://doi.org/10.1177/2049936117713593. 28748089. PMCPMC5507392.

[8] J. Khandare, M. Calderon, N.M. Dagia, R. Haag, Multifunctional dendritic polymers in nanomedicine: opportunities and challenges, Chem. Soc. Rev. 41 (7) (2012) 2824–2848. Epub 2011/12/14 https://doi.org/10.1039/c1cs15242d. 22158998.

[9] S. Gurunathan, M. Qasim, Y. Choi, J.T. Do, C. Park, K. Hong, et al., Antiviral potential of nanoparticles-can nanoparticles fight against coronaviruses? Nanomaterials (Basel) 10 (9) (2020), https://doi.org/10.3390/nano10091645. Epub 2020/08/23 32825737. PMCPMC7557932.

[10] K. Maduray, R. Parboosing, Metal nanoparticles: a promising treatment for viral and arboviral infections, Biol. Trace Elem. Res. (2020), https://doi.org/10.1007/s12011-020-02414-2. Epub 2020/10/09 33029761. PMCPMC7540915.

[11] M.K. Rai, S.D. Deshmukh, A.P. Ingle, A.K. Gade, Silver nanoparticles: the powerful nanoweapon against multidrug-resistant bacteria, J. Appl. Microbiol. 112 (5) (2012) 841–852. Epub 2012/02/14 https://doi.org/10.1111/j.1365-2672.2012.05253.x. 22324439.

[12] M. Rai, S.D. Deshmukh, A.P. Ingle, I.R. Gupta, M. Galdiero, S. Galdiero, Metal nanoparticles: the protective nanoshield against virus infection, Crit. Rev. Microbiol. 42 (1) (2016) 46–56. Epub 2014/04/24 https://doi.org/10.3109/1040841X.2013.879849. 24754250.

[13] B.A. Aderibigbe, Metal-based nanoparticles for the treatment of infectious diseases, Molecules 22 (8) (2017), https://doi.org/10.3390/molecules22081370. Epub 2017/08/19 28820471. PMCPMC6152252.

[14] W. Jiang, B.Y. Kim, J.T. Rutka, W.C. Chan, Nanoparticle-mediated cellular response is size-dependent, Nat. Nanotechnol. 3 (3) (2008) 145–150. Epub 2008/07/26 https://doi.org/10.1038/nnano.2008.30. 18654486.

[15] V.R. Patel, Y.K. Agrawal, Nanosuspension: an approach to enhance solubility of drugs, J. Adv. Pharm. Technol. Res. 2 (2) (2011) 81–87. Epub 2011/12/16 https://doi.org/10.4103/2231-4040.82950. 22171298. PMCPMC3217698.
[16] P. Foroozandeh, A.A. Aziz, Insight into cellular uptake and intracellular trafficking of nanoparticles, Nanoscale Res. Lett. 13 (1) (2018) 339. Epub 2018/10/27 https://doi.org/10.1186/s11671-018-2728-6. 30361809. PMCPMC6202307.
[17] A. Brandelli, A.C. Ritter, F.F. Veras, Antimicrobial activities of metal nanoparticles, in: M. Rai, R. Shegokar (Eds.), Metal Nanoparticles in Pharma, Springer, Cham, 2017, https://doi.org/10.1007/978-3-319-63790-7_15.
[18] S. Galdiero, A. Falanga, M. Vitiello, M. Cantisani, V. Marra, M. Galdiero, Silver nanoparticles as potential antiviral agents, Molecules 16 (10) (2011) 8894–8918. Epub 2011/10/26 https://doi.org/10.3390/molecules16108894. 22024958. PMCPMC6264685.
[19] Y. Cui, Y. Zhao, Y. Tian, W. Zhang, X. Lu, X. Jiang, The molecular mechanism of action of bactericidal gold nanoparticles on Escherichia coli, Biomaterials 33 (7) (2012) 2327–2333. Epub 2011/12/21 https://doi.org/10.1016/j.biomaterials.2011.11.057. 22182745.
[20] E.E. Connor, J. Mwamuka, A. Gole, C.J. Murphy, M.D. Wyatt, Gold nanoparticles are taken up by human cells but do not cause acute cytotoxicity, Small 1 (3) (2005) 325–327. Epub 2006/12/29 https://doi.org/10.1002/smll.200400093. 17193451.
[21] P. Ghosh, G. Han, M. De, C.K. Kim, V.M. Rotello, Gold nanoparticles in delivery applications, Adv. Drug Deliv. Rev. 60 (11) (2008) 1307–1315. Epub 2008/06/17 https://doi.org/10.1016/j.addr.2008.03.016. 18555555.
[22] A.C. Templeton, W.P. Wuelfing, R.W. Murray, Monolayer-protected cluster molecules, Acc. Chem. Res. 33 (1) (2000) 27–36. Epub 2000/03/29 https://doi.org/10.1021/ar9602664. 10639073.
[23] V. Lysenko, V. Lozovski, M. Lokshyn, Y.V. Gomeniuk, A. Dorovskih, N. Rusinchuk, et al., Nanoparticles as antiviral agents against adenoviruses, Adv. Nat. Sci. Nanosci. Nanotechnol. 9 (2018), 025021.
[24] J. Lu, M. Liong, Z. Li, J.I. Zink, F. Tamanoi, Biocompatibility, biodistribution, and drug-delivery efficiency of mesoporous silica nanoparticles for cancer therapy in animals, Small 6 (16) (2010) 1794–1805. Epub 2010/07/14 https://doi.org/10.1002/smll.201000538. 20623530. PMCPMC2952648.
[25] X. Huang, M. Li, Y. Xu, J. Zhang, X. Meng, X. An, et al., Novel gold nanorod-based HR1 peptide inhibitor for middle east respiratory syndrome coronavirus, ACS Appl. Mater. Interfaces 11 (22) (2019) 19799–19807. Epub 2019/05/18 https://doi.org/10.1021/acsami.9b04240. 31099550.
[26] A. Zacheo, J. Hodek, D. Witt, G.F.E. Mangiatordi, Q.K. Ong, O. Kocabiyik, et al., Multi-sulfonated ligands on gold nanoparticles as virucidal antiviral for Dengue virus, Sci. Rep.-Uk 10 (1) (2020), https://doi.org/10.1038/s41598-020-65892-3. PubMed PMID: WOS:000543961800028.
[27] A.M. Paul, Y. Shi, D. Acharya, J.R. Douglas, A. Cooley, J.F. Anderson, et al., Delivery of antiviral small interfering RNA with gold nanoparticles inhibits dengue virus infection in vitro, J. Gen. Virol. 95 (Pt 8) (2014) 1712–1722. Epub 2014/05/16 https://doi.org/10.1099/vir.0.066084-0. 24828333. PMCPMC4103068.
[28] Q.H. Quach, S.K. Ang, J.J. Chu, J.C.Y. Kah, Size-dependent neutralizing activity of gold nanoparticle-based subunit vaccine against dengue virus, Acta Biomater. 78 (2018) 224–235. Epub 2018/08/14 https://doi.org/10.1016/j.actbio.2018.08.011. 30099200.
[29] A. Tavakoli, M.S. Hashemzadeh, Inhibition of herpes simplex virus type 1 by copper oxide nanoparticles, J. Virol. Methods 275 (2020) 113688. Epub 2019/07/05 https://doi.org/10.1016/j.jviromet.2019.113688. 31271792.
[30] A. Tavakoli, A. Ataei-Pirkooh, G. Mm Sadeghi, F. Bokharaei-Salim, P. Sahrapour, S.J. Kiani, et al., Polyethylene glycol-coated zinc oxide nanoparticle: an efficient

nanoweapon to fight against herpes simplex virus type 1, Nanomedicine (London) 13 (21) (2018) 2675–2690. Epub 2018/10/23 https://doi.org/10.2217/nnm-2018-0089. 30346253.

[31] D. Baram-Pinto, S. Shukla, A. Gedanken, R. Sarid, Inhibition of HSV-1 attachment, entry, and cell-to-cell spread by functionalized multivalent gold nanoparticles, Small 6 (9) (2010) 1044–1050. Epub 2010/04/16 https://doi.org/10.1002/smll.200902384. 20394070.

[32] V. Cagno, P. Andreozzi, M. D'Alicarnasso, P. Jacob Silva, M. Mueller, M. Galloux, et al., Broad-spectrum non-toxic antiviral nanoparticles with a virucidal inhibition mechanism, Nat. Mater. (2017), https://doi.org/10.1038/nmat5053. 29251725.

[33] A. Halder, S. Das, D. Ojha, D. Chattopadhyay, A. Mukherjee, Highly monodispersed gold nanoparticles synthesis and inhibition of herpes simplex virus infections, Mater. Sci. Eng. C Mater. Biol. Appl. 89 (2018) 413–421. Epub 2018/05/13 https://doi.org/10.1016/j.msec.2018.04.005. 29752114.

[34] B.M. DeRussy, M.A. Aylward, Z. Fan, P.C. Ray, R. Tandon, Inhibition of cytomegalovirus infection and photothermolysis of infected cells using bioconjugated gold nanoparticles, Sci. Rep.-Uk 4 (2014), https://doi.org/10.1038/srep05550. PubMed PMID: WOS:000338421800002.

[35] C. Feng, P. Fang, Y. Zhou, L. Liu, L. Fang, S. Xiao, et al., Different effects of His-Au NCs and MES-Au NCs on the propagation of pseudorabies virus, Global Chall. 2 (8) (2018) 1800030. Epub 2018/06/25 https://doi.org/10.1002/gch2.201800030. 31565343. PMCPMC6607262.

[36] P.D. Jadzinsky, G. Calero, C.J. Ackerson, D.A. Bushnell, R.D. Kornberg, Structure of a thiol monolayer-protected gold nanoparticle at 1.1 angstrom resolution, Science 318 (5849) (2007) 430–433, https://doi.org/10.1126/science.1148624. PubMed PMID: WOS:000250230400040.

[37] M.C. Bowman, T.E. Ballard, C.J. Ackerson, D.L. Feldheim, D.M. Margolis, C. Melander, Inhibition of HIV fusion with multivalent gold nanoparticles, J. Am. Chem. Soc. 130 (22) (2008) 6896–6897. Epub 2008/05/14 https://doi.org/10.1021/ja710321g. 18473457. PMCPMC2916654.

[38] S. Vijayakumar, S. Ganesan, Gold nanoparticles as an HIV entry inhibitor, Curr. HIV Res. 10 (8) (2012) 643–646, https://doi.org/10.2174/157016212803901383. PubMed PMID: WOS:000311106100003.

[39] C.E. Pena-Gonzalez, P. Garcia-Broncano, M.F. Ottaviani, M. Cangiotti, A. Fattori, M. Hierro-Oliva, et al., Dendronized anionic gold nanoparticles: synthesis, characterization, and antiviral activity, Chemistry 22 (9) (2016) 2987–2999. Epub 2016/02/16 https://doi.org/10.1002/chem.201504262. 26875938.

[40] F. Chiodo, M. Marradi, J. Calvo, E. Yuste, S. Penades, Glycosystems in nanotechnology: gold glyconanoparticles as carrier for anti-HIV prodrugs, Beilstein J. Org. Chem. 10 (2014) 1339–1346, https://doi.org/10.3762/bjoc.10.136. PubMed PMID: WOS: 000338567500001.

[41] M.Y. Lee, J.A. Yang, H.S. Jung, S. Beack, J.E. Choi, W. Hur, et al., Hyaluronic acid-gold nanoparticle/interferon alpha complex for targeted treatment of hepatitis C virus infection, ACS Nano 6 (11) (2012) 9522–9531. Epub 2012/10/25 https://doi.org/10.1021/nn302538y. 23092111.

[42] Z. Wang, H. Liu, S.H. Yang, T. Wang, C. Liu, Y.C. Cao, Nanoparticle-based artificial RNA silencing machinery for antiviral therapy, Proc. Natl. Acad. Sci. U. S. A. 109 (31) (2012) 12387–12392. Epub 2012/07/18 https://doi.org/10.1073/pnas.1207766109. 22802676. PMCPMC3412013.

[43] I. Papp, C. Sieben, K. Ludwig, M. Roskamp, C. Bottcher, S. Schlecht, et al., Inhibition of influenza virus infection by multivalent sialic-acid-functionalized gold nanoparticles, Small 6 (24) (2010) 2900–2906. Epub 2010/11/26 https://doi.org/10.1002/smll.201001349. 21104827.

[44] J. Kim, M. Yeom, T. Lee, H.O. Kim, W. Na, A. Kang, et al., Porous gold nanoparticles for attenuating infectivity of influenza A virus, J. Nanobiotechnol. 18 (1) (2020) 54. Epub 2020/03/27 https://doi.org/10.1186/s12951-020-00611-8. 32209114. PMCPMC7092597.

[45] F. Pfaff, B. Gluck, T. Hoyer, D. Rohlander, A. Sauerbrei, R. Zell, Tungsten carbide nanoparticles show a broad spectrum virucidal activity against enveloped and nonenveloped model viruses using a guideline-standardized in vitro test, Lett. Appl. Microbiol. 69 (4) (2019) 302–309. Epub 2019/08/23 https://doi.org/10.1111/lam.13208. 31436888.

[46] J.J. Broglie, B. Alston, C. Yang, L. Ma, A.F. Adcock, W. Chen, et al., Antiviral activity of gold/copper sulfide core/shell nanoparticles against human norovirus virus-like particles, PLoS One 10 (10) (2015) e0141050. Epub 2015/10/17 https://doi.org/10.1371/journal.pone.0141050. 26474396. PMCPMC4608711.

[47] M.A. Melendez-Villanueva, K. Moran-Santibanez, J.J. Martinez-Sanmiguel, R. Rangel-Lopez, M.A. Garza-Navarro, C. Rodriguez-Padilla, et al., Virucidal activity of gold nanoparticles synthesized by green chemistry using garlic extract, Viruses 11 (12) (2019), https://doi.org/10.3390/v11121111. Epub 2019/12/06 31801280. PMCPMC6950311.

[48] N. Chen, Y. Zheng, J. Yin, X. Li, C. Zheng, Inhibitory effects of silver nanoparticles against adenovirus type 3 in vitro, J. Virol. Methods 193 (2) (2013) 470–477. Epub 2013/07/28 https://doi.org/10.1016/j.jviromet.2013.07.020. 23886562.

[49] T. Du, J.G. Liang, N. Dong, J. Lu, Y.Y. Fu, L.R. Fang, et al., Glutathione-capped Ag2S nanoclusters inhibit coronavirus proliferation through blockage of viral RNA synthesis and budding, ACS Appl. Mater. Interfaces 10 (5) (2018) 4369–4378, https://doi.org/10.1021/acsami.7b13811. PubMed PMID: WOS:000424851600004.

[50] C. Balagna, S. Perero, E. Percivalle, E.V. Nepita, M. Ferraris, Virucidal effect against coronavirus SARS-CoV-2 of a silver nanocluster/silica composite sputtered coating, Open Ceramics 1 (2020) 100006.

[51] A.N. Ben Salem, R. Zyed, M.A. Lassoued, S. Nidhal, S. Sfar, A. Mahjoub, Plant-derived nanoparticles enhance antiviral activity against coxsakievirus B3 by acting on virus particles and vero cells, Dig. J. Nanomater. Bios. 7 (2) (2012) 737–744. PubMed PMID: WOS:000306063200037.

[52] E.G. Haggag, A.M. Elshamy, M.A. Rabeh, N.M. Gabr, M. Salem, K.A. Youssif, et al., Antiviral potential of green synthesized silver nanoparticles of *Lampranthus coccineus* and Malephora lutea, Int. J. Nanomedicine 14 (2019) 6217–6229, https://doi.org/10.2147/Ijn.S214171. PubMed PMID: WOS:000479040500002.

[53] V. Sharma, S. Kaushik, P. Pandit, D. Dhull, J.P. Yadav, S. Kaushik, Green synthesis of silver nanoparticles from medicinal plants and evaluation of their antiviral potential against chikungunya virus, Appl. Microbiol. Biotechnol. 103 (2) (2019) 881–891. Epub 2018/11/11 https://doi.org/10.1007/s00253-018-9488-1. 30413849.

[54] B. Morejon, F. Pilaquinga, F. Domenech, D. Ganchala, A. Debut, M. Neira, Larvicidal activity of silver nanoparticles synthesized using extracts of Ambrosia arborescens (Asteraceae) to control *Aedes aegypti* L. (Diptera: Culicidae), J. Nanotechnol. 2018 (2018), https://doi.org/10.1155/2018/6917938. PubMed PMID: WOS:000446985700001.

[55] A. Aziz, M.A. Alshehri, N.A. Alanazi, C. Panneerselvam, S. Trivedi, F. Maggi, et al., Phytochemical analysis of Rhazya stricta extract and its use in fabrication of silver nanoparticles effective against mosquito vectors and microbial pathogens, Sci. Total Environ. 700 (2020), https://doi.org/10.1016/j.scitotenv.2019.134443. PubMed PMID: WOS:000499480800051.

[56] F. Pilaquinga, B. Morejon, D. Ganchala, J. Morey, N. Pina, A. Debut, et al., Green synthesis of silver nanoparticles using *Solanum mammosum* L. (Solanaceae) fruit extract and their larvicidal activity against *Aedes aegypti* L. (Diptera: Culicidae), PLoS One

14 (10) (2019), https://doi.org/10.1371/journal.pone.0224109. PubMed PMID: WOS: 000532670600024.

[57] K. Murugan, P. Aruna, C. Panneerselvam, P. Madhiyazhagan, M. Paulpandi, J. Subramaniam, et al., Fighting arboviral diseases: low toxicity on mammalian cells, dengue growth inhibition (in vitro), and mosquitocidal activity of Centroceras clavulatum-synthesized silver nanoparticles, Parasitol. Res. 115 (2) (2016) 651–662. Epub 2015/10/16 https://doi.org/10.1007/s00436-015-4783-6. 26462804.

[58] K. Murugan, D. Dinesh, M. Paulpandi, A.D. Althbyani, J. Subramaniam, P. Madhiyazhagan, et al., Erratum to: Nanoparticles in the fight against mosquito-borne diseases: bioactivity of Bruguiera cylindrica-synthesized nanoparticles against dengue virus DEN-2 (in vitro) and its mosquito vector Aedes aegypti (Diptera: Culicidae), Parasitol. Res. 114 (12) (2015) 4743. Epub 2015/09/29 https://doi.org/10.1007/s00436-015-4752-0. 26412056.

[59] K. Murugan, D. Dinesh, M. Paulpandi, A.D. Althbyani, J. Subramaniam, P. Madhiyazhagan, et al., Nanoparticles in the fight against mosquito-borne diseases: bioactivity of Bruguiera cylindrica-synthesized nanoparticles against dengue virus DEN-2 (in vitro) and its mosquito vector Aedes aegypti (Diptera: Culicidae), Parasitol. Res. 114 (12) (2015) 4349–4361. Epub 2015/08/21 https://doi.org/10.1007/s00436-015-4676-8. 26290219.

[60] V. Sujitha, K. Murugan, M. Paulpandi, C. Panneerselvam, U. Suresh, M. Roni, et al., Green-synthesized silver nanoparticles as a novel control tool against dengue virus (DEN-2) and its primary vector *Aedes aegypti*, Parasitol. Res. 114 (9) (2015) 3315–3325, https://doi.org/10.1007/s00436-015-4556-2. PubMed PMID: WOS: 000359865600016.

[61] D. Baram-Pinto, S. Shukla, N. Perkas, A. Gedanken, R. Sarid, Inhibition of herpes simplex virus type 1 infection by silver nanoparticles capped with mercaptoethane sulfonate, Bioconjug. Chem. 20 (8) (2009) 1497–1502.

[62] R.L. Hu, S.R. Li, F.J. Kong, R.J. Hou, X.L. Guan, F. Guo, Inhibition effect of silver nanoparticles on herpes simplex virus 2, Genet. Mol. Res. 13 (3) (2014) 7022–7028, https://doi.org/10.4238/2014.March.19.2. 24682984.

[63] P. Orlowski, E. Tomaszewska, M. Gniadek, P. Baska, J. Nowakowska, J. Sokolowska, et al., Tannic acid modified silver nanoparticles show antiviral activity in herpes simplex virus type 2 infection, PLoS One 9 (8) (2014), https://doi.org/10.1371/journal.pone.0104113. PubMed PMID: WOS:000341230400032.

[64] S. Gaikwad, A. Ingle, A. Gade, M. Rai, A. Falanga, N. Incoronato, et al., Antiviral activity of mycosynthesized silver nanoparticles against herpes simplex virus and human parainfluenza virus type 3, Int. J. Nanomedicine 8 (2013) 4303–4314. Epub 2013/11/16 https://doi.org/10.2147/IJN.S50070. 24235828. PMCPMC3826769.

[65] C.L. Wan, J.H. Tai, J. Zhang, Y. Guo, Q. Zhu, D. Ling, et al., Silver nanoparticles selectively induce human oncogenic gamma-herpesvirus-related cancer cell death through reactivating viral lytic replication, Cell Death Dis. 10 (2019), https://doi.org/10.1038/s41419-019-1624-z. PubMed PMID: WOS:000468676000001.

[66] R.S. El-Mohamady, T.A. Ghattas, M.F. Zawrah, Y.G.M. Abd El-Hafeiz, Inhibitory effect of silver nanoparticles on bovine herpesvirus-1, Int. J. Vet. Sci. Med. 6 (2) (2018) 296–300. Epub 2018/12/20 https://doi.org/10.1016/j.ijvsm.2018.09.002. 30564612. PMCPMC6286414.

[67] J.L. Elechiguerra, J.L. Burt, J.R. Morones, A. Camacho-Bragado, X. Gao, H.H. Lara, et al., Interaction of silver nanoparticles with HIV-1, J. Nanobiotechnol. 3 (2005) 6. Epub 2005/07/01 https://doi.org/10.1186/1477-3155-3-6. 15987516. PMCPMC1190212.

[68] H.H. Lara, N.V. Ayala-Nunez, L. Ixtepan-Turrent, C. Rodriguez-Padilla, Mode of antiviral action of silver nanoparticles against HIV-1, J. Nanobiotechnol. 8 (2010) 1. Epub 2010/02/11 https://doi.org/10.1186/1477-3155-8-1. 20145735. PMCPMC2818642.

[69] L. Lu, R.W. Sun, R. Chen, C.K. Hui, C.M. Ho, J.M. Luk, et al., Silver nanoparticles inhibit hepatitis B virus replication, Antivir. Ther. 13 (2) (2008) 253–262. Epub 2008/05/29 18505176.
[70] D.X. Xiang, Q. Chen, L. Pang, C.L. Zheng, Inhibitory effects of silver nanoparticles on H1N1 influenza A virus in vitro, J. Virol. Methods 178 (1–2) (2011) 137–142. Epub 2011/09/29 https://doi.org/10.1016/j.jviromet.2011.09.003. 21945220.
[71] D. Xiang, Y. Zheng, W. Duan, X. Li, J. Yin, S. Shigdar, et al., Inhibition of A/Human/Hubei/3/2005 (H3N2) influenza virus infection by silver nanoparticles in vitro and in vivo, Int. J. Nanomedicine 8 (2013) 4103–4113. Epub 2013/11/10 https://doi.org/10.2147/IJN.S53622. 24204140. PMCPMC3817021.
[72] Y.H. Li, Z.F. Lin, M.Q. Zhao, T.T. Xu, C.B. Wang, L. Hua, et al., Silver nanoparticle based codelivery of oseltamivir to inhibit the activity of the H1N1 influenza virus through ROS-mediated signaling pathways, ACS Appl. Mater. Interfaces 8 (37) (2016) 24385–24393, https://doi.org/10.1021/acsami.6b06613. PubMed PMID: WOS:000384033600015.
[73] Z.F. Lin, Y.H. Li, M. Guo, T.T. Xu, C.B. Wang, M.Q. Zhao, et al., The inhibition of H1N1 influenza virus-induced apoptosis by silver nanoparticles functionalized with zanamivir, RSC Adv. 7 (2) (2017) 742–750, https://doi.org/10.1039/c6ra25010f. PubMed PMID: WOS:000393743000020.
[74] Z.K. Alghrair, D.G. Fernig, B. Ebrahimi, Enhanced inhibition of influenza virus infection by peptide-noble-metal nanoparticle conjugates, Beilstein J. Nanotechnol. 10 (2019) 1038–1047. Epub 2019/06/06 https://doi.org/10.3762/bjnano.10.104. 31165030. PMCPMC6541353.
[75] A.Z. Bekele, K. Gokulan, K.M. Williams, S. Khare, Dose and size-dependent antiviral effects of silver nanoparticles on feline Calicivirus, a human norovirus surrogate, Foodborne Pathog. Dis. 13 (5) (2016) 239–244. Epub 2016/03/05 https://doi.org/10.1089/fpd.2015.2054. 26938256.
[76] J.L. Castro-Mayorga, W. Randazzo, M.J. Fabra, J.M. Lagaron, R. Aznar, G. Sanchez, Antiviral properties of silver nanoparticles against norovirus surrogates and their efficacy in coated polyhydroxyalkanoates systems, LWT-Food Sci. Technol. 79 (2017) 503–510, https://doi.org/10.1016/j.lwt.2017.01.065. PubMed PMID: WOS:000395603100062.
[77] S. Park, H.H. Park, S.Y. Kim, S.J. Kim, K. Woo, G. Ko, Antiviral properties of silver nanoparticles on a magnetic hybrid colloid, Appl. Environ. Microbiol. 80 (8) (2014) 2343–2350. Epub 2014/02/04 https://doi.org/10.1128/AEM.03427-13. 24487537. PMCPMC3993170.
[78] T.Q. Huy, N.T. Hien Thanh, N.T. Thuy, P.V. Chung, P.N. Hung, A.T. Le, et al., Cytotoxicity and antiviral activity of electrochemical - synthesized silver nanoparticles against poliovirus, J. Virol. Methods 241 (2017) 52–57. Epub 2017/01/04 https://doi.org/10.1016/j.jviromet.2016.12.015. 28040515.
[79] X.X. Yang, C.M. Li, C.Z. Huang, Curcumin modified silver nanoparticles for highly efficient inhibition of respiratory syncytial virus infection, Nanoscale 8 (5) (2016) 3040–3048. Epub 2016/01/20 https://doi.org/10.1039/c5nr07918g. 26781043.
[80] B. Borrego, G. Lorenzo, J.D. Mota-Morales, H. Almanza-Reyes, F. Mateos, E. Lopez-Gil, et al., Potential application of silver nanoparticles to control the infectivity of Rift Valley fever virus in vitro and in vivo, Nanomedicine 12 (5) (2016) 1185–1192. Epub 2016/03/13 https://doi.org/10.1016/j.nano.2016.01.021. 26970026.
[81] J.L. Speshock, R.C. Murdock, L.K. Braydich-Stolle, A.M. Schrand, S.M. Hussain, Interaction of silver nanoparticles with Tacaribe virus, J. Nanobiotechnol. 8 (2010) 19. Epub 2010/08/20 https://doi.org/10.1186/1477-3155-8-19. 20718972. PMCPMC2936366.
[82] J.V. Rogers, C.V. Parkinson, Y.W. Choi, J.L. Speshock, S.M. Hussain, A preliminary assessment of silver nanoparticle inhibition of monkeypox virus plaque formation,

Nanoscale Res. Lett. 3 (4) (2008) 129–133, https://doi.org/10.1007/s11671-008-9128-2. PubMed PMID: WOS:000255572800001.

[83] G. Grass, C. Rensing, M. Solioz, Metallic copper as an antimicrobial surface, Appl. Environ. Microbiol. 77 (5) (2011) 1541–1547, https://doi.org/10.1128/Aem.02766-10. PubMed PMID: WOS:000287700100001.

[84] Y. Fujimori, T. Sato, T. Hayata, T. Nagao, M. Nakayama, T. Nakayama, et al., Novel antiviral characteristics of nanosized copper(I) iodide particles showing inactivation activity against 2009 pandemic H1N1 influenza virus, Appl. Environ. Microbiol. 78 (4) (2012) 951–955. Epub 2011/12/14 https://doi.org/10.1128/AEM.06284-11. 22156433. PMCPMC3272987.

[85] J.O. Noyce, H. Michels, C.W. Keevil, Inactivation of influenza A virus on copper versus stainless steel surfaces, Appl. Environ. Microbiol. 73 (8) (2007) 2748–2750. Epub 2007/01/30 https://doi.org/10.1128/AEM.01139-06. 17259354. PMCPMC1855605.

[86] S.L. Warnes, C.W. Keevil, Inactivation of norovirus on dry copper alloy surfaces, PLoS One 8 (9) (2013) e75017. Epub 2013/09/17 https://doi.org/10.1371/journal.pone.0075017. 24040380. PMCPMC3767632.

[87] G. Borkow, H.H. Lara, C.Y. Covington, A. Nyamathi, J. Gabbay, Deactivation of human immunodeficiency virus type 1 in medium by copper oxide-containing filters, Antimicrob. Agents Chemother. 52 (2) (2008) 518–525. Epub 2007/12/12 https://doi.org/10.1128/AAC.00899-07. 18070974. PMCPMC2224774.

[88] G. Borkow, S.S. Zhou, T. Page, J. Gabbay, A novel anti-influenza copper oxide containing respiratory face mask, PLoS One 5 (6) (2010) e11295. Epub 2010/07/02 https://doi.org/10.1371/journal.pone.0011295. 20592763. PMCPMC2892464.

[89] I.D. Jana, P. Kumbhakar, S. Banerjee, C.C. Gowda, N. Kedia, S.K. Kuila, et al., Copper nanoparticle–graphene composite-based transparent surface coating with antiviral activity against influenza virus, ACS Appl. Nano Mater. 4 (1) (2021) 352–362.

[90] X. Hang, H. Peng, H. Song, Z. Qi, X. Miao, W. Xu, Antiviral activity of cuprous oxide nanoparticles against hepatitis C virus in vitro, J. Virol. Methods 222 (2015) 150–157. Epub 2015/06/28 https://doi.org/10.1016/j.jviromet.2015.06.010. 26116793.

[91] R. Kumar, G. Sahoo, K. Pandey, M.K. Nayak, R. Topno, V. Rabidas, et al., Virostatic potential of zinc oxide (ZnO) nanoparticles on capsid protein of cytoplasmic side of chikungunya virus, Int. J. Infect. Dis. 73 (2018) 368, https://doi.org/10.1016/j.ijid.2018.04.4247. PubMed PMID: WOS:000440345100813.

[92] S. Muthamil Selvan, K. Vijai Anand, K. Govindaraju, S. Tamilselvan, V.G. Kumar, K.S. Subramanian, et al., Green synthesis of copper oxide nanoparticles and mosquito larvicidal activity against dengue, zika and chikungunya causing vector Aedes aegypti, IET Nanobiotechnol. 12 (8) (2018) 1042–1046. Epub 2019/04/10 https://doi.org/10.1049/iet-nbt.2018.5083. 30964011.

[93] S. Agnihothram, L. Mullis, T.A. Townsend, F. Watanabe, T. Mustafa, A. Biris, et al., Titanium dioxide nanoparticles evoke proinflammatory response during murine norovirus infection despite having minimal effects on virus replication, Int. J. Nanotechnol. Eng. Med. 1 (3) (2016) 63–73. Epub 2016/12/01 10.25141/2474-8811-2016-3.0063. 29930994. PMCPMC6007025.

[94] J. Zhong, Y. Xia, L. Hua, X. Liu, M. Xiao, T. Xu, et al., Functionalized selenium nanoparticles enhance the anti-EV71 activity of oseltamivir in human astrocytoma cell model, Artif. Cells Nanomed. Biotechnol. 47 (1) (2019) 3485–3491. Epub 2019/08/20 https://doi.org/10.1080/21691401.2019.1640716. 31422717.

[95] Y.H. Li, T.T. Xu, Z.F. Lin, C.B. Wang, Y. Xia, M. Guo, et al., Inhibition of enterovirus A71 by selenium nanoparticles interferes with JNK signaling pathways, ACS Omega 4 (4) (2019) 6720–6725, https://doi.org/10.1021/acsomega.8b03502. PubMed PMID: WOS:000466552500070.

[96] Z. Lin, Y. Li, G. Gong, Y. Xia, C. Wang, Y. Chen, et al., Restriction of H1N1 influenza virus infection by selenium nanoparticles loaded with ribavirin via resisting caspase-3

apoptotic pathway, Int. J. Nanomedicine 13 (2018) 5787–5797. Epub 2018/10/13 https://doi.org/10.2147/IJN.S177658. 30310281. PMCPMC6165773.

[97] R. Kumar, M. Nayak, G.C. Sahoo, K. Pandey, M.C. Sarkar, Y. Ansari, et al., Iron oxide nanoparticles based antiviral activity of H1N1 influenza A virus, J. Infect. Chemother. 25 (5) (2019) 325–329. Epub 2019/02/17 https://doi.org/10.1016/j.jiac.2018.12.006. 30770182.

[98] B. Ziem, W. Azab, M.F. Gholami, J.P. Rabe, N. Osterrieder, R. Haag, Size-dependent inhibition of herpesvirus cellular entry by polyvalent nanoarchitectures, Nanoscale 9 (11) (2017) 3774–3783, https://doi.org/10.1039/c7nr00611j.

[99] A. Chuvilin, U. Kaiser, E. Bichoutskaia, N.A. Besley, A.N. Khlobystov, Direct transformation of graphene to fullerene, Nat. Chem. 2 (6) (2010) 450–453. Epub 2010/05/22 https://doi.org/10.1038/nchem.644. 20489712.

[100] A.K. Geim, K.S. Novoselov, The rise of graphene, Nat. Mater. 6 (3) (2007) 183–191. Epub 2007/03/03 https://doi.org/10.1038/nmat1849. 17330084.

[101] K.S. Novoselov, A.K. Geim, S.V. Morozov, D. Jiang, Y. Zhang, S.V. Dubonos, et al., Electric field effect in atomically thin carbon films, Science 306 (5696) (2004) 666–669. Epub 2004/10/23 https://doi.org/10.1126/science.1102896. 15499015.

[102] A.E.F. Oliveira, G.B. Braga, C.R.T. Tarley, A.C. Pereira, Thermally reduced graphene oxide: synthesis, studies and characterization, J. Mater. Sci. 53 (17) (2018) 12005–12015, https://doi.org/10.1007/s10853-018-2473-3. PubMed PMID: WOS: 000435357300006.

[103] I. Sengupta, S. Chakraborty, M. Talukdar, S.K. Pal, S. Chakraborty, Thermal reduction of graphene oxide: how temperature influences purity, J. Mater. Res. 33 (23) (2018) 4113–4122, https://doi.org/10.1557/jmr.2018.338. PubMed PMID: WOS: 000454298700024.

[104] W. Gao, L.B. Alemany, L. Ci, P.M. Ajayan, New insights into the structure and reduction of graphite oxide, Nat. Chem. 1 (5) (2009) 403–408. Epub 2009/08/01 https://doi.org/10.1038/nchem.281. 21378895.

[105] J.N. Chen, H. Peng, X.P. Wang, F. Shao, Z.D. Yuan, H.Y. Han, Graphene oxide exhibits broad-spectrum antimicrobial activity against bacterial phytopathogens and fungal conidia by intertwining and membrane perturbation, Nanoscale 6 (3) (2014) 1879–1889, https://doi.org/10.1039/c3nr04941h. PubMed PMID: WOS: 000330041400084.

[106] A.R. Deokar, A.P. Nagvenkar, I. Kalt, L. Shani, Y. Yeshurun, A. Gedanken, et al., Graphene-based "Hot Plate" for the capture and destruction of the herpes simplex virus type 1, Bioconjug. Chem. 28 (4) (2017) 1115–1122, https://doi.org/10.1021/acs.bioconjchem.7b00030. PubMed PMID: WOS:000399965800031.

[107] M. Sametband, I. Kalt, A. Gedanken, R. Sarid, Herpes simplex virus type-1 attachment inhibition by functionalized graphene oxide, ACS Appl. Mater. Interfaces 6 (2) (2014) 1228–1235, https://doi.org/10.1021/am405040z. 24364493.

[108] Z.Y. Song, X.Y. Wang, G.X. Zhu, Q.G. Nian, H.Y. Zhou, D. Yang, et al., Virus capture and destruction by label-free graphene oxide for detection and disinfection applications, Small 11 (9–10) (2015) 1171–1176, https://doi.org/10.1002/smll.201401706. PubMed PMID: WOS:000350981000015.

[109] S. Ye, K. Shao, Z. Li, N. Guo, Y. Zuo, Q. Li, et al., Antiviral activity of graphene oxide: how sharp edged structure and charge matter, ACS Appl. Mater. Interfaces 7 (38) (2015) 21571–21579. Epub 2015/09/16 https://doi.org/10.1021/acsami.5b06876. 26370151.

[110] X.G. Hu, L. Mu, J.P. Wen, Q.X. Zhou, Covalently synthesized graphene oxide-aptamer nanosheets for efficient visible-light photocatalysis of nucleic acids and proteins of viruses, Carbon 50 (8) (2012) 2772–2781, https://doi.org/10.1016/j.carbon.2012.02.038. PubMed PMID: WOS:000303616700012.

[111] O. Akhavan, M. Choobtashani, E. Ghaderi, Protein degradation and RNA efflux of viruses photocatalyzed by graphene-tungsten oxide composite under visible light

irradiation, J. Phys. Chem. C 116 (17) (2012) 9653–9659, https://doi.org/10.1021/jp301707m. PubMed PMID: WOS:000303426500028.
[112] M.F. Gholami, D. Lauster, K. Ludwig, J. Storm, B. Ziem, N. Severin, et al., Functionalized graphene as extracellular matrix mimics: toward well-defined 2D nanomaterials for multivalent virus interactions, Adv. Funct. Mater. 27 (15) (2017), https://doi.org/10.1002/adfm.201606477. PubMed PMID: WOS:000399019000012.
[113] B. Ziem, J. Rahn, I. Donskyi, K. Silberreis, L. Cuellar, J. Dernedde, et al., Polyvalent 2D entry inhibitors for Pseudorabies and African swine fever virus, Macromol. Biosci. 17 (6) (2017), https://doi.org/10.1002/mabi.201600499. PubMed PMID: WOS:000405565300003.
[114] B.J. Reynwar, G. Illya, V.A. Harmandaris, M.M. Muller, K. Kremer, M. Deserno, Aggregation and vesiculation of membrane proteins by curvature-mediated interactions, Nature 447 (7143) (2007) 461–464, https://doi.org/10.1038/nature05840. PubMed PMID: WOS:000246693100044.
[115] I.S. Donskyi, W. Azab, J.L. Cuellar-Camacho, G. Guday, A. Lippitz, W.E.S. Unger, et al., Functionalized nanographene sheets with high antiviral activity through synergistic electrostatic and hydrophobic interactions, Nanoscale 11 (34) (2019) 15804–15809. Epub 2019/08/23 https://doi.org/10.1039/c9nr05273a. 31433428.
[116] S.C. Gupta, S. Patchva, B.B. Aggarwal, Therapeutic roles of curcumin: lessons learned from clinical trials, AAPS J. 15 (1) (2013) 195–218. Epub 2012/11/13 https://doi.org/10.1208/s12248-012-9432-8. 23143785. PMCPMC3535097.
[117] X.X. Yang, C.M. Li, Y.F. Li, J. Wang, C.Z. Huang, Synergistic antiviral effect of curcumin functionalized graphene oxide against respiratory syncytial virus infection, Nanoscale 9 (41) (2017) 16086–16092. Epub 2017/10/17 https://doi.org/10.1039/c7nr06520e. 29034936.
[118] Z. Jendzelovska, R. Jendzelovsky, B. Kucharova, P. Fedorocko, Hypericin in the light and in the dark: two sides of the same coin, Front. Plant Sci. 7 (2016) 560. Epub 2016/05/21 https://doi.org/10.3389/fpls.2016.00560. 27200034. PMCPMC4859072.
[119] X. Du, R. Xiao, H. Fu, Z. Yuan, W. Zhang, L. Yin, et al., Hypericin-loaded graphene oxide protects ducks against a novel duck reovirus, Mater. Sci. Eng. C Mater. Biol. Appl. 105 (2019) 110052. Epub 2019/09/25 https://doi.org/10.1016/j.msec.2019.110052. 31546360.
[120] A. Sharma, J. Das, Small molecules derived carbon dots: synthesis and applications in sensing, catalysis, imaging, and biomedicine, J. Nanobiotechnol. 17 (1) (2019) 92. Epub 2019/08/28 https://doi.org/10.1186/s12951-019-0525-8. 31451110. PMCPMC6709552.
[121] F.M. Lin, Y.W. Bao, F.G. Wu, Carbon dots for sensing and killing microorganisms, C.-J. Carbon Res. (2) (2019) 5, https://doi.org/10.3390/c5020033. PubMed PMID: WOS:000485188900020.
[122] A. Barras, Q. Pagneux, F. Sane, Q. Wang, R. Boukherroub, D. Hober, et al., High efficiency of functional carbon nanodots as entry inhibitors of herpes simplex virus type 1, ACS Appl. Mater. Interfaces 8 (14) (2016) 9004–9013, https://doi.org/10.1021/acsami.6b01681. PubMed PMID: WOS:000374274900019.
[123] M.Z. Fahmi, W. Sukmayani, S.Q. Khairunisa, A.M. Witaningrum, D.W. Indriati, M.Q.Y. Matondang, et al., Design of boronic acid-attributed carbon dots on inhibits HIV-1 entry, RSC Adv. 6 (95) (2016) 92996–93002, https://doi.org/10.1039/c6ra21062g. PubMed PMID: WOS:000385389800102.
[124] T. Du, J.G. Liang, N. Dong, L. Liu, L.R. Fang, S.B. Xiao, et al., Carbon dots as inhibitors of virus by activation of type I interferon response, Carbon 110 (2016) 278–285, https://doi.org/10.1016/j.carbon.2016.09.032. PubMed PMID: WOS:000386402700031.
[125] T. Tong, H. Hu, J. Zhou, S. Deng, X. Zhang, W. Tang, et al., Glycyrrhizic-acid-based carbon dots with high antiviral activity by multisite inhibition mechanisms, Small 16 (13) (2020) e1906206. Epub 2020/02/23 https://doi.org/10.1002/smll.201906206. 32077621. PMCPMC7169479.

[126] X.L. Dong, M.M. Moyer, F. Yang, Y.P. Sun, L.J. Yang, Carbon dots' antiviral functions against noroviruses, Sci. Rep.-Uk 7 (2017), https://doi.org/10.1038/s41598-017-00675-x. PubMed PMID: WOS:000397879800007.

[127] D. Ting, N. Dong, L.R. Fang, J. Lu, J. Bi, S.B. Xiao, et al., Multisite inhibitors for enteric coronavirus: antiviral cationic carbon dots based on curcumin, ACS Appl. Nano Mater. 1 (10) (2018) 5451–5459, https://doi.org/10.1021/acsanm.8b00779. PubMed PMID: WOS:000461401100012.

[128] A. Loczechin, K. Seron, A. Barras, E. Giovanelli, S. Belouzard, Y.T. Chen, et al., Functional carbon quantum dots as medical countermeasures to human coronavirus, ACS Appl. Mater. Interfaces 11 (46) (2019) 42964–42974, https://doi.org/10.1021/acsami.9b15032. PubMed PMID: WOS:000499740300010.

[129] S.M. Huang, J.J. Gu, J. Ye, B. Fang, S.F. Wan, C.Y. Wang, et al., Benzoxazine monomer derived carbon dots as a broad-spectrum agent to block viral infectivity, J. Colloid Interface Sci. 542 (2019) 198–206, https://doi.org/10.1016/j.jcis.2019.02.010. PubMed PMID: WOS:000461536400022.

[130] C.J. Lin, L. Chang, H.W. Chu, H.J. Lin, P.C. Chang, R.Y.L. Wang, et al., High amplification of the antiviral activity of curcumin through transformation into carbon quantum dots, Small 15 (41) (2019), https://doi.org/10.1002/smll.201902641. PubMed PMID: WOS:000484199200001.

[131] D. Iannazzo, A. Pistone, S. Ferro, L. De Luca, A.M. Monforte, R. Romeo, et al., Graphene quantum dots based systems as HIV inhibitors, Bioconjug. Chem. 29 (9) (2018) 3084–3093, https://doi.org/10.1021/acs.bioconjchem.8b00448. PubMed PMID: WOS:000445439500022.

[132] M. Prato, Fullerene chemistry for materials science applications, J. Mater. Chem. 7 (7) (1997) 1097–1109, https://doi.org/10.1039/a700080d. PubMed PMID: WOS:A1997XK01100005.

[133] Sijbesma R., Srdanov G., Wudl F., Castoro J.A., Wilkins C., Friedman S.H., et al. Synthesis of a fullerene derivative for the inhibition of HIV enzymes. J. Am. Chem. Soc. 1993;115(15):6510–2. https://doi.org/10.1021/ja00068a006. PubMed PMID: WOS:A1993LT67300006.

[134] P. Anilkumar, F. Lu, L. Cao, P.G. Luo, J.H. Liu, S. Sahu, et al., Fullerenes for applications in biology and medicine, Curr. Med. Chem. 18 (14) (2011) 2045–2059, https://doi.org/10.2174/092986711795656225. PubMed PMID: WOS:000290612000002.

[135] E. Castro, A.H. Garcia, G. Zavala, L. Echegoyen, Fullerenes in biology and medicine, J. Mater. Chem. B 5 (32) (2017) 6523–6535, https://doi.org/10.1039/c7tb00855d. PubMed PMID: WOS:000407684800008.

[136] J. Luczkowiak, A. Munoz, M. Sanchez-Navarro, R. Ribeiro-Viana, A. Ginieis, B.M. Illescas, et al., Glycofullerenes inhibit viral infection, Biomacromolecules 14 (2) (2013) 431–437, https://doi.org/10.1021/bm3016658. PubMed PMID: WOS:000314908500017.

[137] A. Munoz, D. Sigwalt, B.M. Illescas, J. Luczkowiak, L. Rodriguez-Perez, I. Nierengarten, et al., Synthesis of giant globular multivalent glycofullerenes as potent inhibitors in a model of Ebola virus infection, Nat. Chem. 8 (1) (2016) 50–57, https://doi.org/10.1038/Nchem.2387. PubMed PMID: WOS:000366675400011.

[138] J. Ramos-Soriano, J.J. Reina, B.M. Illescas, N. de la Cruz, L. Rodriguez-Perez, F. Lasala, et al., Synthesis of highly efficient multivalent disaccharide/[60]fullerene nanoballs for emergent viruses, J. Am. Chem. Soc. 141 (38) (2019) 15403–15412, https://doi.org/10.1021/jacs.9b08003. PubMed PMID: WOS:000488322500052.

[139] A.B. Kornev, A.S. Peregudov, V.M. Martynenko, J. Balzarini, B. Hoorelbeke, P.A. Troshin, Synthesis and antiviral activity of highly water-soluble polycarboxylic derivatives of [70]fullerene, Chem. Commun. 47 (29) (2011) 8298–8300, https://doi.org/10.1039/c1cc12209f. PubMed PMID: WOS:000292980900028.

[140] L.B. Piotrovsky, O.I. Kiselev, Fullerenes and viruses, Fuller. Nanotub. Car. N 12 (1–2) (2004) 397–403, https://doi.org/10.1081/Fst-120027198. PubMed PMID: WOS: 000220551600065.

[141] H. Kataoka, T. Ohe, K. Takahashi, S. Nakamura, T. Mashino, Novel fullerene derivatives as dual inhibitors of Hepatitis C virus NS5B polymerase and NS3/4A protease, Bioorg. Med. Chem. Lett. 26 (19) (2016) 4565–4567, https://doi.org/10.1016/j.bmcl.2016.08.086. PubMed PMID: WOS:000383359400002.

[142] N.E. Fedorova, R.R. Klimova, Y.A. Tulenev, E.V. Chichev, A.B. Kornev, P.A. Troshin, et al., Carboxylic fullerene C-60 derivatives: efficient microbicides against herpes simplex virus and cytomegalovirus infections in vitro, Mendeleev Commun. 22 (5) (2012) 254–256, https://doi.org/10.1016/j.mencom.2012.09.009. PubMed PMID: WOS: 000310823600009.

[143] F. Kasermann, C. Kempf, Photodynamic inactivation of enveloped viruses by buckminsterfullerene, Antivir. Res. 34 (1) (1997) 65–70. Epub 1997/03/01 https://doi.org/10.1016/s0166-3542(96)01207-7. 9107386.

[144] A. Moreno-Lanceta, M. Medrano-Bosch, P. Melgar-Lesmes, Single-walled carbon nanohorns as promising nanotube-derived delivery systems to treat cancer, Pharmaceutics 12 (9) (2020), https://doi.org/10.3390/pharmaceutics12090850. PubMed PMID: WOS:000580103100001.

[145] E. Miyako, H. Nagata, K. Hirano, K. Sakamoto, Y. Makita, K. Nakayama, et al., Photoinduced antiviral carbon nanohorns, Nanotechnology 19 (7) (2008), 075106. Epub 2008/02/20 https://doi.org/10.1088/0957-4484/19/7/075106. 21817631.

CHAPTER

5

Antiviral potential of nanomaterials: Novel solutions for emerging challenges

Walid Azab

Institute of Virology, Center for Infection Medicine, Freie Universitaet Berlin, Berlin, Germany

1 Introduction

The antiviral activity of nanoparticles is facilitated by their unique properties that make them attractive tools for viral treatment. The nanoscale size of particles can facilitate cellular uptake and allow reaching anatomically privileged sites. The large surface area-to-volume ratios can ensure decoration with large payloads of functional molecules that can increase the interaction with pathogens, leading to optimized drug dosing and delivery and increasing stability. The surface charge can facilitate binding to infectious agents and subsequently inhibit the initial steps of infection [1,2]. In addition, nanoparticles can contain biomimetic characteristics that enhance their intrinsic antiviral properties as in the case of silver and gold nanoparticles and dendrimers [3,4].

The antiviral mechanisms of nanoparticles can be either direct through exerting virucidal effect that destroys virus particles or indirect by targeting the different steps of virus replication cycle that includes attachment, entry, replication, and egress. Most often, nanoparticles block virus binding to cell surface by the interaction with viral envelope or capsid proteins and thus inhibiting infection. Furthermore, nanoparticles can inhibit viral replication and prevent release of progeny virions from infected cells [5,6].

As a continuation of Chapter 4, this chapter will focus on other types of nanoparticles and recapitulate the mechanism of actions and potential applications of dendrimers, liposomes, and nanobodies for the treatment

https://doi.org/10.1016/B978-0-12-821637-8.00003-1
Copyright © 2023 Elsevier Inc. All rights reserved.

of emerging viral diseases. Finally, perspectives on the potential use of nanoparticles in the future to fight against infectious diseases are discussed.

2 Polymer-based nanogels

Dendritic polymeric nanoparticles derived from polyglycerol (PG) are known to have several characteristic features that make them interesting molecules to be used for broad-spectrum biomedical applications [7]. They have highly biocompatible nature, a good solubility in different solvents, and large number of functional groups at the surface that enable further modifications. Furthermore, the presence of amine groups on the surface is another important feature that proved to be useful for providing an anchoring point in the preparation of protein-resistant surfaces, synthesis of polyglycerol (PG)-based peptide conjugates, drug delivery systems, macromolecular prodrugs, and condensation of nucleic acid drugs [7–10]. Polymeric nanogels are 3D network of cross-linked hydrogel nanoparticles composed of water-soluble and swellable (naturally soft) polymers. Besides the above-mentioned features of dendritic polymers, they possess large surface area that facilitates multivalent interaction and they can be degraded to smaller sized molecules that can be easily removed by renal clearance [11]. Functionalization of polymeric nanogels with certain bioactive ligands may play a crucial role in the field of preventive medicine as they can interfere with pathogen (viral) infection [12]. Given the recent advances in dendritic polyglycerol architecture chemistry, nanogel technology has established itself as a robust platform for creation of functional materials with controlled size and multifunctionality for different fields of biomedical applications.

The mechanism of action of these sulfated nanogels is mainly dependent on the strong multivalent interaction between the nanoparticles and virus particles, which results in virus capture and neutralization before getting in contact with cells. Nanogel-virus particles complexes can afterward be cleared by the immune system. Although some nanogels (based on their size) can enter the cells through endocytic pathways that might be similar to the pathways used by the virus, their inhibitory effect on virus replication is only minimal and needs further investigation [13,14].

To mimic cellular HS, sulfated nanogels (NGs) based on dendritic polyglycerol sulfate (dPGS; 100–200 nm in size) were prepared using the inverse nanoprecipitation technique based on a bioorthogonal click reaction [15]. The flexibility and rigidity (deformation) of these NGs were adjusted by using two different types (dendritic and linear polyglycerols) of spacers. To this end, the efficiency and the mechanism of inhibition of (DNA and RNA) HS-dependent virus (herpesviruses and arteriviruses) infections were investigated [14]. Incubating flexible and rigid NGs with virus particles resulted in a significant inhibition of virus infection; the effect was more potent with flexible NGs. It was clear that the NGs exhibited their inhibiting effects at the level of virus attachment, receptor binding, and/or penetration rather than at downstream events of the virus replication cycle. Interestingly, the presence of NGs in the surrounding media could also inhibit any egressed (new synthesized) virions from cells, preventing infection of new adjacent cells. It was further shown that NGs were noncytotoxic and remain active in the medium for long time [14]. The contact area between soft multivalent architectures and pathogens is an important parameter that determines the efficacy and strength of interaction [15]. The potent inhibitory effect of flexible NGs can be attributed to adaptation of these NGs to the virus surface during the binding process, which leads to higher valent interactions, efficient shielding, and hence reduce the probability of detachment from the viral surface [14].

A series of sialic acid-conjugated polyglycerol-based nanoparticles with different diameters (1–250 nm) were synthesized and tested against influenza A viruses. Stable and biocompatible glycoconjugated nanoparticles using water-soluble hyperbranched dendritic polyglycerol nanoparticles as scaffolds were functionalized with sialic acids (SA). Particle sizes (3, 4, 50, and 70 nm) were varied along with the degree of SA functionalization to match the corresponding virus size and receptor multiplicity to achieve maximum efficiency. The synthesized multivalent nanoparticles could inhibit influenza A virus infection by blocking virus-cell binding and subsequent fusion to the cell membrane. It was evident that the inhibitory activity of the tested polymeric nanoparticles drastically increased with size; particles with similar dimensions to the virus (50–100 nm) were more potent [12]. Flexible and rigid multivalent 3D SA-conjugated nanogels were also tested against influenza A viruses. Although both types of NGs could inhibit virus infection, flexible sialylated NGs (250 nm in diameter) could deform and adapt onto the virus surface via specific ligand-receptor multivalent interactions and efficiently blocked virus adhesion to the cells. It was speculated that SA residues on the NGs surface interact with hemagglutinin spike proteins on virus envelope. As demonstrated through hemagglutination inhibition assay, the high flexibility of sialylated nanogels improves virus inhibition by 400 times as compared to the rigid ones [16]. The concept of "topology-matching design" was facilitated greatly by flexible nanoarchitectures. The concept follows the principle of matching the size and topology to a virion in order to achieve robust binding with the virion and subsequently compete with cell interaction [17]. To this end, rigid nanoparticles with a matching nanotopology to influenza A virus were used to generate nanoinhibitors with increased contact area and strong binding affinities. To achieve these features, a virus-like nanoparticle (VLNP) with nanospikes was first synthesized via a novel single-micelle epitaxial growth approach in a low concentration surfactant oil/water biphase system [18]. The VLNPs were functionalized with linear polyglycerol-sialyllactose (LPG-SAL, to block virus entry) and LPG-zanamivir (LPG-Zan, an NA inhibitor [19]) through copper-free SPAAC click reactions. The synthesized nanoinhibitor could neutralize influenza viruses extracellularly and block attachment and entry to the host cells. Furthermore, virus replication was significantly reduced in the presence of the reverse designed nanoinhibitors. Even when the nanoinhibitors were used 24 hours after infection, more than 99% of inhibition was still achieved indicating that these inhibitors might be a potent antiviral for the prevention and treatment of influenza infection [20].

Synthesis of novel Nucleoside Reverse Transcriptase Inhibitors (NRTIs) prodrugs from cholesteryl-ε-polylysine (CEPL) nanogels by conjugation with NRTI 5′-succinate derivatives (sNRTI) was investigated [21,22]. The new nanodrug is considered attractive due to several properties, which include biodegradability, small particle size, and high NRTI loading (30% by weight) of these carriers. These features can extend drug release due to the slower hydrolysis of ester bonds in the conjugates which would allow a weekly administration schedule, high therapeutic index (>1000) with a lower toxicity compared to free NRTIs, and efficient accumulation in macrophages that are known as carriers for HIV-1 infection. Other nanogel conjugates of Zidovudine (AZT), Lamivudine (3TC), and Abacavir (ABC) were also synthesized and investigated individually and in formulations similar to clinical NRTI cocktails. Interestingly, the nanodrug formulations demonstrated 10-fold suppression of reverse transcriptase activity in HIV-infected macrophages for single, dual, and triple nanodrug cocktails. Nanogel conjugate of Lamivudine was the most effective single nanodrug. Spared intravenous (i.v) injections of PEGylated CEPL-sAZT alone could efficiently suppress HIV-1 RT activity to background level in humanized mouse (hu-PBL) HIV model [21]. Furthermore, these nanogels proved to be efficient carriers for delivery of triphosphorylated NRTIs, which has high anti-HIV activity along with low neurotoxicity [22]. A series of poly(*N*-vinylcaprolactam) (PVCL) nanogels were synthesized by thermo-precipitation in aqueous phase via free radical polymerization in the presence of a surfactant (SDS) to facilitate size control and to reduce the distribution of the generated particles. Among the synthesized NGs, $PVCL80_4$ with a size of ca. 200 nm and 4% of crosslinking agent showed antiviral behavior and inhibited HIV-1 infection in target cells at the maximum nontoxic concentration [23]. It is noteworthy to mention that the synthesized PVCL NGs cannot form anionic moieties, which have been demonstrated to be important for antiviral activity [17,24], therefore, the proposed inhibitory action is speculated to be due to the functional groups of the polymer matrix itself [23].

Conjugation of NGs with natural compounds was also attempted. *Cymbopogon citratus* (DC.) Stapf is a perennial herb originated from India and belongs to Poaceae family. Its volatile oil (VO), with the major compound "citral," which is a natural mixture of the isomers geranial (α-citral) and neral (β-citral), is obtained from fresh leaves and used in food, cosmetics, and pharmaceutical industries [25,26]. Hydrogel containing *C. citratus* Stapf *C. citratus* volatile oil (CcVO) encapsulated in poly(D,L-lactide-*co*-glycolide) nanoparticles was synthesized and incorporated in Carbopol Ultrez hydrophilic gel. The polymeric nanoparticles (with mean diameter of 217 nm and negative zeta potential of −20.5 mV) were produced by solvent emulsification-diffusion method, using poly(D,L-lactide-*co*-glycolide) (PLGA) as wall material, and poly(vinyl alcohol) (PVA) as stabilizer. The synthesized CcVO-conjugated NGs showed a biphasic pattern of volatile oil release with an initial faster release and subsequent sustained release phase, which induced inhibitory effect on HSV infection in vitro at noncytotoxic concentration [27].

3 Liposomes

Liposomes are small spherical artificial vesicles that range from 20 nm to several micrometers in size. They are composed of cholesterol and natural nontoxic hydrated phospholipid bilayer (which can mimic cell membranes) with the polar groups of phospholipids oriented to the inner and outer aqueous phase. Several factors, such as lipid composition, surface charge, size, and the method of preparation, can considerably determine the liposome properties. Furthermore, the composition of bilayer components determines the "rigidity" or "fluidity" and the charge of liposome bilayer. The phospholipid bilayer forms a protective shield that is normally resistant to enzymes, pH, and free radicals, thus protecting the carried cargo from degradation. Furthermore, there are several advantages of liposomes that include low toxicity, high biocompatibility, and biodegradability. Liposomes are widely used as efficient and promising delivery systems for drugs, active molecules, antimicrobials, or antigens, due to their capability to encapsulate bioactive hydrophilic, amphipathic, and lipophilic molecules into the inner aqueous cavity, at their interfaces, or within lipid leaflets, respectively [28–30]. In this chapter, I will focus mainly on the use of liposomes as antiviral agents. All other biomedical applications can be reviewed in one of the several published review articles [28–30].

The mechanism of action described for the antimicrobial effect of liposomes can be due to (i) the "Trojan horse-like mechanism" where liposomes are acting like vehicles carrying antimicrobial active molecules to the target cells, (ii) adjuvant-like mechanism where liposomes can trigger immune responses against conjugated antigens through enhancing antigen processing and presentation by antigen-presenting cells (APCs), and (iii) electrostatic interaction with the microbes leading to blocking of microbe entry and infection of cells [28,31,32].

Targeted drug delivery is one of the main features of liposomes. A unique liposome nanoparticle platform composed of α-tocopherol (vitamin E; VE) and cholesterol-based cationic liposomes (DOTAP-Chol) for systemic delivery of anti-HCV siRNAs to the liver was developed. VE was first attached to cholesterol-based cationic liposome and then coupled with siRNAs. Interestingly, siRNAs were delivered mainly to hepatocytes, both in vitro and in vivo, where they efficiently silenced HCV-gene (NS3 and NS5A) expression and potently suppressed HCV replication. The system is characterized by high delivery efficiency, low injection dose requirements, and low toxicity, which is promising for the clinical development of novel anti-HCV therapeutics [33]. An important approach for the prevention and treatment of RSV infections using exogenous administration of a specific surfactant phospholipid, palmitoyl-oleoyl-phosphatidylglycerol (POPG), was investigated. POPG has been shown to be a potent suppressor of the inflammatory cytokine, interleukin-6 and IL-8, which accompanies RSV infection. The phospholipid POPG could protect cells from RSV infection and subsequent cell death through a direct interaction with the virus, which can interfere with virus-cell binding and entry. Furthermore, POPG could markedly suppress viral spreading in epithelial monolayers preinfected with the virus. Interestingly, administration of POPG to mice could block RSV infection and replication and almost completely eliminated the recovery of virus from the lungs at 3 and 5 days postinfection. Furthermore, it protected the lungs from inflammatory cell infiltration and inflammatory cytokine (IFN-γ) production [34]. Cationic liposomes (of different particle sizes ranging between 100 and 800 nm and even 4 μm) with incorporated stearylamine, but without loading any active antiviral drugs, have been shown to significantly inhibit viral infection in vitro. The synthesized liposomes could inhibit both recombinant baculovirus (BV) and HSV-1 in cell culture. Compared with neutral or negatively charged liposomes, cationic liposomes suppressed the viruses in several mammalian cell lines. Moreover, these antiviral cationic liposomes were not cytotoxic. The mechanism of action was attributed to (i) direct interaction between cationic liposomes and virus particles, which blocked virus binding to cell surfaces and (ii) liposome binding to cell surface, which might decrease cell membrane fluidity and restrict mobility of fatty acyl chains, leading to greater inhibition of virus particle interactions [35].

Direct interaction of differently charged (cationic and anionic) phospholipids with virus particles and their influence on virus infection was shown. Liposomes containing negatively charged phosphatidylserine (PS) or positively charged DOTAP (*N*-[1-(2,3-Dioleoyloxy)propyl]-*N,N,N*-trimethylammonium) could directly interact with herpesviruses, as shown by surface plasmon resonance (SPR) and confocal microscopy, and subsequently inhibited virus entry and infection. However, inhibition of infection with PS was transient and dose dependent. Both cationic and anionic phospholipids can interact with the virus and reduce infectivity, while, presumably, acting through different mechanisms. Exogenous PS liposomes can interact with viral glycoproteins, particularly gH/gL and gB, and block virus entry. This was described also for rhabdoviruses where PS could also interact with glycoprotein G [36]. Furthermore, it was speculated that PS could redistribute the charges on the cell surface and reduce the initial contact between virus particles and cell surface proteoglycans containing heparan sulfate. On the other hand, the strong interaction between virus particles and DOTAP can be due to the electrostatic interaction between the positively charged liposomes and the negatively charged virus particles [31,37].

Liposomes were used as a delivery system for the glycan sialylneolacto-*N*-tetraose-c (LSTc), which is known to form multivalent interactions with influenza viruses. The liposome-based platform allowed spherical display, lateral diffusion, and proper presentation of SA on the surface, which increased the avidity of SA analogs for influenza. The developed LSTc-bearing decoy liposomes could form multivalent, polymer-like, and specific interactions with influenza virus and inhibit infection of host cells in a dose-dependent manner. Furthermore, these decoy liposomes blocked virus spread during multiple rounds of replication in vitro and could extend the survival rate of mice challenged with a lethal dose of influenza virus [38]. Development of multifunctional liposomes for the delivery of antirespiratory syncytial virus-fusion peptide RF-482 to inhibit virus fusion and block infection was described. The peptide, liposomes, and peptide-loaded liposomes were all tested and proved nontoxic when added to cell culture. All three formulations could also inhibit RSV infection of HEp-2 cells. The peptide RF-482 and liposomes alone were capable of inhibiting the RSV; however, when they were combined, the peptide-loaded liposomes had a significant increase in RSV inhibition [39]. Loading of liposomes with viral peptides to induce specific antiviral immunity was tested. Formulations of liposomal vaccine-encapsulated antigenic peptides (GP33 and P13) derived from glycoproteins of the lymphocytic choriomeningitis virus were shown to be highly immunogenic and have elicited protective antiviral immunity when injected intradermally in mice [40]. Liposomal peptide vaccine to prime a MHC class I-restricted T cell response ($CD8^+$ T cell) against 10 different HCV (core, E2, NS3, and NS4) epitopes was able to induce long-lasting strong cytotoxic T cell responses and significant numbers of IFN-γ-secreting cells. Cointegrating CpG oligonucleotides into the liposome further increased the number of IFN-γ-secreting cells for most epitopes tested [41]. The protective efficacy of liposome-encapsulated irradiated Ebola Zaire virus EBO-Z, which contains all of the native EBO-Z proteins, was evaluated in mice. All mice immunized intravenously with these liposomes were protected against virus-induced disease and death after challenge with a uniformly lethal mouse-adapted variant of EBO-Z [42]. Encapsulation of M2e and four highly conserved influenza A virus T and B cell epitope peptides could protect intranasally administered pigs against a virulent zoonotic IAV H1N1 challenge infection. The formulated liposomes elicited robust peptide-specific T cell response that partially helped in virus clearance from the lungs [43]. Similarly, when the peptide nucleoprotein (NP) 366–374 of influenza H3N2 virus was coupled with liposomes and used to immunize mice, peptide-specific $CD8^+$ T cell response was induced and virus replication was successfully suppressed in the lung of challenged mice [44].

The use of liposomes in vaccine development was deeply investigated. Virosomes, viral envelopes incorporating virus-derived proteins but devoid of viral genome and internal proteins, are formed by solubilization and extraction of unilamellar phospholipid membrane vesicle [32]. Virosomes can introduce their antigenic contents into cellular cytoplasm following endocytosis and fusion with the endosomal membrane. Delivery of antigens to the cytosol of antigen-presenting cells (APC) will result in direct access to the MHC-II presentation pathway, which is potentially useful for initiation of a $CD8^+$ T cell response [45,46]. Incorporation of antigens in the virosome can protect them from premature degradation by extracellular proteases. Several virosome-based vaccines are commercially available and several others are currently in clinical development stages [47,48]. The first commercially available nonaluminum adjuvanted human hepatitis A vaccine, Epaxal, consists of inactivated HAV adsorbed to the surface of premade influenza virosomes, containing purified influenza antigens, hemagglutinin, and neuraminidase anchored into a phospholipid bilayer [49]. Epaxal and the follow-up product Epaxal-Junior demonstrated a significantly better tolerability, safety, and a clear trend toward improved immunogenicity through the induction of both T and B cell responses [50–52]. Besides the hepatitis A vaccines, several other influenza vaccines based on virosome technology are currently licensed. Inflexal V and Invivac are trivalent (influenza A H1N1 and H3N2 and influenza B) seasonal influenza vaccines that were successfully developed. The vaccines are a mix of three different virosomes, each assembled separately from an influenza strain [53,54]. In addition, Cervarix, a preventive human papilloma virus (HPV) virosome-based vaccine, is composed of HPV16 and HPV18 L1 virus-like particles (VLPs) formulated in Adjuvant System 04 (AS04) adjuvant. Vaccination with Cervarix has been shown to protect against a high proportion of precursor lesions of cervical cancer caused by HPV16 and HPV18 [55,56]. Several other virosome-based vaccines targeting malaria, HCV, dengue fever virus, tuberculosis, HIV, and *Candida albicans* are currently in clinical development [47,48,57].

Self-amplifying messenger RNA (mRNA) technology has been developed for efficient in situ expression of antigens and has achieved a breakthrough when it was applied as the first approved vaccine against SARS-CoV-2 [58–60]. As a delivery system, cationic liposomes were identified as the more suitable and successful carrier for nonamplifying or self-amplifying mRNA vaccines [61]. It was clear that mRNA encapsulated inside liposomes could enhance both arms (humoral and cellular) of immune responses [62]. mRNA vaccine technology was investigated against Zika virus using a modified mRNA-encoding prM-E gene that produced virus-like particles, which resulted in high neutralizing antibody titers protected against ZIKV infection and conferred sterilizing immunity in mice [63]. Synthetic lipid nanoparticles were engineered to deliver a self-amplifying mRNA encoding seasonal influenza (H7N9) HA antigen and were tested in mouse model. The synthetic mRNA-based vaccine elicited a humoral immune response comparable to that obtained with the licensed influenza subunit vaccine [64]. Another self-amplifying mRNA expressing influenza (H1N1) HA antigen was formulated with a novel oil-in-water cationic nanoemulsion. The vaccine was immunogenic in ferrets and restricted virus replication in the upper respiratory tract. In mice, the synthetic mRNA vaccine induced potent functional neutralizing antibody and cellular immune responses: HA-specific CD4 T helper 1 and CD8 cytotoxic T cells [65]. In another study, the immunogenicity of lipid nanoparticles encapsulating a modified nonreplicating mRNA encoding influenza (H10N8) HA antigen was tested in *rhesus macaques*. The study showed that the vaccine elicited anti-HA protective antibody titers and MHC class II-restricted H10-specific T cell response after two intradermal or intramuscular immunizations. Following the inoculation of mRNA liposomes, dendritic cells (DCs), monocytes, and neutrophils were recruited to the site of immunization and draining lymph nodes (LNs); however, only DCs and monocytes internalized liposomes and upregulated costimulatory molecules [66]. The immunogenicity and the safety of cationic liposomes formulated with the mRNA encoding for a clade C envelope gp140 protein of HIV in *rhesus macaques* were investigated. This formulation has proven to induce stronger cellular and humoral immune responses when compared to self-amplifying mRNA encapsulated in a viral replicon particle or by a recombinant HIV envelope protein formulated with MF59 adjuvant [67]. In another study, vaccination of rhesus macaques with a nucleoside-modified mRNA encapsulated in lipid nanoparticles (mRNA-LNPs) encoding HIV Env protein induced robust Env-specific T follicular helper (Tfh) cells, polyfunctional antigen-specific $CD4^+$ T cell responses, and high numbers of germinal center (GC) B cells and plasma cells that were associated with long-lasting and high-affinity neutralizing antibodies and durable protection [68]. The first two approved COVID-19 vaccine candidates, the Pfizer-BioNTech BNT162b2 and the Moderna mRNA-1273, are both lipid nanoparticle-formulated, nucleoside-modified mRNA encoding the SARS-CoV-2 full-length spike protein, modified with two proline mutations to stabilize the spike protein into a prefusion conformation. Encapsulation into LNPs enabled transfection of the mRNA into host cells after injection. Translation of mRNA into viral protein resulted in the expression of membrane-bound spike protein of SARS-CoV-2, which is recognized by immune cells as a foreign antigen. This elicits both T cell and B cell responses to generate neutralizing antibodies, which may contribute to protection against COVID-19 [69,70].

4 Nanobodies

In general, antibody-based therapeutics are gaining significant clinical success recently. However, the development of antiviral therapeutics based on antibodies has been hampered by high costs of production, limited commercial market, and questionable efficacy due to low tissue accessibility of relatively large sized antibodies [71,72]. The discovery of the unique class of "heavy-chain-only" antibodies (HCAbs; nanobodies) found in the serum of *Camelidae* has solved many of the challenges associated with conventional antibodies. The variable domains of these HCAbs is considered the smallest naturally derived antigen-binding functional fragments with a molecular weight of ca. 15 kDa and 4 nm in length and 2.5 nm in diameter [73]. These nanobodies possess several characteristics: (i) the small size allows rapid and efficient tissue penetration, including the blood-brain barrier; (ii) the longer complementarity-determining regions (CDR3) loop leads to an increased surface area and repertoire that can interact with antigens allowing the nanobodies to access paratopes that would traditionally be difficult to interact with; (iii) high affinities to antigen binding; (iv) overall hydrophilic structure (due to the replacement of the highly conserved hydrophobic amino acids in framework 2 region (FR2) by hydrophilic amino acids) that contributes to high stability, solubility, and resistance to aggregation; (v) exceptional resistance to high temperatures and extreme pH; (vi) easy to engineer; efficient production in bacteria, yeast, mammalian cells, or plant cells; large-scale production at economical costs; (vii) easy administration via intranasal or oral route [73–75]. Due to all these features, nanobodies are considered ideal candidates for developing viable treatment strategies against viruses in harsh (respiratory and gastrointestinal tracts) environments. However, the short half-life of nanobodies, which may not have enough time to cross the endothelial barriers in sufficient amounts to clear out viruses, is one drawback of these structures. Currently, nanobodies are being developed against a number of viruses, including HIV-1, influenza viruses, rotavirus, norovirus, herpesvirus, ebolavirus, HCV, and RSV [73,76,77].

The first human nanobodies, m36, recognized from a human antibody variable domain phage library, targeted a highly conserved sterically restricted domain on HIV gp120 induced by CD4 binding. These developed nanobodies, m36, have higher neutralizing activity and broader cross-reactivity than the conventional anti-gp120 broadly neutralizing antibodies [78]. A panel of broadly neutralizing nanobodies produced using trimeric (trimeric version of gp120 bound to the gp41 ectodomain) gp140, free or bound to a CD4 mimic, as immunogen in llamas have been shown to bind epitopes located at the CD4-binding site (CD4bs) or the coreceptor-binding site (CoRbs) of gp120. These nanobodies exhibited potent neutralizing activity and cross-reactivity against several HIV isolates [79].

Several H5N1 and H5N2 HA-specific nanobodies were isolated from llamas or camels, respectively. The nanobodies had strong neutralizing activities and potently suppressed influenza A virus replication in vivo [80,81]. Furthermore, a panel of nanobodies were generated against H1N1 NP and H3N2 M2 proteins: NP is essential for nuclear trafficking and packaging of the influenza virus genome and M2 is important for virus uncoating, assembly, and egress. The produced nanobodies disrupted virus replication by preventing nuclear import of viral ribonucleoproteins (vRNPs) and showed cross-reactive neutralization of M2 [82,83]. Specific neutralizing nanobodies against RSV (another respiratory virus) fusion (F) protein were identified from immune libraries of llamas. The nanobodies could bind to the antigenic site II epitope on RSV F protein and showed potent cross-reactive neutralization effect in vitro. Furthermore, when the nanobodies were directly delivered prophylactically or therapeutically to the sites of infection, high effective reduction in RSV replication was observed in cotton rat tissues [84]. In addition, in a phase I/IIa trial, RSV viral loads in nasal swabs of hospitalized RSV-infected children were reduced by daily inhalation treatment for three consecutive days [85].

After immunization of alpaca with the conserved HCV E2 glycoprotein, one E2-specific and potent cross-reactive neutralizing nanobodies were identified to recognize a novel conserved epitope overlapping with that of the CD81 binding site. These nanobodies displayed potently neutralizing activity against a panel of HCV pseudo-particles representing different genotypes by impeding E2-CD81 interaction and efficiently inhibited HCV cell-to-cell transmission [86]. Furthermore, a number of anti-NS5B, anti-NS3, and antiprotease nanobodies were produced and have been shown to suppress HCV replication in vitro [87]. For the other hepatitis virus (HBV), nanobodies were produced in llamas after immunization with serum-derived noninfectious HBV-like particles and recognized the HBV S domain (the most abundant viral membrane protein present in the three viral membrane proteins S, M, and L). The S domain-specific nanobodies could prevent secretion of S molecules, were capable of retaining S molecules inside the hepatocytes, and reduced the concentration of the VLPs and HBV virions in the blood of a hydrodynamics-based HBV mouse model [88].

Antirotavirus nanobodies (antirotavirus protein; ARP1) identified from a llama immunized with rhesus monkey group A rotavirus (RVA) strain have been shown to neutralize 11 strains of RVA and exhibited preventive and therapeutic effect against RVA-induced diarrhea in a neonatal mouse model [89]. Further modifications were adopted to develop secreted or anchored antirotavirus nanobodies expressed in *Lactobacillus paracasei* to facilitate direct delivery of therapeutic agents for the treatment of gastroenteritis [90,91]. Clinical trials were performed in infants with RVA-associated diarrhea and showed a significant reduction in diarrhea with no side effects after oral ARP1 application [92]. Antinorovirus (another gastrointestinal tract virus) nanobodies were isolated from alpaca immunized with the genetically distinct genogroup II (GII) norovirus virus-like particles. The nanobodies could bind to the region of protruding (P) domain of the capsid protein and were broadly reactive to several different GII genotypes resulting in particles' disassembly in vitro [93].

Research studies have revealed that camels (the major reservoir host for MERS-CoV) produced a highly efficient neutralizing antibody (both conventional and heavy chain-only antibodies) response to MERS-CoV infection [28]. Bone marrow samples from infected dromedary camels were isolated and screened for possible antibody candidates that are effective in neutralizing the virus in vitro. A candidate anti-MERS-CoV nanobody that was humanized with human Fc was shown to provide complete protection against virus infection in mice and was present at high levels in sera more than a week after treatment [94]. Recently, nanobodies specific for the receptor-binding domain (RBD) of SARS-CoV-2 spike protein were identified after immunization of an alpaca and a llama with the RBD and inactivated virus. Four nanobody candidates (VHHs E, U, V, and W) potently neutralized SARS-CoV-2 and SARS-CoV-2-pseudotyped VSV. Bi- and trivalent nanobodies were further designed and showed improved neutralizing properties. The trivalent VHH EEE was the most potent candidate and did inhibit infection, did not activate fusion, and likely inactivated virions by outcompeting interaction of the virus with cellular receptors; however, evolution experiments revealed emergence of escape mutants in the spike protein with single amino acid changes that were completely insensitive to inhibition by VHH EEE. On the other hand, the bivalent VHH VE neutralized the virus more efficiently than monovalent VHHs and the simultaneous targeting of two independent epitopes largely prevented the emergence of resistant escape mutants in evolution experiments [95]. Several other nanobodies-based therapeutics for emerging coronaviruses can be reviewed in this article [96].

5 Limitations

Biodegradation of nanoparticles is a critical process that determines the sustained drug release, biodistribution profiles, and finally excretion of remaining nanoparticles [97]. Several factors, such as nanoparticle structure, physical and chemical properties, hydrophobicity and hydrophilicity profiles, size, and molecular weight, can affect the fate of nanoparticles in the body. However, the biodegradation and elimination of nanoparticles have been poorly investigated and there is a significant lack of information from in vivo studies. There are two main barriers that nanoparticles need to cross to leave the body: cellular barriers to first leave the cells (exocytosis) and renal barrier to finally leave the body. The process of exocytosis depends on the composition and surface properties of nanoparticles; cationic particles have a slower exocytosis rate due to intracellular agglomeration. Renal clearance depends mainly on the size; nanoparticles $<5\,nm$ may be excreted in the urine while larger particles are often reabsorbed into systemic circulation and degraded in the liver or excreted via digestive system. Some nanoparticles may be too large to be cleared through the kidneys or liver and can accumulate in the body (tissues) since they cannot be degraded [98].

Cell physiology has an important impact on the use of therapeutic agents. The poor permeability of biological membranes can limit the passage of nanoparticles. The lack of cellular machinery necessary for endocytosis or exocytosis can affect the type of used nanoparticles. This can result in the removal of nanoparticles from circulation before reaching the target organs or tissues [99]. The uptake of nanoparticles by reticuloendothelial system and macrophages can impede the efficacy of nanoparticles due to early uptake and clearance from the body [100]. In addition, the nature of nanoparticles besides physical processes that enable contact between nanoparticle surfaces can result in aggregation and clustering of nanoparticles and consequently increasing their size, which results in poor uptake, accumulation in organs, and increased toxicity [101,102].

6 Conclusions and future perspectives

Viral and infectious diseases are emerging since the start of human history and they played an important role in writing the history and reshaping the nations. Several pandemics were existed before and have killed millions of people and animals and could even threaten human existence along the history [103]. However, humans have always searched for, and often found, solutions to mitigate the deadly pandemics starting from the invention of vaccines since the 17th century till the current great advancement in science technologies. As a continuation of the pandemic series, currently, the world is facing a novel (COVID-19) pandemic that, so far, is reshaping human civilization and causing tremendous suffering and deaths.

Coronaviruses represent a large family of viruses that can cause respiratory, hepatic, intestinal, and neurological diseases of variable severity in human and animals alike. There are seven known coronavirus strains that can infect people, with SARS-CoV-2 being the newest member of the family [104]. As mentioned earlier in this chapter, nanotechnology has been proven one of the groundbreaking approaches and could offer promising medical solutions toward fighting the coronaviruses in human and animals. Similarly, nanotechnology can provide a number of strategies to develop diagnostics and highly advanced therapeutics to cope with SARS-CoV-2. Various nanoarchitectures (graphene sheets, nanogels, liposomes, silver, gold, and copper nanoparticles, and nanobodies) show intrinsic antiviral properties and can act as potential therapeutics against the disease. Furthermore, nanoparticles can have a potential role as efficient drug delivery vehicles for potential therapeutics, immunomodulators, or vaccines against COVID-19 that ensure targeted delivery and optimized concentration while minimizing side effects and undesirable degradation enhancing efficacy [105,106].

Nanoparticles possess remarkable antiviral activities and play a critical role in both viral disease diagnosis and therapeutics [107]. They represent a branch of novel antiviral agents that are suitable, alternative, biosafe, and biocompatible to prevent the spread of infections. With the continuous emergence of viral disease and deadly infections, nanotechnology has become a focal point of research and various types of nanoarchitectures have been explored and evaluated for their prophylactic and therapeutic activity against different viruses. Although promising, many challenges and barriers still exist to achieve the full potential of nanoparticles to be successfully translated from the laboratory to the clinical setting. Future research on nanoparticle-based antiviral platforms needs to focus on (i) functionalization with specific molecules to achieve effective targeting, potent inhibitory effect, and broad-spectrum nature and (ii) improvement of bioavailability and biodegradability and reducing cytotoxicity and side effects in in vivo clinical studies.

References

[1] S.E. McNeil, Unique benefits of nanotechnology to drug delivery and diagnostics, Methods Mol. Biol. 697 (2011) 3–8. Epub 2010/12/01 https://doi.org/10.1007/978-1-60327-198-1_1. PubMed PMID: 21116949.

[2] N. Sanvicens, M.P. Marco, Multifunctional nanoparticles—properties and prospects for their use in human medicine, Trends Biotechnol. 26 (8) (2008) 425–433. Epub 2008/06/03 https://doi.org/10.1016/j.tibtech.2008.04.005. PubMed PMID: 18514941.

[3] R.A. Petros, J.M. DeSimone, Strategies in the design of nanoparticles for therapeutic applications, Nat. Rev. Drug Discov. 9 (8) (2010) 615–627. Epub 2010/07/10 https://doi.org/10.1038/nrd2591. PubMed PMID: 20616808.

[4] M. Gagliardi, Biomimetic and bioinspired nanoparticles for targeted drug delivery, Ther. Deliv. 8 (5) (2017) 289–299. Epub 2017/04/01 https://doi.org/10.4155/tde-2017-0013. PubMed PMID: 28361608.

[5] S. Gurunathan, M. Qasim, Y. Choi, J.T. Do, C. Park, K. Hong, et al., Antiviral potential of nanoparticles—can nanoparticles fight against coronaviruses? Nanomaterials (Basel) 10 (9) (2020), https://doi.org/10.3390/nano10091645. Epub 2020/08/23. PubMed PMID: 32825737; PubMed Central PMCID: PMCPMC7557932.

[6] K. Maduray, R. Parboosing, Metal nanoparticles: a promising treatment for viral and arboviral infections, Biol. Trace Elem. Res. (2020), https://doi.org/10.1007/s12011-020-02414-2. Epub 2020/10/09. PubMed PMID: 33029761; PubMed Central PMCID: PMCPMC7540915.

[7] A.R. Menjoge, R.M. Kannan, D.A. Tomalia, Dendrimer-based drug and imaging conjugates: design considerations for nanomedical applications, Drug Discov. Today 15 (5–6) (2010) 171–185. Epub 2010/02/02 https://doi.org/10.1016/j.drudis.2010.01.009. PubMed PMID: 20116448.

[8] J. Khandare, M. Calderon, N.M. Dagia, R. Haag, Multifunctional dendritic polymers in nanomedicine: opportunities and challenges, Chem. Soc. Rev. 41 (7) (2012) 2824–2848. Epub 2011/12/14 https://doi.org/10.1039/c1cs15242d. PubMed PMID: 22158998.

[9] A. Tschiche, S. Malhotra, R. Haag, Nonviral gene delivery with dendritic self-assembling architectures, Nanomedicine (Lond.) 9 (5) (2014) 667–693. Epub 2014/05/16 https://doi.org/10.2217/nnm.14.32. PubMed PMID: 24827843.

[10] A. Boreham, R. Brodwolf, K. Walker, R. Haag, U. Alexiev, Time-resolved fluorescence spectroscopy and fluorescence lifetime imaging microscopy for characterization of dendritic polymer nanoparticles and applications in nanomedicine, Molecules 22 (1) (2016), https://doi.org/10.3390/molecules22010017. Epub 2016/12/29. PubMed PMID: 28029135; PubMed Central PMCID: PMCPMC6155873.

[11] D. Steinhilber, A.L. Sisson, D. Mangoldt, P. Welker, K. Licha, R. Haag, Synthesis, reductive cleavage, and cellular interaction studies of biodegradable, polyglycerol nanogels, Adv. Funct. Mater. 20 (23) (2010) 4133–4138, https://doi.org/10.1002/adfm.201000410.

[12] I. Papp, C. Sieben, A.L. Sisson, J. Kostka, C. Böttcher, K. Ludwig, et al., Inhibition of influenza virus activity by multivalent glycoarchitectures with matched sizes, ChemBioChem 12 (6) (2011) 887–895. Epub 2011/03/09 https://doi.org/10.1002/cbic.201000776. PubMed PMID: 21384484.

[13] K. Achazi, R. Haag, M. Ballauff, J. Dernedde, J.N. Kizhakkedathu, D. Maysinger, et al., Understanding the interaction of polyelectrolyte architectures with proteins and biosystems, Angew. Chem. Int. Ed. Engl. (2020), https://doi.org/10.1002/anie.202006457. Epub 2020/06/27. PubMed PMID: 32589355.

[14] P. Dey, T. Bergmann, J.L. Cuellar-Camacho, S. Ehrmann, M.S. Chowdhury, M. Zhang, et al., Multivalent flexible nanogels exhibit broad-spectrum antiviral activity by blocking virus entry, ACS Nano 12 (7) (2018) 6429–6442. Epub 2018/06/13 https://doi.org/10.1021/acsnano.8b01616. PubMed PMID: 29894156.

[15] J. Vonnemann, S. Liese, C. Kuehne, K. Ludwig, J. Dernedde, C. Böttcher, et al., Size dependence of steric shielding and multivalency effects for globular binding inhibitors, J. Am. Chem. Soc. 137 (7) (2015) 2572–2579, https://doi.org/10.1021/ja5114084.

[16] S. Bhatia, M. Hilsch, J.L. Cuellar-Camacho, K. Ludwig, C.X. Nie, B. Parshad, et al., Adaptive flexible sialylated nanogels as highly potent influenza A virus inhibitors, Angew. Chem. Int. Edit. 59 (30) (2020) 12417–12422, https://doi.org/10.1002/anie.202006145. PubMed PMID: WOS:000545431600001.

[17] S. Bhatia, L.C. Camacho, R. Haag, Pathogen inhibition by multivalent ligand architectures, J. Am. Chem. Soc. 138 (28) (2016) 8654–8666, https://doi.org/10.1021/jacs.5b12950.

[18] W. Wang, P. Wang, X. Tang, A.A. Elzatahry, S. Wang, D. Al-Dahyan, et al., Facile synthesis of uniform virus-like mesoporous silica nanoparticles for enhanced cellular internalization, ACS Cent. Sci. 3 (8) (2017) 839–846. Epub 2017/08/31 https://doi.org/10.1021/acscentsci.7b00257. PubMed PMID: 28852697; PubMed Central PMCID: PMCPMC5571464.

[19] J.J. Shie, J.M. Fang, S.Y. Wang, K.C. Tsai, Y.S. Cheng, A.S. Yang, et al., Synthesis of tamiflu and its phosphonate congeners possessing potent anti-influenza activity, J. Am. Chem. Soc. 129 (39) (2007) 11892–11893. Epub 2007/09/14 https://doi.org/10.1021/ja073992i. PubMed PMID: 17850083.

[20] C. Nie, B. Parshad, S. Bhatia, C. Cheng, M. Stadtmuller, A. Oehrl, et al., Topology-matching design of an influenza-neutralizing spiky nanoparticle-based inhibitor with a dual mode of action, Angew. Chem. Int. Ed. Engl. 59 (36) (2020) 15532–15536. Epub 2020/05/19 https://doi.org/10.1002/anie.202004832. PubMed PMID: 32421225; PubMed Central PMCID: PMCPMC7497169.

[21] T.H. Senanayake, S. Gorantla, E. Makarov, Y. Lu, G. Warren, S.V. Vinogradov, Nanogel-conjugated reverse transcriptase inhibitors and their combinations as novel antiviral agents with increased efficacy against HIV-1 infection, Mol. Pharm. 12 (12) (2015) 4226–4236, https://doi.org/10.1021/acs.molpharmaceut.5b00424. PubMed PMID: WOS:000366151500005.

[22] G. Warren, E. Makarov, Y. Lu, T. Senanayake, K. Rivera, S. Gorantla, et al., Amphiphilic cationic nanogels as brain-targeted carriers for activated nucleoside reverse transcriptase inhibitors, J. Neuroimmune Pharmacol. 10 (1) (2015) 88–101. Epub 2015/01/07 https://doi.org/10.1007/s11481-014-9576-7. PubMed PMID: 25559020; PubMed Central PMCID: PMCPMC4336642.

[23] M.A. Macchione, C. Guerrero-Beltran, A.P. Rosso, E.M. Euti, M. Martinelli, M.C. Strumia, et al., Poly(N-vinylcaprolactam) nanogels with antiviral behavior against HIV-1 infection, Sci. Rep. 9 (1) (2019) 5732. Epub 2019/04/07 https://doi.org/10.1038/s41598-019-42150-9. PubMed PMID: 30952921; PubMed Central PMCID: PMCPMC6450967.

[24] C.E. Pena-Gonzalez, P. Garcia-Broncano, M.F. Ottaviani, M. Cangiotti, A. Fattori, M. Hierro-Oliva, et al., Dendronized anionic gold nanoparticles: synthesis, characterization, and antiviral activity, Chemistry 22 (9) (2016) 2987–2999. Epub 2016/02/16 https://doi.org/10.1002/chem.201504262. PubMed PMID: 26875938.

[25] G.F. Santoro, M.G. Cardoso, L.G. Guimaraes, J.M. Freire, M.J. Soares, Anti-proliferative effect of the essential oil of *Cymbopogon citratus* (DC) Stapf (lemongrass) on intracellular amastigotes, bloodstream trypomastigotes and culture epimastigotes of *Trypanosoma cruzi* (Protozoa: Kinetoplastida), Parasitology 134 (Pt 11) (2007) 1649–1656. Epub 2007/08/10 https://doi.org/10.1017/S0031182007002958. PubMed PMID: 17686189.

[26] M. Minami, M. Kita, T. Nakaya, T. Yamamoto, H. Kuriyama, J. Imanishi, The inhibitory effect of essential oils on herpes simplex virus type-1 replication in vitro, Microbiol.

Immunol. 47 (9) (2003) 681–684. Epub 2003/10/31 https://doi.org/10.1111/j.1348-0421.2003.tb03431.x. PubMed PMID: 14584615.
[27] K.B. Almeida, J.L. Araujo, J.F. Cavalcanti, M.T.V. Romanos, S.C. Mourão, A.C.F. Amaral, et al., In vitro release and anti-herpetic activity of *Cymbopogon citratus* volatile oil-loaded nanogel, Rev. Bras Farmacogn. 28 (4) (2018) 495–502.
[28] A. Akbarzadeh, R. Rezaei-Sadabady, S. Davaran, S.W. Joo, N. Zarghami, Y. Hanifehpour, et al., Liposome: classification, preparation, and applications, Nanoscale Res. Lett. 8 (1) (2013) 102, https://doi.org/10.1186/1556-276X-8-102. PubMed PMID: 23432972; PubMed Central PMCID: PMC3599573.
[29] R. Nisini, N. Poerio, S. Mariotti, F. De Santis, M. Fraziano, The multirole of liposomes in therapy and prevention of infectious diseases, Front. Immunol. 9 (2018) 155. Epub 2018/02/21 https://doi.org/10.3389/fimmu.2018.00155. PubMed PMID: 29459867; PubMed Central PMCID: PMCPMC5807682.
[30] H. Daraee, A. Etemadi, M. Kouhi, S. Alimirzalu, A. Akbarzadeh, Application of liposomes in medicine and drug delivery, Artif. Cells Nanomed. Biotechnol. 44 (1) (2016) 381–391. Epub 2014/09/16 https://doi.org/10.3109/21691401.2014.953633. PubMed PMID: 25222036.
[31] O. Kolyvushko, J. Latzke, I. Dahmani, N. Osterrieder, S. Chiantia, W. Azab, Differentially-charged liposomes interact with alphaherpesviruses and interfere with virus entry, Pathogens 9 (5) (2020), https://doi.org/10.3390/pathogens9050359. Epub 2020/05/14. PubMed PMID: 32397270; PubMed Central PMCID: PMCPMC7281745.
[32] M.J. Copland, T. Rades, N.M. Davies, M.A. Baird, Lipid based particulate formulations for the delivery of antigen, Immunol. Cell Biol. 83 (2) (2005) 97–105. Epub 2005/03/08 https://doi.org/10.1111/j.1440-1711.2005.01315.x. PubMed PMID: 15748206.
[33] L. Duan, Y. Yan, J. Liu, B. Wang, P. Li, Q. Hu, et al., Target delivery of small interfering RNAs with vitamin E-coupled nanoparticles for treating hepatitis C, Sci. Rep. 6 (2016) 24867. Epub 2016/04/27 https://doi.org/10.1038/srep24867. PubMed PMID: 27113197; PubMed Central PMCID: PMCPMC4845054.
[34] M. Numata, H.W. Chu, A. Dakhama, D.R. Voelker, Pulmonary surfactant phosphatidylglycerol inhibits respiratory syncytial virus-induced inflammation and infection, Proc. Natl. Acad. Sci. U. S. A. 107 (1) (2010) 320–325, https://doi.org/10.1073/pnas.0909361107. PubMed PMID: WOS:000273559200056.
[35] K. Tahara, M. Kobayashi, S. Yoshida, R. Onodera, N. Inoue, H. Takeuchi, Effects of cationic liposomes with stearylamine against virus infection, Int. J. Pharm. 543 (1–2) (2018) 311–317. Epub 2018/04/07 https://doi.org/10.1016/j.ijpharm.2018.04.001. PubMed PMID: 29625169.
[36] A. Estepa, J.M. Coll, Phosphatidylserine binding to solid-phase rhabdoviral peptides: a new method to study phospholipid/viral protein interactions, J. Virol. Methods 61 (1–2) (1996) 37–45. Epub 1996/09/01 https://doi.org/10.1016/0166-0934(96)02067-8. PubMed PMID: 8882935.
[37] J. Kreuger, D. Spillmann, J.P. Li, U. Lindahl, Interactions between heparan sulfate and proteins: the concept of specificity, J. Cell Biol. 174 (3) (2006) 323–327, https://doi.org/10.1083/jcb.200604035. PubMed PMID: WOS:000239422800001.
[38] G.L. Hendricks, K.L. Weirich, K. Viswanathan, J. Li, Z.H. Shriver, J. Ashour, et al., Sialylneolacto-N-tetraose c (LSTc)-bearing liposomal decoys capture influenza A virus, J. Biol. Chem. 288 (12) (2013) 8061–8073. Epub 2013/01/31 https://doi.org/10.1074/jbc.M112.437202. PubMed PMID: 23362274; PubMed Central PMCID: PMCPMC3605625.
[39] S. Joshi, A.A. Chaudhari, V. Dennis, D.J. Kirby, Y. Perrie, S.R. Singh, Anti-RSV peptide-loaded liposomes for the inhibition of respiratory syncytial virus, Bioengineering (Basel) 5 (2) (2018), https://doi.org/10.3390/bioengineering5020037. Epub 2018/05/12. PubMed PMID: 29747393; PubMed Central PMCID: PMCPMC6027156.

[40] B. Ludewig, F. Barchiesi, M. Pericin, R.M. Zinkernagel, H. Hengartner, R.A. Schwendener, In vivo antigen loading and activation of dendritic cells via a liposomal peptide vaccine mediates protective antiviral and anti-tumour immunity, Vaccine 19 (1) (2000) 23–32. Epub 2000/08/05 https://doi.org/10.1016/s0264-410x(00)00163-8. PubMed PMID: 10924783.
[41] O.B. Engler, R.A. Schwendener, W.J. Dai, B. Wolk, W. Pichler, D. Moradpour, et al., A liposomal peptide vaccine inducing CD8(+) T cells in HLA-A2.1 transgenic mice, which recognise human cells encoding hepatitis C virus (HCV) proteins, Vaccine 23 (1) (2004) 58–68, https://doi.org/10.1016/j.vaccine.2004.05.009. PubMed PMID: WOS:000225200900009.
[42] M. Rao, M. Bray, C.R. Alving, P. Jahrling, G.R. Matyas, Induction of immune responses in mice and monkeys to Ebola virus after immunization with liposome-encapsulated irradiated Ebola virus: protection in mice requires CD4(+) T cells, J. Virol. 76 (18) (2002) 9176–9185. Epub 2002/08/21 https://doi.org/10.1128/jvi.76.18.9176-9185. 2002. PubMed PMID: 12186901; PubMed Central PMCID: PMCPMC136452.
[43] S. Dhakal, X. Cheng, J. Salcido, S. Renu, K. Bondra, Y.S. Lakshmanappa, et al., Liposomal nanoparticle-based conserved peptide influenza vaccine and monosodium urate crystal adjuvant elicit protective immune response in pigs, Int. J. Nanomedicine 13 (2018) 6699–6715. Epub 2018/11/15 https://doi.org/10.2147/IJN.S178809. PubMed PMID: 30425484; PubMed Central PMCID: PMCPMC6205527.
[44] T. Nagata, T. Toyota, H. Ishigaki, T. Ichihashi, K. Kajino, Y. Kashima, et al., Peptides coupled to the surface of a kind of liposome protect infection of influenza viruses, Vaccine 25 (26) (2007) 4914–4921. Epub 2007/05/29 https://doi.org/10.1016/j.vaccine. 2007.04.010. PubMed PMID: 17531358.
[45] L. Bungener, K. Serre, L. Bijl, L. Leserman, J. Wilschut, T. Daemen, et al., Virosome-mediated delivery of protein antigens to dendritic cells, Vaccine 20 (17–18) (2002) 2287–2295. Epub 2002/05/16 https://doi.org/10.1016/s0264-410x(02)00103-2. PubMed PMID: 12009284.
[46] C. Moser, I.C. Metcalfe, J.F. Viret, Virosomal adjuvanted antigen delivery systems, Expert Rev. Vaccines 2 (2) (2003) 189–196. Epub 2003/08/06 https://doi.org/10. 1586/14760584.2.2.189. PubMed PMID: 12899570.
[47] C. Moser, M. Amacker, R. Zurbriggen, Influenza virosomes as a vaccine adjuvant and carrier system, Expert Rev. Vaccines 10 (4) (2011) 437–446. Epub 2011/04/22 https:// doi.org/10.1586/erv.11.15. PubMed PMID: 21506642.
[48] C. Moser, M. Muller, M.D. Kaeser, U. Weydemann, M. Amacker, Influenza virosomes as vaccine adjuvant and carrier system, Expert Rev. Vaccines 12 (7) (2013) 779–791. Epub 2013/07/28 https://doi.org/10.1586/14760584.2013.811195. PubMed PMID: 23885823.
[49] R. Gluck, E. Walti, Biophysical validation of Epaxal Berna, a hepatitis A vaccine adjuvanted with immunopotentiating reconstituted influenza virosomes (IRIV), Dev. Biol. (Basel) 103 (2000) 189–197. Epub 2001/02/24. PubMed PMID: 11214236.
[50] P.A. Bovier, Recent advances with a virosomal hepatitis A vaccine, Expert. Opin. Biol. Ther. 8 (8) (2008) 1177–1185. Epub 2008/07/11 https://doi.org/10.1517/14712598.8.8. 1177. PubMed PMID: 18613769.
[51] B.R. Holzer, C. Hatz, D. Schmidt-Sissolak, R. Gluck, B. Althaus, M. Egger, Immunogenicity and adverse effects of inactivated virosome versus alum-adsorbed hepatitis A vaccine: a randomized controlled trial, Vaccine 14 (10) (1996) 982–986. Epub 1996/07/01 https://doi.org/10.1016/0264-410x(96)00042-4. PubMed PMID: 8873392.
[52] P.A. Bovier, T. Farinelli, L. Loutan, Interchangeability and tolerability of a virosomal and an aluminum-adsorbed hepatitis A vaccine, Vaccine 23 (19) (2005) 2424–2429. Epub 2005/03/09 https://doi.org/10.1016/j.vaccine.2004.11.024. PubMed PMID: 15752828.
[53] I.A. de Bruijn, J. Nauta, L. Gerez, A.M. Palache, The virosomal influenza vaccine Invivac®: immunogenicity and tolerability compared to an adjuvanted influenza vaccine

(Fluad®) in elderly subjects, Vaccine 24 (44–46) (2006) 6629–6631. Epub 2006/08/12 https://doi.org/10.1016/j.vaccine.2006.05.035. PubMed PMID: 16901593.

[54] R. Mischler, I.C. Metcalfe, Inflexal V a trivalent virosome subunit influenza vaccine: production, Vaccine 20 (Suppl. 5) (2002) B17–B23. Epub 2002/12/13 https://doi.org/10.1016/s0264-410x(02)00512-1. PubMed PMID: 12477413.

[55] S.J. Keam, D.M. Harper, Human papillomavirus types 16 and 18 vaccine (recombinant, AS04 adjuvanted, adsorbed) [Cervarix], Drugs 68 (3) (2008) 359–372. Epub 2008/02/09 https://doi.org/10.2165/00003495-200868030-00007. PubMed PMID: 18257611.

[56] A. Monie, C.F. Hung, R. Roden, T.C. Wu, Cervarix: a vaccine for the prevention of HPV 16, 18-associated cervical cancer, Biologics 2 (1) (2008) 97–105. Epub 2008/03/01. PubMed PMID: 19707432; PubMed Central PMCID: PMCPMC2727782.

[57] V. Bernasconi, K. Norling, M. Bally, F. Hook, N.Y. Lycke, Mucosal vaccine development based on liposome technology, J. Immunol. Res. 2016 (2016) 5482087. Epub 2017/01/28 https://doi.org/10.1155/2016/5482087. PubMed PMID: 28127567; PubMed Central PMCID: PMCPMC5227169.

[58] L.A. Brito, S. Kommareddy, D. Maione, Y. Uematsu, C. Giovani, F. Berlanda Scorza, et al., Self-amplifying mRNA vaccines, Adv. Genet. 89 (2015) 179–233. Epub 2015/01/27 https://doi.org/10.1016/bs.adgen.2014.10.005. PubMed PMID: 25620012.

[59] N. Pardi, M.J. Hogan, F.W. Porter, D. Weissman, mRNA vaccines—a new era in vaccinology, Nat. Rev. Drug Discov. 17 (4) (2018) 261–279, https://doi.org/10.1038/nrd.2017.243. PubMed PMID: WOS:000428524100014.

[60] B. Faiman, Approved mRNA vaccines for SARS-CoV-2, J. Adv. Pract. Oncol. 12 (1) (2021) 17–18. Epub 2021/02/09 https://doi.org/10.6004/jadpro.2021.12.1.1. PubMed PMID: 33552658; PubMed Central PMCID: PMCPMC7844189.

[61] T. Schlake, A. Thess, M. Fotin-Mleczek, K.J. Kallen, Developing mRNA-vaccine technologies, RNA Biol. 9 (11) (2012) 1319–1330. Epub 2012/10/16 https://doi.org/10.4161/rna.22269. PubMed PMID. 23064118; PubMed Central PMCID: PMCPMC3597572.

[62] K. Kariko, H.P. Ni, J. Capodici, M. Lamphier, D. Weissman, mRNA is an endogenous ligand for Toll-like receptor 3, J. Biol. Chem. 279 (13) (2004) 12542–12550, https://doi.org/10.1074/jbc.M310175200. PubMed PMID: WOS:000220334900066.

[63] J.M. Richner, S. Himansu, K.A. Dowd, S.L. Butler, V. Salazar, J.M. Fox, et al., Modified mRNA vaccines protect against zika virus infection, Cell 169 (1) (2017) 176. Epub 2017/03/25 https://doi.org/10.1016/j.cell.2017.03.016. PubMed PMID: 28340344.

[64] A. Hekele, S. Bertholet, J. Archer, D.G. Gibson, G. Palladino, L.A. Brito, et al., Rapidly produced SAM (R) vaccine against H7N9 influenza is immunogenic in mice, Emerg. Microbes Infect. 2 (2013), https://doi.org/10.1038/emi.2013.54, e52. PubMed PMID: WOS:000339188500001.

[65] M. Brazzoli, D. Magini, A. Bonci, S. Buccato, C. Giovani, R. Kratzer, et al., Induction of broad-based immunity and protective efficacy by self-amplifying mRNA vaccines encoding influenza virus hemagglutinin, J. Virol. 90 (1) (2016) 332–344. Epub 2015/10/16 https://doi.org/10.1128/JVI.01786-15. PubMed PMID: 26468547; PubMed Central PMCID: PMCPMC4702536.

[66] F. Liang, G. Lindgren, A. Lin, E.A. Thompson, S. Ols, J. Rohss, et al., Efficient targeting and activation of antigen-presenting cells in vivo after modified mRNA vaccine administration in rhesus macaques, Mol. Ther. 25 (12) (2017) 2635–2647. Epub 2017/09/30 https://doi.org/10.1016/j.ymthe.2017.08.006. PubMed PMID: 28958578; PubMed Central PMCID: PMCPMC5768558.

[67] W.M. Bogers, H. Oostermeijer, P. Mooij, G. Koopman, E.J. Verschoor, D. Davis, et al., Potent immune responses in rhesus macaques induced by nonviral delivery of a self-amplifying RNA vaccine expressing HIV type 1 envelope with a cationic nanoemulsion, J. Infect. Dis. 211 (6) (2015) 947–955. Epub 2014/09/23 https://doi.org/10.1093/infdis/jiu522. PubMed PMID: 25234719; PubMed Central PMCID: PMCPMC4416123.

[68] N. Pardi, M.J. Hogan, M.S. Naradikian, K. Parkhouse, D.W. Cain, L. Jones, et al., Nucleoside-modified mRNA vaccines induce potent T follicular helper and germinal center B cell responses, J. Exp. Med. 215 (6) (2018) 1571–1588, https://doi.org/10.1084/jem.20171450. PubMed PMID: WOS:000440821300007.
[69] F.P. Polack, S.J. Thomas, N. Kitchin, J. Absalon, A. Gurtman, S. Lockhart, et al., Safety and efficacy of the BNT162b2 mRNA Covid-19 vaccine, N. Engl. J. Med. 383 (27) (2020) 2603–2615. Epub 2020/12/11 https://doi.org/10.1056/NEJMoa2034577. PubMed PMID: 33301246; PubMed Central PMCID: PMCPMC7745181.
[70] Z. Wang, F. Schmidt, Y. Weisblum, F. Muecksch, C.O. Barnes, S. Finkin, et al., mRNA vaccine-elicited antibodies to SARS-CoV-2 and circulating variants, Nature (2021), https://doi.org/10.1038/s41586-021-03324-6. Epub 2021/02/11. PubMed PMID: 33567448.
[71] J.M. Reichert, Antibodies to watch in 2017, MAbs 9 (2) (2017) 167–181. Epub 2016/12/15 https://doi.org/10.1080/19420862.2016.1269580. PubMed PMID: 27960628; PubMed Central PMCID: PMCPMC5297518.
[72] H. Samaranayake, T. Wirth, D. Schenkwein, J.K. Raty, S. Yla-Herttuala, Challenges in monoclonal antibody-based therapies, Ann. Med. 41 (5) (2009) 322–331. Epub 2009/02/24 https://doi.org/10.1080/07853890802698842. PubMed PMID: 19234897.
[73] Y.L. Wu, S.B. Jiang, T.L. Ying, Single-domain antibodies as therapeutics against human viral diseases, Front. Immunol. 8 (2017), https://doi.org/10.3389/fimmu.2017.01802, 1802. PubMed PMID: WOS:000417809000001.
[74] S. Muyldermans, Nanobodies: natural single-domain antibodies, Annu. Rev. Biochem. 82 (2013) 775–797, https://doi.org/10.1146/annurev-biochem-063011-092449. PubMed PMID: WOS:000321698100028.
[75] S. Muyldermans, T. Atarhouch, J. Saldanha, J.A.R.G. Barbosa, R. Hamers, Sequence and structure of V-H domain from naturally-occurring camel heavy-chain immunoglobulins lacking light-chains, Protein Eng. 7 (9) (1994) 1129–1135, https://doi.org/10.1093/protein/7.9.1129. PubMed PMID: WOS:A1994PG50700011.
[76] P. Vanlandschoot, C. Stortelers, E. Beirnaert, L.I. Ibanez, B. Schepens, E. Depla, et al., Nanobodies (R): new ammunition to battle viruses, Antivir. Res. 92 (3) (2011) 389–407, https://doi.org/10.1016/j.antiviral.2011.09.002. PubMed PMID: WOS:000298127100001.
[77] R.A. Weiss, C.T. Verrips, Nanobodies that neutralize HIV, Vaccines (Basel) 7 (3) (2019), https://doi.org/10.3390/vaccines7030077, 77. PubMed PMID: WOS:000487982000036.
[78] W.Z. Chen, Z.Y. Zhu, Y. Feng, D.S. Dimitrov, Human domain antibodies to conserved sterically restricted regions on gp120 as exceptionally potent cross-reactive HIV-1 neutralizers, Proc. Natl. Acad. Sci. U. S. A. 105 (44) (2008) 17121–17126, https://doi.org/10.1073/pnas.0805297105. PubMed PMID: WOS:000260913800059.
[79] J. Matz, P. Kessler, J. Bouchet, O. Combes, O.H.P. Ramos, F. Barin, et al., Straightforward selection of broadly neutralizing single-domain antibodies targeting the conserved CD4 and coreceptor binding sites of HIV-1 gp120, J. Virol. 87 (2) (2013) 1137–1149, https://doi.org/10.1128/Jvi.00461-12. PubMed PMID: WOS:000312934400040.
[80] L.I. Ibanez, M. De Filette, A. Hultberg, T. Verrips, N. Temperton, R.A. Weiss, et al., Nanobodies with in vitro neutralizing activity protect mice against H5N1 influenza virus infection, J. Infect. Dis. 203 (8) (2011) 1063–1072, https://doi.org/10.1093/infdis/jiq168. PubMed PMID: WOS:000289168500005.
[81] S.V. Tillib, T.I. Ivanova, L.A. Vasilev, M.V. Rutovskaya, S.A. Saakyan, I.Y. Gribova, et al., Formatted single-domain antibodies can protect mice against infection with influenza virus (H5N2), Antivir. Res. 97 (3) (2013) 245–254, https://doi.org/10.1016/j.antiviral.2012.12.014. PubMed PMID: WOS:000317709400005.
[82] J. Ashour, F.I. Schmidt, L. Hanke, J. Cragnolini, M. Cavallari, A. Altenburg, et al., Intracellular expression of camelid single-domain antibodies specific for influenza virus nucleoprotein uncovers distinct features of its nuclear localization, J. Virol.

89 (5) (2015) 2792–2800, https://doi.org/10.1128/Jvi.02693-14. PubMed PMID: WOS:000350207400030.

[83] G.W. Wei, W.X. Meng, H.J. Guo, W.Q. Pan, J.S. Liu, T. Peng, et al., Potent neutralization of influenza A virus by a single-domain antibody blocking M2 ion channel protein, PLoS One 6 (12) (2011), https://doi.org/10.1371/journal.pone.0028309, e28309. PubMed PMID: WOS:000298171400067.

[84] L. Detalle, T. Stohr, C. Palomo, P.A. Piedra, B.E. Gilbert, V. Mas, et al., Generation and characterization of ALX-0171, a potent novel therapeutic nanobody for the treatment of respiratory syncytial virus infection, Antimicrob. Agents Chemother. 60 (1) (2016) 6–13, https://doi.org/10.1128/Aac.01802-15. PubMed PMID: WOS:000369154600002.

[85] G. Van Heeke, K. Allosery, V. De Brabandere, T. De Smedt, L. Detalle, A. de Fougerolles, Nanobodies (R)dagger as inhaled biotherapeutics for lung diseases, Pharmacol. Ther. 169 (2017) 47–56, https://doi.org/10.1016/j.pharmthera.2016.06.012. PubMed PMID: WOS:000393012000004.

[86] A.W. Tarr, P. Lafaye, L. Meredith, L. Damier-Piolle, R.A. Urbanowicz, A. Meola, et al., An alpaca nanobody inhibits hepatitis C virus entry and cell-to-cell transmission, Hepatology 58 (3) (2013) 932–939. Epub 2013/04/05 https://doi.org/10.1002/hep.26430. PubMed PMID: 23553604.

[87] P. Sroga, D. Safronetz, D.R. Stein, Nanobodies: a new approach for the diagnosis and treatment of viral infectious diseases, Futur. Virol. 15 (3) (2020) 195–205, https://doi.org/10.2217/fvl-2019-0167. PubMed PMID: WOS:000532019700006.

[88] B. Serruys, F. Van Houtte, P. Verbrugghe, G. Leroux-Roels, P. Vanlandschoot, Llama-derived single-domain intrabodies inhibit secretion of hepatitis B virions in mice, Hepatology 49 (1) (2009) 39–49, https://doi.org/10.1002/hep.22609. PubMed PMID: WOS:000262127400008.

[89] J.M. van der Vaart, N. Pant, D. Wolvers, S. Bezemer, P.W. Hermans, K. Bellamy, et al., Reduction in morbidity of rotavirus induced diarrhoea in mice by yeast produced monovalent llama-derived antibody fragments, Vaccine 24 (19) (2006) 4130–4137. Epub 2006/04/18 https://doi.org/10.1016/j.vaccine.2006.02.045. PubMed PMID: 16616802.

[90] N. Pant, A. Hultberg, Y. Zhao, L. Svensson, Q. Pan-Hammarstrom, K. Johansen, et al., Lactobacilli expressing variable domain of llama heavy-chain antibody fragments (lactobodies) confer protection against rotavirus-induced diarrhea, J. Infect. Dis. 194 (11) (2006) 1580–1588. Epub 2006/11/04 https://doi.org/10.1086/508747. PubMed PMID: 17083044.

[91] N. Pant, H. Marcotte, P. Hermans, S. Bezemer, L. Frenken, K. Johansen, et al., Lactobacilli producing bispecific llama-derived anti-rotavirus proteins in vivo for rotavirus-induced diarrhea, Future Microbiol. 6 (5) (2011) 583–593. Epub 2011/05/19 https://doi.org/10.2217/fmb.11.32. PubMed PMID: 21585264.

[92] S.A. Sarker, M. Jakel, S. Sultana, N.H. Alam, P.K. Bardhan, M.J. Chisti, et al., Anti-rotavirus protein reduces stool output in infants with diarrhea: a randomized placebo-controlled trial, Gastroenterology 145 (4) (2013) 740–748. e8. Epub 2013/07/09 https://doi.org/10.1053/j.gastro.2013.06.053. PubMed PMID: 23831050.

[93] A.D. Koromyslova, G.S. Hansman, Nanobody binding to a conserved epitope promotes norovirus particle disassembly, J. Virol. 89 (5) (2015) 2718–2730. Epub 2014/12/19 https://doi.org/10.1128/JVI.03176-14. PubMed PMID: 25520510; PubMed Central PMCID: PMCPMC4325747.

[94] V. Stalin Raj, N.M.A. Okba, J. Gutierrez-Alvarez, D. Drabek, B. van Dieren, W. Widagdo, et al., Chimeric camel/human heavy-chain antibodies protect against MERS-CoV infection, Sci. Adv. 4 (8) (2018) eaas9667. Epub 2018/08/14 https://doi.org/10.1126/sciadv.aas9667. PubMed PMID: 30101189; PubMed Central PMCID: PMCPMC6082650.

[95] P.A. Koenig, H. Das, H. Liu, B.M. Kummerer, F.N. Gohr, L.M. Jenster, et al., Structure-guided multivalent nanobodies block SARS-CoV-2 infection and suppress mutational

escape, Science 371 (6530) (2021), https://doi.org/10.1126/science.abe6230. Epub 2021/01/14. PubMed PMID: 33436526.

[96] S. Bessalah, S. Jebahi, N. Mejri, I. Salhi, T. Khorchani, M. Hammadi, Perspective on therapeutic and diagnostic potential of camel nanobodies for coronavirus disease-19 (COVID-19), 3 Biotech 11 (2) (2021) 89. Epub 2021/01/28 https://doi.org/10.1007/s13205-021-02647-5. PubMed PMID: 33500874; PubMed Central PMCID: PMCPMC7820838.

[97] E. Sulheim, H. Baghirov, E. von Haartman, A. Boe, A.K. Aslund, Y. Morch, et al., Cellular uptake and intracellular degradation of poly(alkyl cyanoacrylate) nanoparticles, J. Nanobiotechnol. 14 (1) (2016), https://doi.org/10.1186/s12951-015-0156-7. Epub 2016/01/09. PubMed PMID: 26743777; PubMed Central PMCID: PMCPMC4705582.

[98] N. Bertrand, J.C. Leroux, The journey of a drug-carrier in the body: an anatomo-physiological perspective, J. Control. Release 161 (2) (2012) 152–163. Epub 2011/10/18 https://doi.org/10.1016/j.jconrel.2011.09.098. PubMed PMID: 22001607.

[99] S.S. Yu, C.M. Lau, S.N. Thomas, W.G. Jerome, D.J. Maron, J.H. Dickerson, et al., Size- and charge-dependent non-specific uptake of PEGylated nanoparticles by macrophages, Int. J. Nanomedicine 7 (2012) 799–813. Epub 2012/02/24 https://doi.org/10.2147/IJN.S28531. PubMed PMID: 22359457; PubMed Central PMCID: PMCPMC3284223.

[100] D.A. Kuhn, D. Vanhecke, B. Michen, F. Blank, P. Gehr, A. Petri-Fink, et al., Different endocytotic uptake mechanisms for nanoparticles in epithelial cells and macrophages, Beilstein J. Nanotechnol. 5 (2014) 1625–1636. Epub 2014/11/11 https://doi.org/10.3762/bjnano.5.174. PubMed PMID: 25383275; PubMed Central PMCID: PMCPMC4222452.

[101] A. Albanese, W.C. Chan, Effect of gold nanoparticle aggregation on cell uptake and toxicity, ACS Nano 5 (7) (2011) 5478–5489. Epub 2011/06/23 https://doi.org/10.1021/nn2007496. PubMed PMID: 21692495.

[102] J.M. Zook, R.I. Maccuspie, L.E. Locascio, M.D. Halter, J.T. Elliott, Stable nanoparticle aggregates/agglomerates of different sizes and the effect of their size on hemolytic cytotoxicity, Nanotoxicology 5 (4) (2011) 517–530. Epub 2010/12/15 https://doi.org/10.3109/17435390.2010.536615. PubMed PMID: 21142841.

[103] D.M. Morens, G.K. Folkers, A.S. Fauci, Emerging infections: a perpetual challenge, Lancet Infect. Dis. 8 (11) (2008) 710–719. Epub 2008/11/11 https://doi.org/10.1016/S1473-3099(08)70256-1. PubMed PMID: 18992407; PubMed Central PMCID: PMCPMC2599922.

[104] E. de Wit, N. van Doremalen, D. Falzarano, V.J. Munster, SARS and MERS: recent insights into emerging coronaviruses, Nat. Rev. Microbiol. 14 (8) (2016) 523–534, https://doi.org/10.1038/nrmicro.2016.81. PubMed PMID: 27344959.

[105] T. Dube, A. Ghosh, J. Mishra, U.B. Kompella, J.J. Panda, Repurposed drugs, molecular vaccines, immune-modulators, and nanotherapeutics to treat and prevent COVID-19 associated with SARS-CoV-2, a deadly nanovector, Adv. Ther. (Weinh.) (2020) 2000172. Epub 2020/11/12 https://doi.org/10.1002/adtp.202000172. PubMed PMID: 33173808; PubMed Central PMCID: PMCPMC7645867.

[106] R. Itani, M. Tobaiqy, F.A. Al, Optimizing use of theranostic nanoparticles as a life-saving strategy for treating COVID-19 patients, Theranostics 10 (13) (2020) 5932–5942. Epub 2020/06/03 https://doi.org/10.7150/thno.46691. PubMed PMID: 32483428; PubMed Central PMCID: PMCPMC7254986.

[107] L. Singh, H.G. Kruger, G.E.M. Maguire, T. Govender, R. Parboosing, The role of nanotechnology in the treatment of viral infections, Ther. Adv. Infect. Dis. 4 (4) (2017) 105–131. Epub 2017/07/28 https://doi.org/10.1177/2049936117713593. PubMed PMID: 28748089; PubMed Central PMCID: PMCPMC5507392.

CHAPTER

6

Antimicrobial nanoparticles: Synthesis, mechanism of actions

Harleen Kaur, Protima Rauwel, and Erwan Rauwel

Institute of Technology, Estonian University of Life Science, Tartu, Estonia

1 Introduction

The intensive and broad spectrum utilization of antibiotic drugs during the past century has brought about antimicrobial resistance (AMR) against them [1]. The enhancement in the resilience of the microbial strains toward existing antibiotics, in the event of intensive care and other medical procedures, is presently a colossal concern. AMR is usually based on microbial evolutionary processes that take place during the antibiotic treatment leading to inheritable resistance [2]. For these reasons, over the past few decades, researchers have investigated nanoparticles that have the potential to fight AMR [3]. In fact, metals and more particularly silver have been used as antimicrobial agents since antiquity. For this reason, there have been tremendous advancements in the development of nanoparticles with variable morphologies, such as nanoflowers, nanobars, nanorods, nanospheres, nanoplatelets as they hold promise for potential biomedical applications [4,5]. Their nanosize results in generating novel properties and makes them capable of penetrating into cell bodies. Their large surface-to-volume ratio is of substantial interest in promoting an efficient antimicrobial interaction with bacteria and biofilms. The antimicrobial modes of actions are still not well understood and stem from multiple origins that include the shape, size, zeta potential, surface area, and surface curvature, which enable tuning their antimicrobial activity for specific applications dealing with microbial threat [6]. Metal oxides and metals have different modes of action. In fact, studies show that metals act by provoking injuries to microbial cells that result in oxidative stress, protein dysfunction, or membrane damage, followed by death of the bacteria [7]. Lemire et al. reviewed the different antimicrobial mechanisms of

https://doi.org/10.1016/B978-0-12-821637-8.00008-0
Copyright © 2023 Elsevier Inc. All rights reserved.

metals and described the chemical and toxicological principles that underlie their antimicrobial activity (Fig. 1) [8]. Fig. 1 describes all the antimicrobial mechanisms that take place with metal species, including in their nanoparticle form. For this reason, inorganic nanoparticles are now considered as the most promising solution to battle AMR [9].

Nanomaterials are already being used in antibacterial coatings for implantable devices and medical equipment in order to prevent

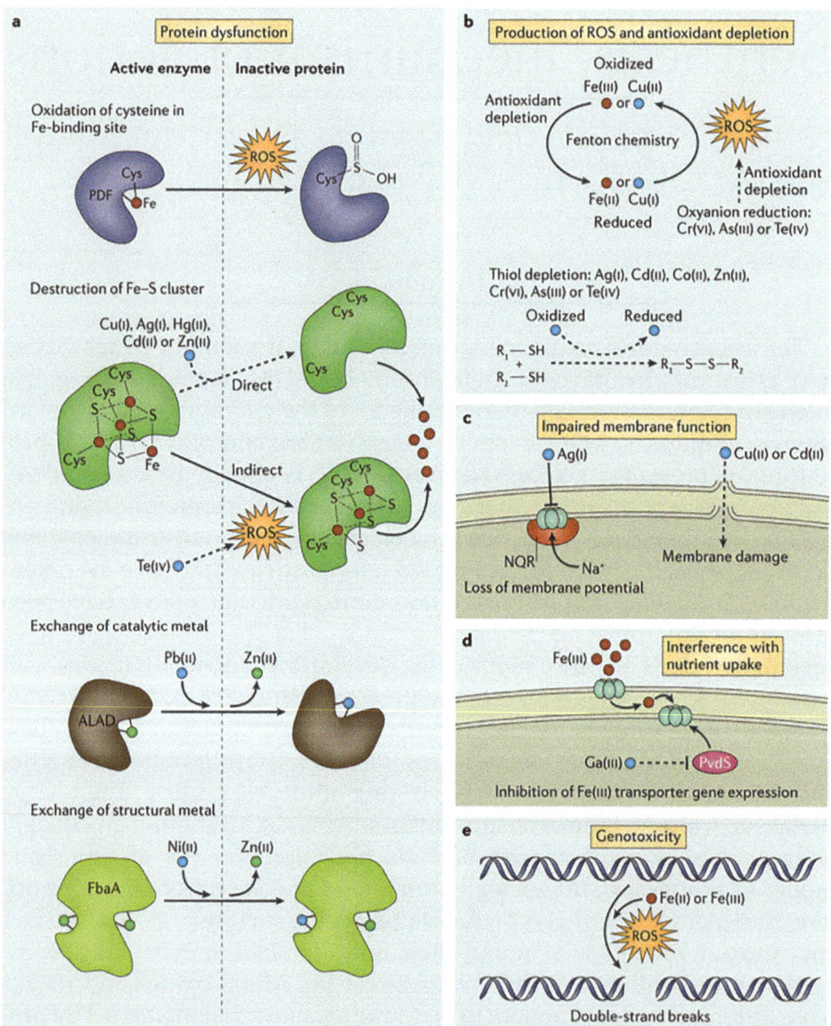

Nature Reviews | Microbiology

FIG. 1 Toxicity mechanisms of different metal species [8].

nosocomial infection and promote faster wound healing. They can also be used as carriers of conventional antibiotics leading to an improvement of the antibacterial activity compared to conventional antibiotics [3]. Some nanoparticles show enhanced antibacterial activities and improved biocidal properties compared to conventional disinfectants, when combined with active antimicrobial molecules. Besinis et al., for example, compared the toxicity of silver (Ag), titanium dioxide, and chlorhexidine against the oral pathogenic species of Streptococcus and demonstrated that Ag NPs exhibit higher biocidal properties [10]. Till now, only three methods have been known to fight against antimicrobial resistance, i.e., synthesis and development of new drugs, increased dose of already available drugs, and combination of multiple existing drugs. Nanotechnology gives a promising alternative to current antibiotic-based approaches as they can overcome antimicrobial resistance by following distinct approaches in immunization, designing and delivery of antimicrobial drugs, and fighting against bacterial infections [11,12].

Metal-based nanoparticles easily bind in a nonspecific manner to microorganisms, which expand the spectrum of antimicrobial action. Metal and metal oxide nanoparticles are known to show unique antimicrobial properties that are different in many ways as compared to their bulk form. For example, iron compounds exhibit antimicrobial properties in nanosize form but not in bulk form. This property has paved the path for the development of "nanoantibiotics" during the last decade [13]. Several metal-based nanoparticles such as silver, gold, copper, iron, and gallium and metal oxide nanoparticles such as titanium dioxide (TiO_2), silicon oxide (SiO_2), iron oxide, copper oxide, zinc oxide (ZnO), cobalt oxide, calcium oxide (CaO), cerium oxide (CeO_2), and magnesium oxide (MgO) are known to be effective against a wide spectrum of microorganisms [14–16]. The antimicrobial properties and their respective toxicity have been widely investigated [17]. These nanomaterials are intensively studied for applications in antibacterial coating. These coatings involving nanomaterials can be applied to implantable prostheses, medical tools and materials to prevent infection, as well as promote wound healing [18]. TiO_2 and Ag nanocompounds are already integrated into antimicrobial coatings in commercially available panels [19].

Organic-inorganic hybrid nanoparticles refer to materials, where an inorganic component is infused with an organic one, typically polymeric components, ligands, biomolecules, and pharmaceutical substances [20]. The antimicrobial properties of these molecules such as a combination of lipoic acid organic nanoparticles (ONPs) with Ag and Au have been explored [21]. Polymeric nanoparticles composed mainly of linear polymers and amphiphilic block polymers are used to deliver antimicrobial agents to treat infections. Coworkers have conducted various studies on metal nanoparticles embedded in polymer matrices demonstrating the efficiency of such integrated methods in blocking the bacterial/viral transmissions. For example, Cu NPs mixed by melt blending with polypropylene showed excellent antimicrobial properties; hence, it can be incorporated in plastic materials [22,23].

Apart from metal-based nanomaterials, the advancements in carbon-based nanomaterials (graphene, carbon nanotubes, fullerene, etc.) have shown compelling results against antimicrobial resistance [24,25]. In a recent study, carbon nanotubes exhibit higher effectiveness than fullerenes, depending on the diameter, length, electronic structure, and surface energy of the former [26]. These nanostructures exhibit strong antimicrobial activities due to the simultaneous effect of cell membrane disruption and reactive oxygen species (ROS) generation [27,28].

In this book chapter, the multiple mechanisms that are involved in their antimicrobial processes are described and discussed. These mechanisms are different for metal oxide and metal nanoparticles; therefore, the two categories are treated separately. At first, the recent works involving metal oxide nanoparticles will be described and discussed with specific examples (ZnO, Fe_3O_4, TiO_2, MgO, CuO, CeO_2, and Co_3O_4). The antimicrobial properties of metal nanoparticles will be then addressed in the second part with the most studied compounds (Ag, Au, Cu, and Ga). The different mechanisms of action and their present applications will also be discussed.

2 Synthesis

There are multiple approaches toward the synthesis of nanomaterials and more specifically antimicrobial nanoparticles as shown in Fig. 2. Physical methods are based on a top-down approach in which a bulk structure

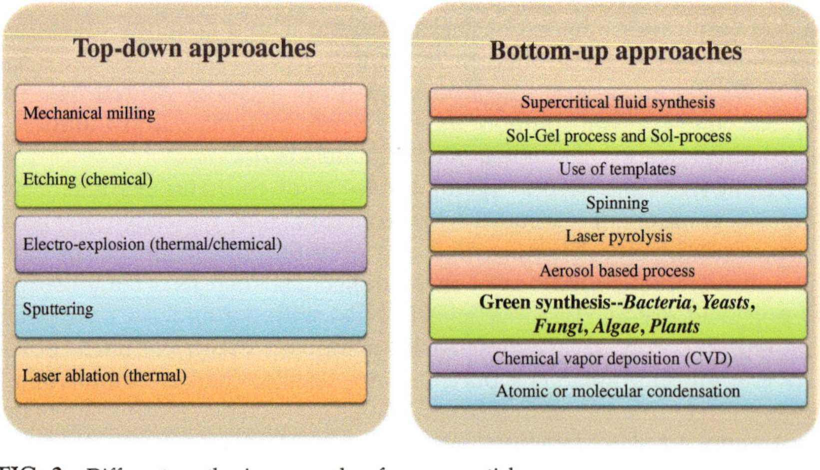

FIG. 2 Different synthesis approaches for nanoparticles.

is reduced to nanoscale by numerous methods that include mechanical milling, etching, electro-explosion (thermal/chemical), sputtering, and laser ablation among others. However, these approaches including mechanical milling do not allow an accurate control of the size, shape, size distribution during the synthesis. In addition, pulsed laser deposition that enables an accurate control of the previously mentioned nanoparticle characteristics during the synthesis is too specialized and expensive. The bottom-up approaches enable the synthesis of nanoparticles with specific size, shape, structure, and size distribution at lower costs; they are therefore preferred to top-down approaches for the synthesis of antimicrobial nanoparticles. Bottom-up approaches are mostly classified as chemical methods *viz.* supercritical fluid synthesis, sol-gel and sol processes, use of templates, spinning, laser pyrolysis, aerosol-based process, green synthesis, chemical vapor deposition (CVD), and atomic or molecular condensation. However, chemical approaches usually involve the production of undesired chemical by-products. Green chemistry was then developed to minimize the use of hazardous substances and eliminate the production of toxic by-products to make the chemistry, and more particularly syntheses, more environmental friendly through 12 principles [29]. Green chemistry includes biochemical methods that use bacteria, yeasts, fungi, algae, or plant extracts to synthesize nanoparticles [30]. The utilization of plant extracts for the synthesis of nanomaterials [30,31] also allows combining the properties of the plant extract (i.e., antioxidant properties, antibacterial properties, antifungal properties, etc.) [32] with the intrinsic properties of the nanomaterials through a synergistic effect [33]. Along with the cost-effectiveness of these methods, the synthesis of metal and metal oxide nanoparticles using plant extracts has been the focus of numerous investigations in the field of antimicrobial applications. Antimicrobial nanomaterials can be applied to biocidal coating (painting, medical tool, etc.), antimicrobial bandage, antimicrobial clothing (socks), and in antimicrobial treatment methods, among others.

3 Mechanism

The mechanism behind the antimicrobial activity of nanoparticles does not only depend on the intrinsic properties of the material (chemistry) but also depends on the physical properties of the nanomaterial, such as size, shape, solubility, surface charge, and agglomeration, for example. Few studies have shown the effect of the morphology on the antimicrobial activity. Wang et al. demonstrated that sharp edges are capable of mechanically damaging the cell wall of bacteria [34,35]. Raza et al. compared the antimicrobial properties of spherical and triangular Ag nanoparticles of different sizes against *P. aeruginosa* and *E. coli*. In their study, they showed that smaller sized spherical nanoparticles were more efficient in destroying bacteria than triangular nanoparticles. They demonstrated that the nanoparticle size is a critical parameter and the infiltration of nanoparticles inside the bacteria kills them more efficiently than direct contact with sharp edges [36]. Morones et al. studied the "electronic effects" or zeta potential against Gram-negative bacteria (*E. coli, V. cholera, P. aeruginosa,* and *S. typhus*) and showed that electrostatic forces increase bactericidal activity [15,37]. This is due to the attachment of the nanoparticles to the bacterial membrane by electrostatic interaction or Van der Walls forces that disrupt the membrane [38]. In their investigation on silver nanoparticles, Morones et al. have identified three antimicrobial ways against Gram-negative bacteria. The smaller nanoparticles (1–10nm) attach to the cell membrane and disturb the permeability and respiration, whereupon inducing bacterial death. The small nanoparticles are also able to invade into the bacteria and damage the DNA, and finally the Ag NPs also release silver ions that kill bacteria by binding with their DNA. A positive zeta potential results in stronger interactions between the nanoparticles and negatively charged bacteria (Gram negative) which causes cell membrane disruption. The same was observed for Gram-positive bacteria (*B. subtilis* and *S. aureus*) that are more sensitive to negatively charged nanoparticles [39,40]. This enables a targeted antimicrobial treatment through the selection of the suitable nanomaterials. Even though the specific mechanism is not thoroughly identified nor understood, for each type of nanoparticles, it is however widely accepted that there are multiple mechanisms involved in the antimicrobial activities.

The possible antimicrobial mechanisms of metal-based nanoparticles investigated and revealed from numerous studies are described in Fig. 3. Fig. 3 summarizes the antimicrobial mechanisms that are related to reactive oxygen species (ROS) production and interaction of the bacteria with NP. The combination of antimicrobial mechanisms via generation of reactive oxygen species (ROS) that have a strong redox potential, disruption of cell membranes by electrostatic static interaction, disturbance in metal/metal ion homeostasis, genotoxicity, and induction of intracellular antibacterial effects, including interactions with DNA and proteins, are the possible modes of action leading to bacterial death. The penetration of nanoparticles into the bacteria through the cell wall induces the generation of ROS like superoxide anion (O_2^-), hydroxyl radicals (OH•), hydrogen peroxide (H_2O_2), oxygen (O_2), and organic hydroperoxides. These ROS are in fact toxic to bacteria and provoke their death through protein denaturation, DNA damages, and enzyme disruption. However, the four types of ROS do not have the same impact on bacteria metabolism and H_2O_2 can be neutralized by endogenous antioxidants produced by the bacteria. On the other hand, (OH•) and (O_2) are more efficient and are able to induce acute microbial death. Oxidative stress has also been identified

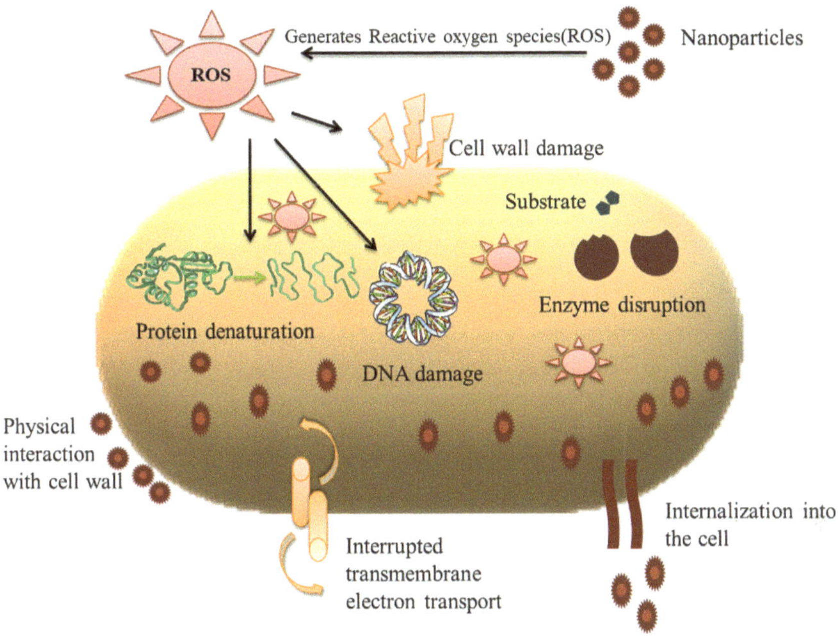

FIG. 3 Various processes underlying the antimicrobial mechanism.

as responsible for the modification of cell membrane permeability, which results in membrane damage. Active redox cycling, functional group modifications on nanoparticle surfaces, and particle-cell interactions play a vital role in the ROS production. The multiple solutions that involve nanomaterials to combat microbial resistance are summarized as follows.

- Nanoparticles with multiple simultaneous mechanisms of action against microbes,
- Multiple antimicrobial agents packaged within the same nanoparticle,
- Nanoparticles can overcome decreased uptake and increased efflux of drug by bacteria,
- Nanoparticles prevent or overcome biofilm formation,
- Nanoparticles can combat intracellular bacteria,
- Nanoparticles target antimicrobial agents to the site of infection.

4 Metal oxide nanoparticles

Metal oxide nanoparticles (MONP) have attracted attention of researchers, as they are usually easier to synthesize than their metallic counterpart. MONP nanoparticles can be synthesized by various methods that determine their size distribution, shape, roughness, surface energy, and enable their utilization in targeted applications. Over the past two decades, MONPs have spurred a lot of interest as an attractive alternative to combat microbes and antimicrobial resistance [41]. MONP nanoparticles use multiple mechanisms to overcome microbial resistance to antimicrobial agents and offer promising results when tested against a wide range of microorganisms. In this section, we discuss about the synthesis, mechanism, and antimicrobial applications of MONP.

4.1 Zinc oxide nanoparticles (ZnO)

Zinc oxide nanoparticles (ZnO NP) that play a vital role in the development of several antimicrobial agents were preliminary studied by Brainler et al. [42] and Jones et al. [43] against numerous pathogens. Out of the known zinc compounds, ZnO has been recognized as 'safe' by the U.S. Food and Drug Administration [44]. Nanosized ZnO is investigated widely due to its numerous intrinsic properties (i.e., photocatalyst, photoluminescence, antimicrobial, etc.) [45–47]. Among all the other metal oxide nanoparticles, ZnO NP are known to display strong antimicrobial effects and ZnO NP are already used in many applications [48] such as food preservation, packaging materials, [49] paints, ceramics, batteries, solar cells, textiles and biosensors, [50] piezoelectric devices, photodiode devices, sun screens, antireflection coatings, and photocatalysis. The band gap of 3.37 eV is responsible for the sensitivity of ZnO toward shorter wavelengths [45,51]. The antimicrobial properties of ZnO nanomaterials are linked to the structure [52]. ZnO NP toxicity not only depends on the size, [53] modification of the surface, and intrinsic physicochemical properties of ZnO [54] but also on the medium used to disperse the ZnO NP [52]. Raghupathi et al. showed that antimicrobial properties of ZnO NP are inversely proportional to the size in the case of *S. aureus* and not due to Zn^{2+} released [53]. However, Pasquet et al. showed that antimicrobial properties of ZnO NP depend on the intrinsic physicochemical properties and showed that ZnO NP with smaller sizes and higher porosities exhibit higher antimicrobial activity, [54] which therefore implies that no consensus has been reached on the same. Reyes et al. suggested that ZnO toxicity partially originates from membrane damage and disruption of ion homeostasis [55]. In the antimicrobial photocatalytic activity of ZnO, when the surface of ZnO is exposed to light, electron-hole pairs are created. These excitons interact with water molecules and produce ROS (mainly OH•, H_2O_2, O_2^-), known to be toxic toward bacteria. Their mechanism consists of entering into the bacterial cell and disrupting lipids and proteins of the bacterial cell membrane, which results in the leakage of intracellular contents, finally leading to death of the bacteria. The antimicrobial activity is highly dependent on the ZnO NP size with smaller NP penetrating more easily into the bacteria, thereupon enhancing the antimicrobial activity. [56] ZnO NP have shown strong potential to generate ROS inside the bacteria, which effectively damage the cell wall, disrupt DNA replication, and kill the bacteria. In addition, in the NP assembly, the antimicrobial activity is also higher due to the high surface-to-volume ratio that accelerates the process of ROS generation. Moreover, antimicrobial activity of nanomaterials is also dependent on bacterial species (i.e., Gram-positive and Gram-negative bacteria). Azam et al. showed that ZnO NP are efficient against both Gram-positive and Gram-negative bacteria with higher efficiencies against Gram positive because of differences in the membrane's thickness and ROS sensitivity, more specifically when comparing *S. aureus* to *E. coli* [57]. However, contrary to Fe_2O_3 NP that were only efficient against Gram-positive bacteria, ZnO was efficient in killing both bacteria. In a recent study, the nano-ZnO multilayer deposited on cotton fabrics showed excellent antibacterial activity against *S. aureus* (Fig. 4) [58]. Fig. 4 shows SEM images of cotton fabric covered by ZnO NP that prevent the development of bacteria on its surface.

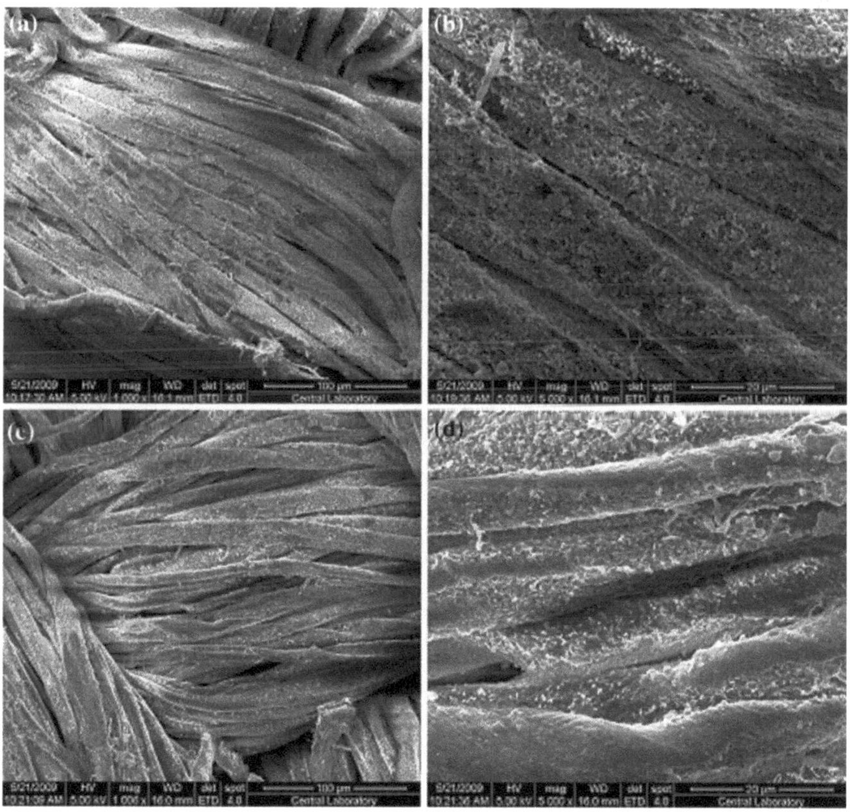

FIG. 4 SEM images of cationized cotton fabrics deposited with ZnO/ZnO10 layer at (A) low magnification (B) and ZnO/ZnO 16 layer (C) at low magnification and (D) at high magnification [58].

McCarthy et al. studied the possible synergy between Zn ions and antioxidant molecules and found a possible synergy between the Zn^{2+} and these molecules [59]. It is possible to obtain such synergistic effects during the synthesis itself of nanomaterials like ZnO by using green synthesis techniques based on plant extracts [31,60], which also provide more environmental friendly solutions. The synthesis of ZnO NP using plant extracts for antimicrobial applications is presently the main focus to not only battle AMR and other clinical pathogens [61], but also to combat other infections that affect plants and harvests [62]. In a recent study, Nasser et al. showed that ZnO synthesized with medicinal plants like *Cassia fistula* and *Melia azedarach* improves its antimicrobial properties through a synergistic effect [63]. These nanomaterials exhibit higher antimicrobial properties than conventional antibiotic drugs. Ahmad et al. studied the biocidal properties of ZnO NP synthesized using *Euphorbia hirta* leaf extract. They tested ZnO NP against bacteria (*Streptococcus aureus* and *Escherichia coli*) and fungi (*Aspergillus fumigatus* and *Aspergillus niger*) and showed that it is possible to produce antimicrobial ZnO NP at a large scale in a rapid, cost-effective, and environmental friendly manner [64]. Hence, ZnO NP synthesized using plant extracts are seen as a potential candidate for different medicinal and biological applications, but the antimicrobial mechanisms that occur in such combinations need to be further studied. In general, both antimicrobial mechanisms for ZnO are accepted, i.e., photocatalytic effect as well as nanosize effects, which change their surface states. The latter has also been known to engender visible light activation in ZnO, which provides cost-effective solutions to photocatalytic antimicrobial nanoparticles (Table 1).

TABLE 1 Zinc oxide nanoparticles.

Co_3O_4			
Method of synthesis	**Biocidal mechanism**	**Microorganisms tested**	**Ref.**
Sol-gel combustion route	production of ROS, accumulation of NP inside bacteria	*E. coli, P. aeruginosa, B. subtilis, S. aureus*	[57]
Chemical synthesis (hydrolysis and precipitation)	Membrane damage due to particle abrasiveness	*E. coli*	[56]
Purchased	Accumulation inside bacteria	Methicillin sensitive *S. aureus*(MSSA), *S. epidermis, S. pyogenes, E. faecalis, B. subtilis, E. coli*	[43]
Hydrolysis of zinc acetate	Membrane damage	*S. agalactiae, S. aureus*	[65]

TABLE 1 Zinc oxide nanoparticles—cont'd

Co_3O_4			
Method of synthesis	**Biocidal mechanism**	**Microorganisms tested**	**Ref.**
Sol-gel synthesis (zinc acetate dihydrate)	Multiple mechanisms	Methicillin-sensitive *S. aureus*, Methicillin-resistant *S. aureus* and *S. epidermidis*	[66]
Room temperature synthesis and solvothermal synthesis	ROS production, accumulation of NP inside and outside bacteria	Gram positive (Methicillin-sensitive *S. aureus, S. epidermidis, S. pyogenes, E. faecalis, B. subtilis, B. cereus*) and Gram negative (*E. coli, P. vulgaris, S. typhimurium, S. flexneri, P. alcaligenes, E. aerogenes*)	[53]
Biosynthesis using plant extracts (*Cassia fistula* and *Melia azedarach*)	Multiple mechanisms	*E. coli, S. aureus*	[63]
Biosynthesis using *Euphorbia hirta* leaf extract	ROS production, accumulation of NP inside and outside bacteria	*Streptococcus mutans and aureus, C. absonum, E. coli and P. mirabilis, Arthogrophis cuboida, Aspergillus fumigates and Aspergillus niger*	[64]
Hydrolysis of zinc acetate	ROS production, electrostatic interaction with bacterial membrane (disruption and permeability)	*P. aeruginosa* strains (A−B−, A+B−, A−B+, and A+B+)	[67]
Biosynthesis using *Glycosmis pentaphylla* leaf extract	Bacterial membrane disruption, (respiratory and cytoplasmic content damage)	*B. cereus, S. aureus, S. dysenteriae, S. paratyphi, Candida albicans, Aspergillus niger*	[68]
Bacterial growth (*Pseudomonas putida* broth culture)	ROS production	*P. otitidis, P. oleovorans, A. baumannii, B. cereus, E. faecalis*	[69]
Biosynthesis using *Cinnamomum tamala* leaf extract	Bacterial membrane damage and leakage of intracellular proteins and cellular contents	*S. aureus*	[70]
Biosynthesis using *Coptidis Rhizoma*	ROS production (hydrogen peroxide), bacterial membrane damage, and intracellular content leakage	*B. megaterium, B. pumilus, B. cereus, E. coli*	[71]

4.2 Titanium dioxide nanoparticles (TiO_2)

TiO_2 is abundant on Earth and is a well-known photocatalyst [72,73]. It is already used in self-cleaning windows [74] as well as in self-cleaning and UV protection clothing [75]. In addition, TiO_2 NP exhibit quantum size effects, long-term stability in water and air, and are already used as antimicrobial agents for a large range of pathogenic microorganisms [76–78]. The small size of TiO_2 NP increases the specific surface and hence enhances the photocatalytic and photoelectrical chemical conversion processes due to higher surface-to-volume ratio. The antimicrobial properties of TiO_2 NP depend on photocatalytic properties that are also dependent on the TiO_2 structure (anatase, rutile, brookite). In fact, anatase and rutile have different bandgaps for ROS production and depend on the photoexcitation. For this reason, studies show that the anatase structure exhibits higher antimicrobial properties than rutile and amorphous TiO_2 [79]. Under UV light excitation, three ROS are generated mainly (OH•, O_2^{-}•, and O_2^{-}), which engender bacterial cell membrane disruption that kills bacteria. Foster et al. showed that TiO_2 NP can attach to bacterial membranes and produce ROS that will damage the membrane itself, resulting in the death of bacteria [80]. A further study reported that the cell membrane's complexity and density also affect the antimicrobial efficiency of TiO_2 NP. Kühn et al. showed that it is possible to disinfect surfaces consisting of a TiO_2 stimulated by indirect UVA. They then compared the antimicrobial efficiency against *E. coli, P. aeruginosa, S. aureus, E. faecium, C. albicans,* in that order [81]. The combination of TiO_2 NP with conventional antibiotics (beta lactams, cephalosporins, aminoglycosides, glycopeptides, macrolides, lincosamides, and tetracyclines) was studied by Roy et al., in the aim to combine the photocatalytic properties of TiO_2 (ROS production) with toxicity of antibiotics against methicillin-resistant *S. aureus*. They showed that it is possible to obtain a synergy and improve the antimicrobial effect with such combinations [82]. In another study, the authors revealed the role of zeta potential in TiO_2 NP's invasiveness toward the cell. The smaller sized TiO_2 NP with more positive zeta potential exhibit more significant antimicrobial effect when tested against bacteria. Carré et al. studied the impact of TiO_2 photocatalyst under UVA excitation on *E. coli* survival. They observed that antibacterial photocatalytic activity is accompanied by lipid peroxidation that induces membrane fluidity and disrupts the bacterial cell integrity [83]. TiO_2 NP were also studied for applications in water purification systems and their antimicrobial properties were compared with conventional disinfectants. TiO_2 NP are seen as promising alternatives to traditional chemical disinfectants, but the barrier remains the production scalability, cost-effectiveness and the need for UV sources, feasibility for direct application, and finally the safety [84]. TiO_2 NP have also been investigated for

TABLE 2 Titanium dioxide nanoparticles.

TiO_2			
Method of synthesis	**Biocidal mechanism**	**Microorganisms tested**	**Ref.**
Purchased	Cell wall damages through OH radicals that lead to punctures	*E. coli, P. aeruginosa, S. aureus, E. faecium, Candida albicans*	[81]
Chemical synthesis (surfactant)	"Electromagnetic" attraction between NP and bacterial membrane and ROS production	*Methicillin-resistant S. aureus*	[22]
Purchased	ROS production (lipid peroxidation)	*E. coli*	[83]
Purchased	ROS production (lipid peroxidation) and loss of respiratory activity (membrane)	*E. coli, S. aureus, E. hirae, P. aeruginosa, B. fragilis*	[91]
Sol-gel methods	Multiple mechanisms	*S. aureus, K. pneumoniae*	[92]
Laser pyrolysis	Impairment of membrane integrity	*E. coli*	[93]
Sol-gel method	Damage of the bacterial membranes (ROS production)	*S. aureus, P. aeruginosa, E. coli*	[79]
Purchased	ROS production and breakdown of the bacterial membrane	*P. aeruginosa, E. hire, E. coli, S. aureus, B. fragilis*	[94]
Purchased	Cellular uptake, ROS production	*S. typhimurium*	[95]

application in medical implants [85] and orthodontic appliances like dental implants, screws, and tools due to their antibacterial activity against *Lactobacillus acidophilus* [86,87]. The FDA has certified the use of TiO_2 in food industry, [75] cosmetics, drugs, and clothing [88], which has therefore broadened the spectrum of various applications [89]. However, studies highlight that the utilization of TiO_2 NP and ZnO NP needs safe disposal protocols as their release into surface or ground waters will have detrimental effects on the environment [90] (Table 2).

4.3 Magnesium oxide nanoparticles (MgO)

Magnesium is one of the essential nutrients for the human body as it regulates metabolism and cardiorespiratory endurance [96]. The low toxicity of MgO NP against human cells is well known and allows their utilization in dietary supplements and medication. He et al. showed that MgO NP can prevent the growth of food-borne pathogens such as *S. enteritidis, E. coli,* and *C. jejuni* by multiple mechanisms and can be incorporated into food packaging materials in order to prevent food spoilage [97]. Even if the antimicrobial action of MgO NP is still not well understood, three possible mechanisms were identified *viz.,* the direct interaction of nanoparticles with bacteria that subsequently damage the bacterial cell, the formation of ROS, and an alkaline effect [98,99]. Alkaline effect results in water molecules adsorption on the entire surface of MgO NP. Sawai et al. demonstrated that due to a higher pH of this thin surface water layer, direct contact with the bacterial membrane will damage it and consequently induce bacterial death [100]. When exposed to light, MgO NP generate ROS (mainly H_2O_2) that attach themselves to the bacterial membrane and induce lipid peroxidation and oxidative stress on bacterial organelles. Leung et al. conducted a study revealing antimicrobial activity of MgO NP (~20 nm) against *E. coli* even without UV illumination. Under UV illumination MgO NP are more toxic to *E. coli* when compared to its activity in dark, due to the formation of ROS. Nevertheless, unilluminated MgO NP still exhibited antibacterial activity and therefore confirms the presence of other antimicrobial mechanisms [99]. Sellik et al. demonstrated that size, structure, and arrangement of MgO NP affect the antimicrobial properties by testing MgO-325 mesh and MgO nanopowders. MgO nanosheets (~11 nm) exhibited higher antimicrobial activity against *E. coli* and *S. aureus* compared to MgO nanopowder(~ 25 nm). In fact, smaller NPs arranged into sheets with more surface area increase ROS production [101]. In another study, Al-Hazmi et al. showed that one-dimensional MgO nanowires are able to damage the membrane wall of bacteria and higher concentration of MgO nanowires increases the antibacterial activity [102]. MgO NP can be combined with other antimicrobial agents to promote a synergistic effect between the components. For example, Jin et al. reported a comparative study of antimicrobial effect of MgO NP (~20 nm) alone and when combined either with ZnO (~20 nm) or nisin (a polycyclic antibacterial peptide) against *E. coli* and *Salmonella* [103]. The combination of MgO and ZnO did not show significant improvement of the antimicrobial properties. However, when combined with nisin, MgO NP exhibited higher antimicrobial activity. Nisin is known to show antibacterial activity against both Gram-positive and Gram-negative species. The interaction of nisin with phospholipid components of the cell membrane causes the formation of pores within the membrane, allowing MgO NP to enter more easily into the bacterial cell and produce ROS. The overall combination of MgO NP with these antimicrobial agents increased visibly the antimicrobial effect [104,105]. For similar reasons, the synthesis of MgO NPs using selected plant extracts has also been studied [106–108]. MgO NP exhibit antimicrobial properties due to their very high surface area and enhanced surface reactivity. When treated with halogen (Cl or

TABLE 3 Magnesium oxide nanoparticles.

MgO			
Method of synthesis	**Biocidal mechanism**	**Microorganisms tested**	**Ref.**
Biosynthesis using *Emblica officinalis* extract and sol-gel method	ROS production followed by decay of bacterial outer membranes and phospholipid peroxidation	*S. aureus, E. coli*	[106]
Microwave hydrothermal synthesis	ROS production	*E. coli, B.* spp	[102]
Purchased	Synergistic effect	*Salmonella, E. coli*	[103]
Purchased	Cell membrane damage, even without ROS production	*E. coli*	[99]
Biosynthesis using *Rhizophora lamarckii* leaf extract	Destruction of the cytoplasmic membrane of the bacteria	*S. aureus, S. pneumoniae, E. coli, S. typhimurium*	[111]
Purchased	ROS production and disruption of bacterial membrane	*C. jejuni, E. coli, S. enteritidis*	[97]

Br), the nanocrystals adsorb and carry a high load of active halogens that will be supplied all around the bacteria in high concentrations. MgO NP combined with halogen (MgO/X_2 (X = Cl, Br) have shown promising antimicrobial properties against both Gram-positive and Gram-negative bacteria. MgO/X_2 nanoparticles also generate ROS, cause lipid peroxidation, penetrate the cell, and disrupt the bacterial cell membrane [109]. MgO/X_2 were tested against *E. coli, B. megaterium* and *B. subtilis* endospores and showed enhanced antimicrobial activity. In a similar study, Lellouche et al. studied the antimicrobial properties of magnesium fluorides (MgF_2) and observed that they killed both *E. coli* and *B. megaterium* in 20 min. The probable killing mechanism is the attachment and penetration of MgF_2 NP into the cells that causes a disruption in membrane potential, enhanced lipid peroxidation, and DNA binding. This study demonstrated the high antimicrobial potential of metal fluoride NP for the design of sterile surface coatings [110] (Table 3).

4.4 Copper oxide nanoparticles (CuO)

Copper oxide (CuO) has attracted a lot of attention owing to its applications in solar cells, lithium ion batteries, catalysis, and more particularly antimicrobial materials [112]. CuO NP are known to exhibit stronger antimicrobial behavior than metallic copper nanoparticles and organic antimicrobial agents [113]. The main reason is the release of Cu^{2+} ions combined with electrostatic interactions, cell wall damage, rupture of the plasma membrane, and disruptive biochemical processes [114]. Ren et al. demonstrated that CuO NP prepared by plasma processes have potential to kill pathogenic bacteria involved in hospital-acquired infections [115]. CuO NP synthesized by chemical methods or by using a biotemplate also exhibit similar antimicrobial properties against *S. aureus* and *E. coli* than antibiotics [116,117]. Jaramillo et al. compared the antimicrobial properties of nanocomposites combining poly(butylene adipate-*co*-terephthalate) (PBAT) and Cu nanoparticles, Cu/Cu_2O nanoparticles, and copper sulfate ($CuSO_4$). They showed that PBAT does not exhibit any antimicrobial properties, but when combined with copper nanomaterials or $CuSO_4$, the nanocomposite prevented bacterial development [118]. They showed that the release of copper ions was the main factor that kills bacteria. CuO NP exhibit high biocidal properties and are also cheaper when compared to other metal oxide nanoparticles, which make them good candidates for applications like wound dressings, antimicrobial plastics, textile industry, face masks (Fig. 5), and biomedical applications [119,120]. Fig. 5 shows SEM images of polymer fibers covered with CuO NP that provides protection against bacteria. CuO NP should be well anchored on the surface of the fibers. These types of mask can be adapted to other threats including viruses. Although the cytotoxicity of CuO NP is

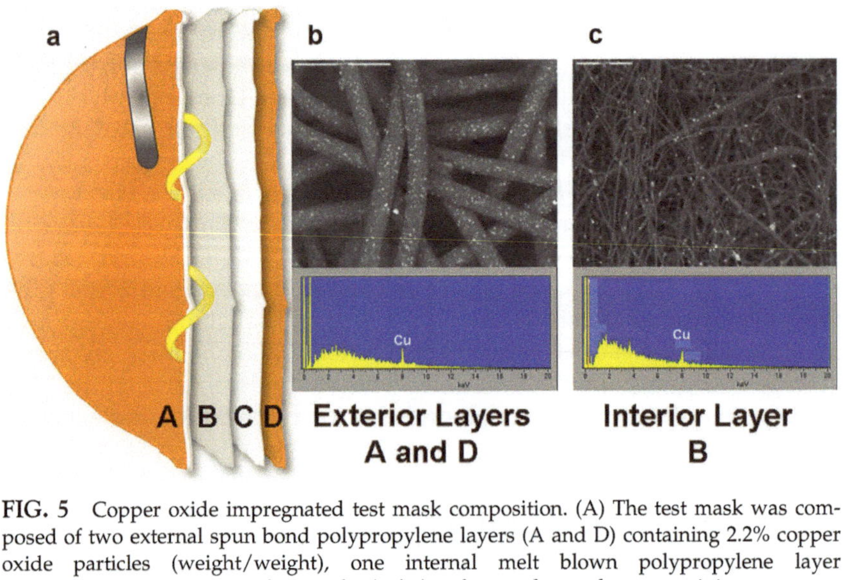

FIG. 5 Copper oxide impregnated test mask composition. (A) The test mask was composed of two external spun bond polypropylene layers (A and D) containing 2.2% copper oxide particles (weight/weight), one internal melt blown polypropylene layer (B) containing 2% copper oxide particles (*w*/w) and one polyester layer containing no copper oxide particles. (B) Scanning electronic microscope picture and X-ray analysis of external layer A. (C) Scanning electronic microscope picture and X-ray photoelectron spectrum analysis of internal layer B [120].

higher than other MONP, when functionalized they tend to exhibit lower toxicities. This has spurred interest in synthesizing CuO NP with plant molecules to simultaneously decrease their toxicity and enhance their antimicrobial properties. Kumar et al. showed that CuO NP synthesized using *Aloe vera* leaf extract exhibit higher antimicrobial properties against three bacterial fish pathogens—*Aeromonas hydrophila, Pseudomonas fluorescens,* and *Flavobacterium branchiophilum* [121]. Other studies show that CuO NP synthesized from brown algae (*Bifurcaria bifurcata*) or ultrasound-assisted biosynthesis using brown algae *Cystoseira trinodis* exhibit both antimicrobial activity against *E. aerogenes* (Gram negative) and *S. aureus* (Gram positive) [122,123]. Environmental friendly syntheses are also investigated and colloidal CuO NP were prepared via microwave-assisted synthesis using glucose, starch, and $CuCl_2$. Their antibacterial properties were successfully assessed against *E. coli* (Gram negative), *S. epidermis* (Gram positive), methicillin-resistant *S. aureus* (superbug MRSA) isolate, and the spore-forming *Bacillus megatherium* [124] (Table 4).

TABLE 4 Copper oxide nanoparticles.

CuO			
Method of synthesis	**Biocidal mechanism**	**Microorganisms tested**	**Ref.**
Sonochemical synthesis	ROS production	*E. coli, S. aureus*	[114]
Colloid-thermal synthesis	ROS production	*E. coli, S. aureus*	[117]
Biosynthesis using *Aloe vera* leaf extract	Multiple mechanisms	*A. hydrophila, P. fluorescens, F. branchiophilum*	[121]
Microwave-assisted synthesis	ROS production, interaction of CuO NP with bacterial membrane	*E. coli, S. epidermidis,* methicillin-resistant *S. aureus* (MRSA)	[124]
Biosynthesis using brown alga extract (*Bifurcaria bifurcata*)	Synergistic effect	*E. aerogenes, S. aureus*	[122]
Biosynthesis using *Pterocarpus marsupium* extract	Cell membrane disruption	*E. coli, P. vulgaris, K. pneumoniae, S. aureus, S. epidermis, B. cereus*	[125]
Electrochemical deposition	Contact-based mechanical damage, ROS production	*E. coli, S. typhi, S. aureus, B. subtilis*	[126]
Biosynthesis using *Cystoseira trinodis* extract	ROS production	*E. coli, E. faecalis, S. typhimurium, S. aureus, B. subtilis, S. faecalis*	[123]

4.5 Cerium oxide nanoparticles

Cerium oxide (CeO_2) is a cubic-fluorite type of oxide, one of the most used among the rare earth oxide family. CeO_2 is a semiconductor with bandgap energy (3.19 eV) used in applications such as solar cells, catalytic converters, biosensors, drug delivery, and biomedical applications [127]. CeO_2 NP are biologically relevant because of their catalytic and multienzyme mimetic properties [128]. For these reasons, CeO_2 NP can be infused into sunscreens as they are good for absorbing UV light. Although CeO_2 NP can be synthesized via precipitation, hydrothermal, solvothermal, spray pyrolysis methods, green methods however (plant mediated, fungus mediated, polymer mediated, and nutrient mediated) tend to produce less toxic and more stable nanoparticles. [129] The antimicrobial mechanism of cerium oxide NPs is not well explored, but they induce oxidative stress resulting in cell membrane leakage. CeO_2 NP enter the cell and produce ROS, which deactivate the enzymes, damage the organic molecules, and ultimately cause cell death. In addition, protein denaturation occurs because of strong electrostatic potential when CeO_2 NP interact with thiol groups present in membrane proteins. The antimicrobial mechanism of CeO_2 NP was first studied by Thill et al. [38] Along with plant-based methods, mycosynthesis method also appears simple, rapid, water miscible, and nontoxic and provides higher yield even on a large scale. [130] Spherical CeO_2 NP were synthesized from *Acalypha indica* leaf extract and showed excellent antibacterial property against both Gram-positive and Gram-negative microorganisms such as *S. aureus* and *E. coli* [131]. CeO_2 NP synthesized from *Leucas aspera* extract have shown positive results when tested against *K. aerogenes, E. coli, P. desmolyticum,* and *S. aureus* [132]. In another investigation, CeO_2 NP synthesized using *Aspergillus niger* culture filtrate showed high antibacterial activity against Gram-positive bacteria (*S. pneumoniae, B. subtilis*) and Gram-negative bacteria (*P. vulgaris, E. coli*) [133]. In Fig. 6, the TEM images provide an overview of the mycosynthesized CeO_2. Even though the size distribution

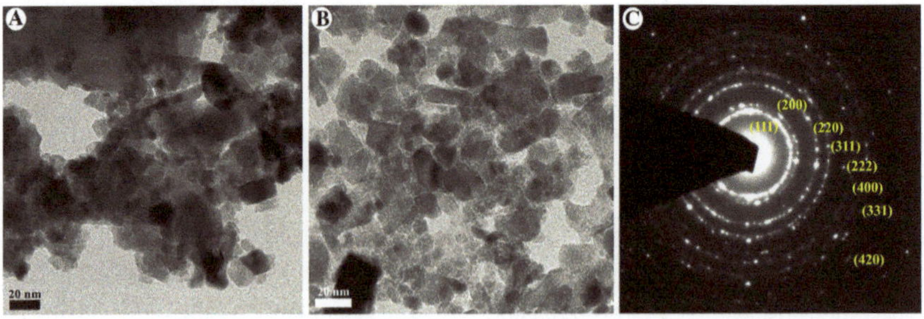

FIG. 6 (A), (B) TEM images of mycosynthesized CeO_2 NP calcined at 400°C, (C) selected area electron diffraction of CeO_2 NP [133].

TABLE 5 Cerium oxide nanoparticles.

CeO_2			
Method of synthesis	**Biocidal mechanism**	**Microorganisms tested**	**Ref.**
Biosynthesis using *Acalypha indica* leaf extract	Penetration into the bacteria (enzymes deactivation, ROS production)	*E. coli, S. aureus*	[131]
Biosynthesis using *Leucas aspera* leaf extract	Cell membrane disruption, ROS production	*E. coli, S. aureus, K. aerogenes, P. desmolyticum*	[132]
Biosynthesis using *Aspergillus niger* culture filtrate	Interference with bacteria membrane (binding), disruption of cellular respiration and DNA replication, ROS production	*S. pneumoniae, B. subtilis, P. vulgaris, E. coli*	[133]
Biosynthesis using *Gloriosa superba* leaf extract	ROS production, intracellular functional disruption	*E. coli, S. dysenteriae, P. aeruginosa, P. vulgaris, K. pneumonia, S. pneumoniae*	[134]
Surfactant and template-free synthesis	ROS production	*E. coli, S. oneidensis, B. subtilis*	[135]

is not uniform, the larger nanoparticles appear facetted, while the smaller nanoparticles are round. This implies that the larger and smaller NP could have different antimicrobial mechanisms based on their morphologies, as explained previously. Similarly, the antibacterial activity of CeO_2 NP synthesized using *Gloriosa superba* leaf extract was studied by disc diffusion method against *S. aureus, S. pneumoniae, E. coli, P. aeruginosa, P. vulgaris, K. pneumoniae,* and *S. dysenteriae* [134] (Table 5).

4.6 Iron oxide nanoparticles (IO NP)

Among all the known metal oxides nanoparticles, iron oxide nanoparticles (IO NP), particularly maghemite (Fe_2O_3) and magnetite (Fe_3O_4), are given importance in theranostics for diagnostic applications and in drug delivery and anticancer treatment. IO NP exhibit unique properties at nanoscale due to their superparamagnetic properties, surface chemistry, and their unique oxidation state-dependent properties. IO NP have triggered a lot of interest for applications in the biomedical field due to their magnetic properties, biocompatibility, antimicrobial properties and, more importantly their superparamagnetic properties that can be used for drug delivery, hyperthermia therapy, and cancer treatment [136]. IO NP are also promising candidates for antimicrobial activities, similarly to ZnO and TiO_2, as they produce ROS that can kill pathogenic bacteria. IO NP can be synthesized using many methods, but conventional chemical syntheses involve toxic reagents that are not environmental friendly. Therefore biosynthesis using algae, fungi, bacteria, or plant extracts is seen as better alternatives [137]. Durmus et al. demonstrated that superparamagnetic iron oxide nanoparticles (SPION) exhibit biocidal properties against antibiotic-resistant biofilms produced by methicillin-resistant *S. aureus* (MRSA). They also showed that IO NP can be used alone or conjugated with other antibacterial metals like ZnO to increase their antibiofilm properties. They discovered that the process is facilitated by the addition of fructose as a supplement to IO NP uptake by biofilms [138]. However, only a few studies are available of IO NP as antibacterial agents. Tran et al. revealed that IO NP synthesized using a polyvinyl matrix have antimicrobial properties against *S. aureus*, but only at a high dose ($3\,mg\,mL^{-1}$) [139]. This study also demonstrates that the antimicrobial activity of IO NP increases with their concentration. The mode of action behind bacterial death is an increase in oxidative stress and cell membrane interference.

Geilich et al. showed that SPION combined with encapsulated hydrophilic antibiotic methicillin are able to overcome antibiotic-resistant biofilms and kill the bacteria inside the biofilm (Fig. 7). They obtained a synergistic effect between SPIONs and methicillin that are unable to damage the biofilm and kill bacteria if taken separately [140]. Arokiyaraj et al. demonstrated that treating IO NPs with *Argemone mexicana* extract increases the antibacterial properties, and the resulting antimicrobial properties against *E. coli, P. mirabilis,* and *B. subtilis* are higher than the plant extract itself or untreated IO NP. IO NP were also synthesized using *Ruellia tuberosa* leaf extract to coat cotton fabrics. The composite shows antimicrobial properties against both Gram-negative (*E. coli, K. pneumonia*) and Gram-positive (*S. aureus*) pathogens, with a higher activity against Gram-negative bacteria [141] (Table 6).

4.7 Cobalt oxide nanoparticles (Co_3O_4)

Metal ions are known to exhibit antimicrobial activity and cobalt ion complexes have also been studied to that end [144]. The biocompatibility and cytotoxicity of the cobalt oxide nanoparticles (CONP) toward human cells are still under investigation. However, despite a probable

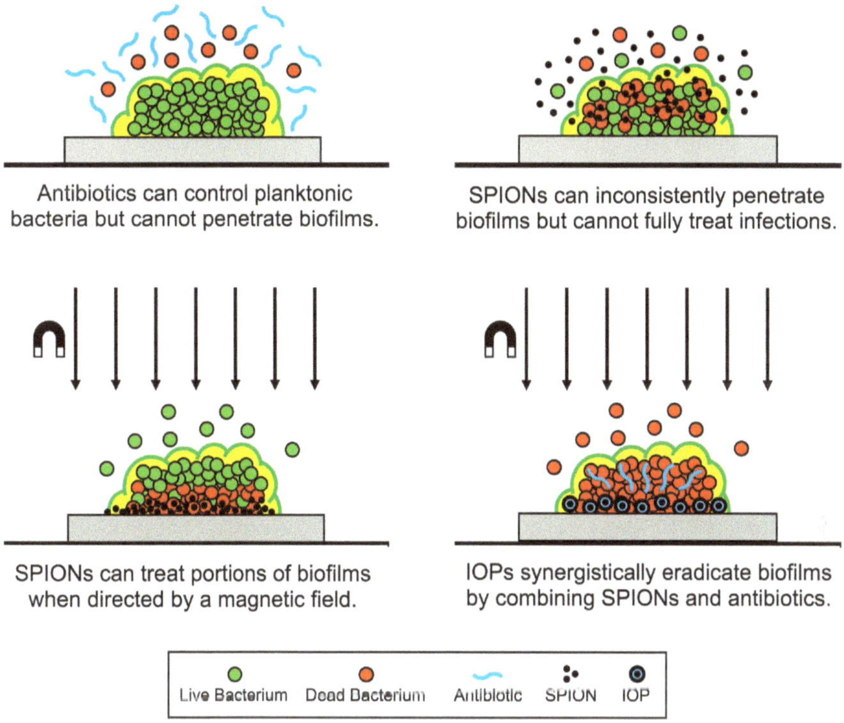

FIG. 7 Current strategies for biofilm treatment using SPIONs and/or antimicrobials. IOP = iron oxide-encapsulating polymersome; SPION = superparamagnetic iron oxide nanoparticle [140].

TABLE 6 Iron oxide nanoparticles.

Iron oxide			
Method of synthesis	**Biocidal mechanism**	**Microorganisms tested**	**Ref.**
Biosynthesis *using Argemone mexicana L.* leaf extract	ROS production	*P. mirabilis, E. coli, B. subtilis*	[142]
In situ synthesis (matrix mediated)	ROS production	*S. aureus*	[139]
Biocompatible amphiphilic diblock copolymer	Synergistic effect	*Methicillin-resistant S. epidermidis*	[140]
Biosynthesis using *R. tuberosa* leaf extract	Cell wall disruption, ROS production	*E. coli, K. pneumoniae, S. aureus*	[119]
Wet chemistry method	ROS production, membrane disruption	*S. epidermidis*	[143]

moderate toxicity of Co_3O_4 nanomaterials, [145] the antimicrobial properties of CONP and their nanocomposites combining carbon-based nanomaterials like graphene oxide(GO) [146] or nanodiamond [147] were investigated. Alsharaeh et al. showed that combining Co_3O_4 with GO(8%) generates antimicrobial properties, whereas Co_3O_4 and GO taken separately did not exhibit any antimicrobial properties [99]. In fact, CONP synthesized using conventional chemical methods like simple urea-based thermal decomposition method do not exhibit high antimicrobial properties [148]. So, other methods were investigated in order to improve these properties and Omran et al. reported that CONP synthesized using *Aspergillus brasiliensis* exhibit high antimicrobial properties [149]. They showed that the attachment of biomolecules present in *A. brasiliensis* enhanced these properties in addition to stabilize the cobalt oxide nanoparticle. Similarly, the utilization of plant extracts for the synthesis of nanomaterials enhances the antimicrobial properties of the nanomaterials and might make them more biocompatible [150]. For these reasons, in most of the studies that report investigations of antimicrobial properties of Co_3O_4 nanoparticles, the nanoparticles were synthesized using selected plant extracts. CONP synthesized using leaf extract of *Sageretia thea* showed varying degree of antimicrobial properties against different pathogens (Gram positive: *S. aureus, S. epidermis, B. subtilis* and Gram negative: *K. pneumonia, P. aeruginosa, E. coli*). Similar results, with similar bacteria, were recently found by Hafeez et al., with CONP synthesized using *Populus ciliata* leaf extract with higher antimicrobial activity against Gram-positive bacteria [151]. This result is attributed to the porous nature of the cell wall of Gram-positive bacteria that present a higher permeability and facilitate the penetration of the CONP inside the bacteria [152]. In another study, *Vijaya* et al. showed that CONP synthesized using *A. indica* plant extract exhibit higher antimicrobial properties than chloramphenicol against *S. aureus, B. subtilis, P. aeruginosa,* and *E. coli* [153]. Khalil et al. investigated the antimicrobial properties of CONP synthesized using *Sageretia thea*. They also assessed the cytotoxicity of CONP against Brine shrimp larvae and the biocompatibility of CONP. They found that CONP can be used for disease treatment, but further studies on cytotoxicity and biocompatibility are still necessary [150]. However, contradictory scientific reports claim that CONP of 10nm of diameter synthesized using *Calotropis procera* plant extract do not exhibit antimicrobial properties against pathogens like *E. coli* or these properties are less notable than antibiotic ampicillin used as control [154]. Further investigations are then needed to better understand the antimicrobial mechanism. Nevertheless, the mechanism seems to be related to bacterial cell penetration followed by ROS production (Table 7).

TABLE 7 Cobalt oxide nanoparticles.

Co_3O_4			
Method of synthesis	**Biocidal mechanism**	**Microorganisms tested**	**Ref.**
Biosynthesis using *Aspergillus brasiliensis*	Cell penetration, ROS, mycelial coating	*B. subtilis, S. aureus, P. aeruginosa, E. coli*	[149]
Biosynthesis using *Sageretia thea*	Cell penetration, ions release, ROS	*B. subtilis, S. aureus, S. epidermis, P. aeruginosa, E. coli, K. pneumonia*	[150]
Biosynthesis using *Populus ciliata* leaf extract	Cell penetration, ROS	*B. subtilis, B. lichernifermia, P. aeruginosa, E. coli*	[151]
Biosynthesis using *Azadirachta indica* leaf extract	Bacterial cell damage due to different oxidation state, ROS	*B. subtilis, S. aureus, P. aeruginosa, E. coli*	[153]

5 Metal nanoparticles

Metal nanoparticles (MNP) are the most interesting nanomaterials for many applications. They have been more particularly investigated for their intrinsic antimicrobial properties and their possible use as antimicrobial agents during the last two decades [14]. MNP are highly toxic to microorganism and more specifically bacteria with a low or zero toxicity to human cells depending on the metal. Hence, they can be employed in various applications like medicinal drugs, food industry, and coating. During the last decade, investigations were focused on MNP synthesized using biosynthesis and selected plant extracts. Due to the additional antimicrobial properties arising from the plant extracts used during the synthesis, these MNP exhibit higher antimicrobial properties with lower cytotoxicity to human cells than their counterpart synthesized using conventional methods [155]. Ongoing investigations are focusing on understanding the combined mechanisms underlying antimicrobial activity of biosynthesized nanoparticles.

5.1 Silver nanoparticles

Since antiquity, silver has been used as coins and cutlery owing to its resistance to corrosion and to its antibacterial properties. Silver has multiple modes of antibacterial actions; Chernousova et al. have reported from a chemical viewpoint that the utilization of silver in applications depends on its assemblies and compositions, i.e., nanoparticles, salt, or metal [156]. Contrary to salt and metal that have their own mode of action, the nanoparticle form of silver combines all modes of antibacterial actions in addition to its own modes of action *viz.*, mechanical membrane damage and bacterial cell penetration [157]. Silver metal nanoparticles (Ag MNP) have probably been the most investigated nanomaterials during the last decade and are now being employed in many fields such as biocidal coatings, [158] photocatalysis, lithography, microelectronics, cosmetics, pharmaceutical drugs, wastewater treatment, wound dressing, [159] treatment for various diseases, fabric coatings, dental tools, and food industry [160,161]. Ag MNP exhibit unique antibacterial, antifungal, and antiviral properties that facilitate their mode of action against a wide range of infectious pathogens. In addition, Ag MNP can be used as nanocarriers and combined with antibiotics (amoxicillin, penicillin G, vancomycin, and erythromycin) to target multidrug-resistant bacterial strains. [162] Most of the publications about antimicrobial properties of Ag MNP published during the last decades report studies about plant-based synthesized Ag NMP due to their enhanced antimicrobial activity [60,160,163]. Ag MNP interact with the bacterial cell membrane, simultaneously releasing biochemically active ions. These ions cause "pit" formation, which damage the cellular membrane leading to leakage of cellular contents, obstruction of the vital transport processes, denaturation of the proteins, blockage of DNA functions, and alteration of enzymatic processes, all of which contribute to kill bacteria [157,164]. Morones et al. used high angle annular dark field scanning transmission electron microscopy to highlight that antimicrobial properties linked to direct interaction are size dependent and only Ag MNPs of 1–10 nm of diameter present a direct interaction with the Gram-negative bacteria (*E. coli, S. typhus, P. aeruginosa*, and *V. cholerae*) [37]. Ag MNPs attach to the surface of the cell membrane, disturbing permeability and respiration and then pervade inside the bacteria. When Ag MNP break the outer membrane, it results in the leakage of cellular materials and affects the cell functions such as DNA damage, protein denaturation, and enzyme deactivation. Ag MNP then disable respiratory chain dehydrogenases, which inhibit the respiration and growth of cells [164]. In fact, Ag MNP have shown stronger interactions toward sulfur-containing proteins present on the cell wall, [165] which further hinder the enzymatic function of affected proteins and inhibit the growth of bacteria. This results in the collapse of the bacterial membrane that results in cell decomposition and death of the bacteria. The other antimicrobial effect is the silver ion release inside the bacteria, [166] inducing the interruption of intracellular O_2 reduction, which induces ROS production and the death of bacteria [167]. The disruption mechanism of bacterial membrane followed by loss of membrane integrity through ROS generation in the case of Ag MNP combined with clay was studied by Su et al. [168] Ag MNP have the ability to act as modulators of signal transduction in microbial cells. Jena et al. have investigated the ability of gold-silver MNP to disrupt the bacterial actin cytoskeletal network, which mediates the death of bacteria. Their results revealed that MNP ruptured the cells by affecting actin cytoskeleton MreB which alters cell morphology and increases the fluidity in the cell membrane [169]. Zafar et al. compared the antimicrobial properties of Ag MNP synthesized via chemical processes and laser ablation against Gram-positive (*B. subtilis, S. aureus*) and Gram-negative (*Salmonella, E. coli, K. pneumonia*) bacteria. They showed that antimicrobial properties are size dependent, and the smaller Ag MNP synthesized by laser ablation showed the best minimum inhibitory concentration (MIC) [170]. Another investigation revealed that combining tannic acid with Ag MNP can avert bacterial infections that occur during wound healing processes [159]. Ag MNP are more efficient against Gram-negative bacteria than Gram-positive bacteria due to electrostatic attraction between negatively charged cell membrane and positively charged silver ions present on the surface of Ag MNP that promote surface attachment [171]. For this reason, negatively charged Ag MNP with reduced toxicity were developed for pharmaceutical preparations [40]. Srikar et al. summarized 120 publications reporting the antimicrobial study of Ag MNP synthesized using plant extracts [163]. Antimicrobial mechanism of Ag MNP also involves ROS production. Liao et al. studied the antibacterial activity of Ag MNP against multidrug-resistant bacteria (*P. aeruginosa*) [172]. When treated with Ag MNP, *P. aeruginosa* presented very high levels of superoxide dismutase (SOD), catalase (CAT), and peroxidase (POD) and confirmed the excessive production of ROS. They highlighted that in the case of multidrug-resistant bacteria the antimicrobial mechanism is a disequilibrium of oxidation and antioxidation processes and the failure to eliminate excess ROS (Table 8).

TABLE 8 Silver nanoparticles.

Ag			
Method of synthesis	**Biocidal mechanism**	**Microorganisms tested**	**Ref.**
Chemical reduction method	Scavenging of free radicals and reactive oxygen species (ROS)	*P. aeruginosa, S. aureus, E. coli*	[159]
Purchased	Damage bacterial membrane structure and permeability	*E. coli*	[164]
Hydrothermal synthesis	Cell membrane disruption, ROS, upregulation of expression of thioredoxin and glutaredoxin	*methicillin-resistant S. aureus, E. coli*	[166]

Continued

TABLE 8 Silver nanoparticles—cont'd

Ag			
Method of synthesis	**Biocidal mechanism**	**Microorganisms tested**	**Ref.**
Chemical reduction method	Cell membrane disruption, ROS production	*S. aureus, P. aeruginosa, S. pyogenes, methicillin/ oxacillin-resistant S. aureus*	[168]
Laser ablation method	Cell membrane disruption	*E. coli, S. aureus, B. subtilis, Salmonella*	[170]
Biosynthesis using *Chrysanthemum indicum* extract	Deactivate the cellular enzymes, causing disruptions in membrane permeability, enzymatic dysfunction, Ros production	*B. subtilis, S. aureus, S. epidermidis, E. coli, K. pneumoniae, P. aeruginosa*	[171]
Chemical synthesis	Membrane disruption, release of Ag ions (interference with DNA replication)	*E. coli, V. cholerae, P. aeruginosa, S. typhus*	[37]
Chemical synthesis	Cell permeability, release of Ag ions (interference with DNA replication)	*S. aureus, E. coli*	[165]

5.2 Gold nanoparticles

Despite being less popular and more expensive to produce than Ag MNP, gold metal nanoparticles (Au MNP) have also gained attention owing to their antimicrobial properties [173]. Due to their lower toxicity and higher stability than Ag MNP, Au MNP have been preferred for targeted applications such as drug delivery, hyperthermia therapy, catalysis, data storage, biosensing, cancer treatment, and bioimaging. As in the case of Ag MNP, the modifications of Au MNP surface using different coatings result in changes in their properties. For this reason, Au MNP synthesized by green synthesis have been more particularly studied for antimicrobial applications [174]. The combination of antibiotics with Au MNP enables enhancing the antimicrobial efficiency against multidrug-resistant bacteria. Grace et al. showed that Au MNP.

capped with antibiotics like streptomycin, gentamycin, or neomycin show higher antibacterial activity against *S. aureus, M. luteus, E. coli,* and *P. aeruginosa* [175]. As compared to free ampicillin a twofold increase in antimicrobial activity was shown by chitosan-capped Au MNP coupled with ampicillin [176]. Au MNP are hence promising adjuvants for treating bacterial infections at lower doses of antibiotics. Photothermal therapy also intends to improve the antimicrobial potential of Au MNP [177]. The antimicrobial mechanism of Au NP is similar to Ag MNP and is related to cell membrane disruption as well as ROS production causing oxidative stress that impairs basic cell function, leading to the death of bacteria. However, ion release is less predominant in Au MNP compared to Ag MNP, and Cui et al. observed in their study that Au MNP do not induce the production of ROS [178]. In their study, the origin of the antimicrobial activity is the modification of the membrane potential, which results in the inhibition of ATP synthase activities, indicating a general decline in metabolism. The other origin is the collapse of biological processes through the inhibition of the subunit of ribosome for tRNA binding. They showed that such a mechanism that does not involve ROS production would allow the development of antibacterial agents that are harmful for the host while killing bacteria. Conventional synthesis methods of Au MNP are known to be less effective as compared to biosynthesized nanoparticles that combine other active antimicrobial molecules. Das et al. found that Au MNP (~10 nm) synthesized via in situ reduction of $HAuCl_4$ on the surface of a fungal strain (*Rhizopus oryzae*, MTCC 262) exhibit high antimicrobial activity against several Gram-negative and Gram-positive pathogenic bacteria (*E. coli, P. aeruginosa, B. subtilis, S. aureus*, and *S. cerevisiae*) [179]. Mishra et al. have reported similar results for Au MNP (~50–70 nm) and Ag MNP (10–20 nm) synthesized using yeast *Candida guilliermondii* via an extracellular synthesis method [180]. The highest efficiency for both Au MNP and Ag MNP was observed against *S. aureus*. Other studies report the use of plant extracts like *Mentha piperita, T. decandra* root extract, *Helianthus annuus* flower extract, and dried flower extract of *Carthamus tinctorius* for synthesis and study of the antibacterial activity of Au MNP and Ag MNP. [181–183] The authors compared the antimicrobial properties against Gram-negative (*E. coli*) and Gram-positive (*S. aureus*) bacteria finding similar mechanisms and toxicities in both types of bacteria (Table 9).

TABLE 9 Gold nanoparticles.

Au			
Method of synthesis	**Biocidal mechanism**	**Microorganisms tested**	**Ref.**
Chemical synthesis ($HAuCl_4{\cdot}3H_2O$)	Cell membrane disruption	*S. aureus, M. luteus, E. coli, P. aeruginosa*	[175]
Wet precipitation method	Multiple mechanisms	*E coli, S. aureus, K. mobilis*	[176]

Continued

TABLE 9 Gold nanoparticles—cont'd

Au			
Method of synthesis	**Biocidal mechanism**	**Microorganisms tested**	**Ref.**
Chemical synthesis ($HAuCl_4{\cdot}3H_2O$)	Collapse membrane potential, inhibiting ATPase activities; inhibit ribosome binding to tRNA	*E. coli*	[178]
Microbial synthesis using *R. oryzae* mycelia	Cell membrane disruption	*P. aeruginosa, E. coli, B. subtilis, S. aureus, Salmonella* spp, *S. cerevisiae, Candida albicans*	[179]
Biosynthesis using *C. guilliermondii* (yeast)	Complex mechanism	*E. coli, S. aureus, B. cereus, K. pneumoniae, S. typhimurium*	[180]
Biosynthesis using *Mentha piperita* extract	Complex mechanism	*E. coli, S. aureus*	[181]
Biosynthesis using *Trianthema decandra* extract	Cell membrane disruption	*S. aureus, S. faecalis, E. faecalis, E. coli, P. aeruginosa, P. vulgaris, B. subtilis, Y. enterocolitica, K. pneumoniae, Candida albicans*	[182]
Biosynthesis using *Hibiscus cannabinus* extract	Cell membrane penetration/ disruption	*P. aeruginosa, S. aureus*	[183]
Biosynthesis using *Alternanthera bettzickiana* leaf extract	Cellular damage	*B. subtilis, S. aureus, S. typhi, P. aeruginosa, M. luteus, E. aerogenes*	[184]
Biosynthesis using *Deinococcus radiodurans* (bacteria)	Cell penetration inducing disruption of the cytoplasmic membrane	*S. aureus, E. coli*	[185]
Biosynthesis using *Stoechospermum marginatum* (algae)	Complex mechanism	*P. aeruginosa, V. cholerae, V. parahaemolyticus, S. enterica, P. vulgaris, S. typhi, K. pneumoniae, K. oxytoca, E. faecalis*	[186]

5.3 Copper nanoparticles

Since Roman times and before the advent of antibiotics, Cu-based compounds were used to protect against infections. Copper metal nanoparticles (Cu MNP) are already integrated in diverse applications like catalysis [187], antimicrobial coatings, solar cells, and fuel cells. Cu MNP exhibit properties similar to other noble metals, but are cost effective and more reactive. Their biocidal properties make them suitable candidates for application in medical devices [188], plastics, textiles, coatings, and protection against nosocomial pathogens. [189] Cu MNP can be synthesized by various physical and chemical methods [190]; however, green synthesis methods and more particularly methods based on plant extract are more complicated to apply to Cu MNP synthesis. In most of the cases, plant extract syntheses only enable the synthesis of CuO NP due to the high sensitivity of copper to oxygen. Conventionally synthesized Cu MNP are air sensitive as they oxidize easily, however, if well-controlled, bio-based synthesis methods can prevent the oxidation by capping the Cu MNPs with biomolecules during the synthesis. Hence, green chemistry-based Cu MNP synthesis that involves biological entities like plants, fungi, bacteria, actinomycetes, yeast, and viruses in the synthesis has high potential for antimicrobial applications [191]. For example, Cu MNP synthesized using aqueous extract of *Eryngium caucasicum* Trautv leaves exhibit antimicrobial properties against *E. coli, S. typhimurium, B. cereus,* and *S. aureus* [192]. Cu MNPs can also be synthesized using bacteria, fungi, or yeast. Biosynthesized Cu MNP using *Shewanella loihica* PV-4 exhibit high antibacterial activity against *E. coli* and the author showed that cell damages are attributed to chemical oxidation due to ROS production. They observed disabling mechanisms accompanied with cell membrane damage, lipid peroxidation, and DNA damage [193]. The antimicrobial mechanisms of Cu MNP are found to be similar to that of Ag MNP. The mechanism is based on the interaction of Cu MNP with bacteria that results in protein complexation, ROS generation, mitochondrial membrane damage, electron chain dysfunction, and finally death of bacteria. Similarly to Ag MNP, Cu MNP also interact with thiol groups and inactivate proteins. Studies show that Cu MNP easily associate to bacterial cell membrane to generate ROS response and also enter the cells intact and reach the intracellular DNA, resulting in complete degradation of the plasmid DNA. Cu MNP were also combined with other molecules to enhance their antimicrobial properties. Usman et al. reported the antimicrobial activity of Cu-chitosan nanoparticles against several microorganisms, including methicillin-resistant *S. aureus, B. subtilis, P. aeruginosa, S. choleraesuis,* and *C. albicans* [194]. They also showed that polymers like chitosan will prevent the agglomeration of Cu MNP, which tend to occur due to the high surface energy. In addition, the chitosan capping did not affect

TABLE 10 Copper nanoparticles.

Cu			
Method of synthesis	**Biocidal mechanism**	**Microorganisms tested**	**Ref.**
Biosynthesis using *Eryngium caucasicum Trautv* extract	Accumulation inside bacteria, membrane permeability change, ROS production, release of Cu ions (reduced intracellular ATP levels and impaired DNA replication)	*E. coli*, *S. typhimurium*, *B. cereus*, *S. aureus*	[192]
Biosynthesis using *Shewanella loihica* PV-4 (bacteria)	ROS, cell membrane disruption	*E. coli*	[193]
Chemical synthesis (with chitosan stabilizer)	Synergistic effect	*S. aureus*, *B. subtilis*, *P. aeruginosa*, *S. choleraesuis*, *C. albicans*	[194]
Electrochemical synthesis	Synergistic effect	*E. coli*, *S. aureus*, *L. monocytogenes*	[195]
Purchased	Protein inactivation via thiol interaction	*E. coli*, *B. subtilis*	[196]

the biocidal properties and chitosan-stabilized Cu MNP exhibit antibacterial activity against Gram-positive and Gram-negative bacteria. These results are promising in the potential use of Cu MNP as antimicrobial agents. Cu MNP antimicrobial release systems based on modified biological polymers will contribute to Cu MNP stabilization and synergistic antimicrobial effects. Cu MNP modified with polymer thin films have demonstrated antimicrobial activity against bacterial species (*E. coli*, *S. aureus*, and *L. monocytogenes*) and yeast species (*S. cerevisiae*) [190,195] (Table 10).

5.4 Gallium nanoparticles

Over a decade ago, Gallium(Ga) was proposed as a possible antimicrobial agent to treat various infections. Ga has no particular natural function in the body, but its ferro-mimicking properties are the origin of its antimicrobial properties. Ga can either be used alone or in combination with other materials against microbes; hence they can be employed for a variety

TABLE 11 Gallium nanoparticles.

Ga			
Method of synthesis	**Biocidal mechanism**	**Microorganisms tested**	**Ref.**
Hydrogels	Multiple mechanisms	*S. aureus,* *S. epidermidis,* *P. aeruginosa,* *A. johnsonii*	[198]
Chemical synthesis using $Ga(NO_3)_3$	Decreasing bacterial Fe uptake and by interfering with Fe signaling by the transcriptional regulator pvdS	*P. aeruginosa*	[197]
Double emulsification by sonication	Multitargeting mechanism	*HIV(human immunodeficiency virus)-mycobacterium*	[199]

of applications. Iron is one of the most important elements for reproductive, metabolic, and growth processes of bacteria. However, bacteria cannot distinguish between Ga and Fe, which makes Ga a suitable antimicrobial agent for iron-dependent growth of multiresistant bacteria. The reduction of Fe^{3+} to Fe^{2+} facilitates the vital cellular pathways; however, Ga cannot be reduced, so it disrupts the enzymatic processes of enzymes, damages cell DNA, obstructs nutrient/iron release, promotes starvation, eventually leading to cell death [197]. Richter et al. reported that the combination of Deferiprone and Ga-protoporphyrin into a hydrogel controls the kinetics to enhance the antimicrobial activity against iron-metabolized bacteria and biofilms [198]. In similar studies, Ga tetraphenyl porphyrin (Ga-TP) nanoparticles (Ga MNP) (305 ± 0.29 nm) were synthesized, stained using p24, and tested to inhibit Human immunodeficiency virus (HIV) and the growth of *M. tuberculosis*. Ga MNP are the first single drug to show complete inhibition of HIV development as well as coinfection of HIV- *M. tuberculosis* [199,200] (Table 11).

6 Discussion and conclusion

Microbial infectious diseases are one of the leading causes of mortality in our world, hence a global threat to humans. Researchers are developing effective and safe antimicrobial agents that show promising results in combating various infectious diseases. Narrow spectrum antibiotics have been used explicitly to treat specific type of diseases. Unfortunately, the irrational use and misuse of antibiotics has brought about antimicrobial resistance and the emergence of multidrug-resistant bacterial strains. These "superbugs" are resistant to almost all types of existing antibiotics because they carry a specific gene. The latter can make bacteria impervious to action modes of antibiotics that target more specifically translational machinery, DNA replication, and cell wall synthesis. The simultaneous formation of biofilms that protect bacteria proliferation is also part of the origin of this AMR. The onset of AMR has directed research to elaborate new antimicrobial solutions that can prevent multiresistance and affect the biofilm formation. The utilization of nanomaterials appears to be the most promising solution because it combines simultaneously multiple antimicrobial mechanisms, i.e., oxidative stress, mechanical stress, metal ion release, nonoxidative mechanism. In fact, as opposed to conventional antibiotics, nanomaterials use simultaneously multiple antibacterial mechanisms to neutralize bacteria. The main advantage of such multimechanism actions is that bacteria will need to have multiple mutated genes to develop a resistance, which is unlikely. In addition, nanomaterials can be combined with existing antibiotics in order to enhance their antibacterial activity. The mechanisms mentioned in the previous parts that describe the different antimicrobial effects of nanomaterials can be summarized as follows: generation of ROS, damage of the bacterial cell membrane, penetration of the nanomaterial into the bacteria, and interaction with the bacterial intracellular material (proteins, DNA, etc.). However, the predominant modes of antibacterial actions will depend on the type of nanomaterial, i.e., oxide, metal, and on the type of bacteria (i.e., Gram negative or Gram positive). The antimicrobial mechanism is first linked to the contact between the NP and the bacteria through electrostatic attraction in the case of polar NP, receptor-ligand interaction, van der Walls forces, or hydrophilic interactions followed by the penetration through the bacterial membrane inside the bacteria and interaction with the bacterial cell components (enzymes, ribosome, lysosome, DNA). NP have greater antimicrobial activity against Gram-positive bacteria because their cell wall is thin and composed of peptidoglycan and teichoic acid with numerous pores that allow molecules and NP to penetrate. In the case of Gram-negative bacteria, the membrane is composed of peptidoglycan, lipoproteins, and phospholipids that only allows the penetration of selected macromolecules, which make them more difficult to penetrate and induce internal damage and death of bacteria. The NP size has also an influence on the antimicrobial activity; the larger specific surface of smaller NP enables a higher probability of contact with bacteria combined with an easier penetration inside the bacterial cell. Azam et al. compared the minimum bacterial concentration (MBC) of ZnO, CuO, and Fe_2O_3 NP and they found that the smaller NP exhibit the highest antibacterial properties with following order ZnO > CuO > Fe_2O_3 [57]. However, they did not consider differences in antimicrobial mechanisms that exist between these MONP. They also observe that for all MONP types, Gram-positive bacteria are more sensitive to MONP than Gram-negative bacteria. In contrast, another study from *Arokiyaraj* et al. showed that Gram-negative bacteria are more sensitive to Ag MNP synthesized using *Chrysanthemum indicum* than Gram-positive bacteria [171]. This is more probably related to the surface nature and properties of the NP itself. In fact, zeta potential of NP should also be considered in the probability of contact, because positively charged NP will have more probability to be adsorbed on the bacterial membrane that is negatively charged.

Another particularity of NP is their ability to destroy the biofilm that bacteria can produce. This biofilm is usually composed of extracellular polymeric substances from bacteria that enable bacteria to multiply rapidly inside the biofilm. At the same time, due to its composition, the biofilm promotes genetic material exchange and protects bacteria from external threats like antibiotics and host defenses, which result in systemic chronic infection. Contrary to antibiotics, due to their physical properties and their intrinsic polarity, nanomaterials can pass through the biofilm formed by bacteria and inhibit its regeneration. Like bacteria, the biofilm will be affected by the antimicrobial action of the nanomaterial (mechanical, ROS production, ion release) and damage it, which will result in its destruction without the possibility of regeneration. However, the type of ROS produced also depends on the type of NP. Whereas CuO NP can produce the four types of ROS, ZnO NP only produce (H_2O_2) and ($OH\bullet$) and MgO only (O_2^-). The intrinsic properties of NP play an important role in the antimicrobial mechanism and ZnO NP are able to split H_2O into H^+ ion that will then react with dissolved oxygen to produce H_2O_2. However, in the case of TiO_2 NP, their photocatalytic properties generate the production of ROS. The other mechanism involves dissolved metallic ions that are released by the NP. These ions will penetrate the bacterial membrane and interact with the nucleic acids and proteins inside the bacterial cells, which will affect the internal physiological bacterial processes. However, the antimicrobial effect of dissolved ions is lower than ROS, which also makes antimicrobial treatment using only metallic ions less efficient than NP.

The method of NP synthesis should be wisely selected and the synthesis needs to be reliable and well controlled in order to obtain the suitable

NP that will exhibit the desired antimicrobial properties. However, environmental friendly methods of synthesis and more particularly plant extract syntheses should be preferred, as they employ greener routes and allow a synergistic effect between the NP properties and the intrinsic properties of the plant during synthesis. The impact of synthesis, but also of the NP on the environment, should be understood and estimated. NP can also affect biological organisms in the same way they affect bacteria and their cytotoxicity needs to be evaluated. Considering these facts, the synthesis of NP using plant extracts appears to be the most suitable route in addition to the possibility of scaling-up the synthesis for production [30]. The cost-effectiveness in addition to the possible toxicity of antimicrobial agent should also be considered. The cost of production of Au MNPs is higher than Ag MNP, which makes it the preferred element for antimicrobial coating and for this reason most of the scientific literature reports studies on Ag MNP. Despite a lower stability of Ag MNP compared to Au NP, the stability of Ag is sufficiently high to remain attractive. Copper is more abundant on Earth and Cu MNPs and CuO NP are cheaper than noble metals; however, they are known to exhibit higher cytotoxicity than Ag MNP and Au MNP. Despite their good antimicrobial properties, Cu MNP oxidize easily and are less stable than Au MNP and Ag MNP, which make noble metals and more particularly Ag MNP more attractive for application. These are the main reasons, in addition to the synergistic effect, for using environmental friendly methods of synthesis and more particularly plant extract syntheses appear to be the most suitable for such applications. The selection of the most suitable candidate for each application is still difficult due to the lack of information about the antimicrobial mechanism and exact cytotoxicity. The different modes of antimicrobial mechanism and action of the NP described in the book chapter are summarized in Fig. 8, where each type of MONP and NP has their specific mode of action against bacteria depending on their intrinsic properties, size, shape, and functionalization. In addition to their specific antimicrobial properties, it is also possible to combine these different NPs to exhibit enhanced antibacterial properties. Such combinations would be fatal for bacteria, with lower chances of their developing antimicrobial resistance. Numerous types of nanomaterials (Ag, Au, Cu, CuO, TiO_2, ZnO, CeO_2, MgO, etc.) have been investigated for antimicrobial applications and further investigations are ongoing [17].

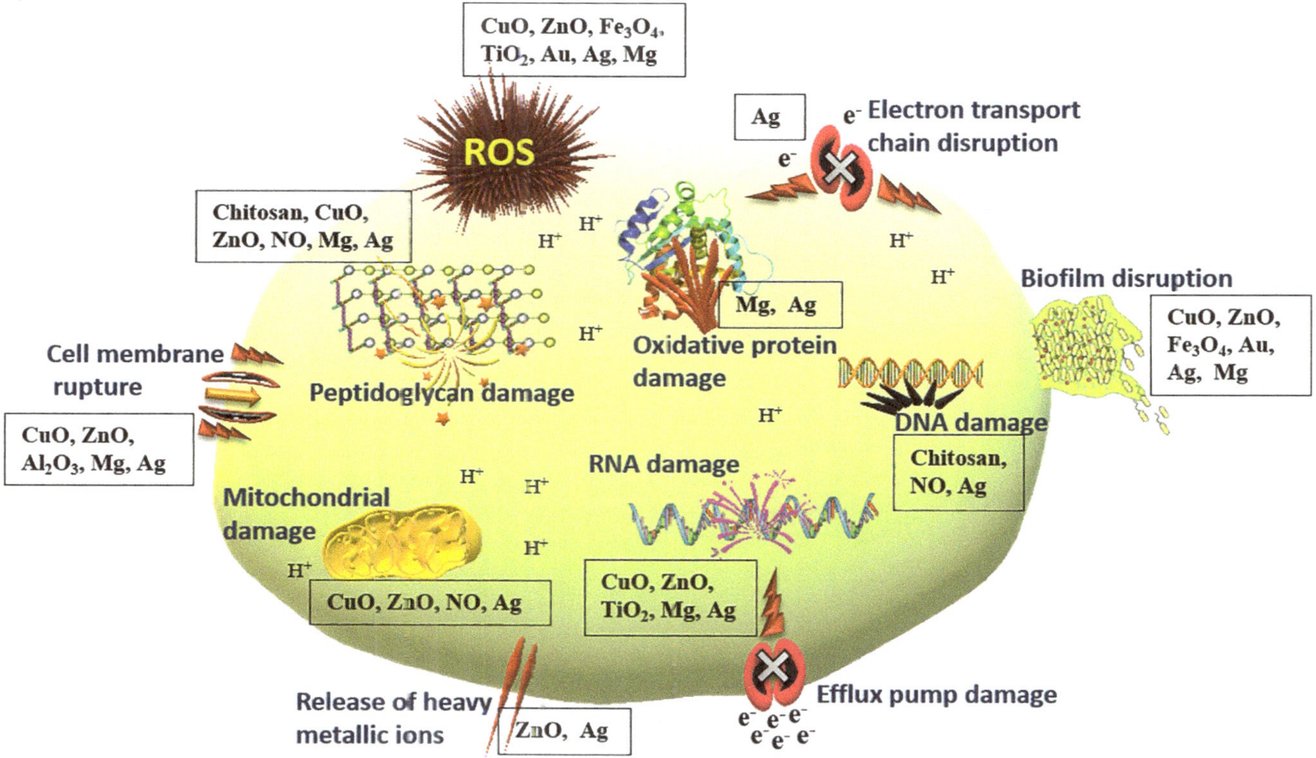

FIG. 8 Summary of the different antimicrobial mechanisms of nanomaterials.

Acknowledgment

The authors acknowledge the Estonian Center of Excellence project EQUiTANT number TK134(F180175TIBT) and Bridge funding (P200030TIBT) from EMÜ for financial support.

References

[1] S.B. Levy, B. Marshall, Antibacterial resistance worldwide: causes, challenges and responses, Nat. Med. 10 (2004) S122–S129, https://doi.org/10.1038/nm1145.

[2] J. Davies, D. Davies, Origins and evolution of antibiotic resistance, J. Microbiol. Mol. Biol. Rev. 74 (2010) 417–433, https://doi.org/10.1128/MMBR.00016-10.

[3] K. Blecher, A. Nasir, A. Friedman, The growing role of nanotechnology in combating infectious disease, Virulence 2 (2011) 395–401, https://doi.org/10.4161/viru.2.5.17035.

[4] P. Yuan, X. Ding, Y.Y. Yang, Q.-H. Xu, Metal nanoparticles for diagnosis and therapy of bacterial infection, Adv. Healthc. Mater. 7 (2018) 1701392, https://doi.org/10.1002/adhm.201701392.

[5] I. Seil, T. Webster, Antimicrobial applications of nanotechnology: methods and literature, Int. J. Nanomedicine 7 (2012) 2767–2781, https://doi.org/10.2147/IJN.S24805.

[6] M.J. Hajipour, K.M. Fromm, A. Akbar Ashkarran, D. Jimenez de Aberasturi, I.R.D. Larramendi, T. Rojo, V. Serpooshan, W.J. Parak, M. Mahmoudi, Antibacterial properties of nanoparticles, Trends Biotechnol. 30 (2012) 499–511, https://doi.org/10.1016/j.tibtech.2012.06.004.

[7] Y.N. Slavin, J. Asnis, U.O. Häfeli, H. Bach, Metal nanoparticles: understanding the mechanisms behind antibacterial activity, J. Nanobiotechnol. 15 (2017) 65, https://doi.org/10.1186/s12951-017-0308-z.

[8] J.A. Lemire, J.J. Harrison, R.J. Turner, Antimicrobial activity of metals: mechanisms, molecular targets and applications, Nat. Rev. Microbiol. 11 (2013) 371–384, https://doi.org/10.1038/nrmicro3028.

[9] M. Boboc, F. Curti, A.M. Fleacă, M.L. Jianu, A.-M. Roşu, C. Curutiu, V. Lazar, M.C. Chifiriuc, A.M. Grumezescu, Chapter 14—preparation and antimicrobial activity of inorganic nanoparticles: promising solutions to fight antibiotic resistance, in: A. Ficai, A.-M. Grumezescu (Eds.), Nanostructures for Antimicrobial Therapy, Elsevier, 2017, pp. 325–340, https://doi.org/10.1016/B978-0-323-46152-8.00014-7pp.

[10] A. Besinis, T. De Peralta, R.D. Handy, The antibacterial effects of silver, titanium dioxide and silica dioxide nanoparticles compared to the dental disinfectant chlorhexidine on Streptococcus mutans using a suite of bioassays, Nanotoxicology 8 (2014) 1–16, https://doi.org/10.3109/17435390.2012.742935.

[11] R.Y. Pelgrift, A.J. Friedman, Nanotechnology as a therapeutic tool to combat microbial resistance, Adv. Drug Deliv. Rev. 65 (2013) 1803–1815, https://doi.org/10.1016/j.addr.2013.07.011.

[12] M.C. Sportelli, R.A. Picca, N. Cioffi, Nano-antimicrobials based on metals, in: D.A. Phoenix, F. Harris, S.R. Dennison (Eds.), Novel Antimicrobial Agents and Strategies, 2014, pp. 181–218, https://doi.org/10.1002/9783527676132.ch8pp.

[13] A.J. Huh, Y.J. Kwon, "Nanoantibiotics": a new paradigm for treating infectious diseases using nanomaterials in the antibiotics resistant era, J. Control. Release 156 (2011) 128–145, https://doi.org/10.1016/j.jconrel.2011.07.002.

[14] E. Sánchez-López, D. Gomes, G. Esteruelas, L. Bonilla, A.L. Lopez-Machado, R. Galindo, A. Cano, M. Espina, M. Ettcheto, A. Camins, et al., Metal-based nanoparticles as antimicrobial agents: an overview, Nanomaterials 10 (2020) 292, https://doi.org/10.3390/nano10020292.

[15] K. Gold, B. Slay, M. Knackstedt, A.K. Gaharwar, Antimicrobial activity of metal and metal-oxide based nanoparticles, Adv. Ther. 1 (2018) 1700033, https://doi.org/10.1002/adtp.201700033.
[16] S.M. Dizaj, F. Lotfipour, M. Barzegar-Jalali, M.H. Zarrintan, K. Adibkia, Antimicrobial activity of the metals and metal oxide nanoparticles, Mater. Sci. Eng. C 44 (2014) 278–284, https://doi.org/10.1016/j.msec.2014.08.031.
[17] G.V. Vimbela, S.M. Ngo, C. Fraze, L. Yang, D.A. Stout, Antibacterial properties and toxicity from metallic nanomaterials, Int. J. Nanomedicine 12 (2017) 3941–3965, https://doi.org/10.2147/IJN.S134526.
[18] L.L. Wang, C. Hu, L.Q. Shao, The antimicrobial activity of nanoparticles: present situation and prospects for the future, Int. J. Nanomedicine 12 (2017) 1227–1249, https://doi.org/10.2147/IJN.S121956.
[19] J. Molling, J. Seezink, B. Teunissen, I. Muijrers-Chen, P. Borm, Comparative performance of a panel of commercially available antimicrobial nanocoatings in Europe, Nanotechnol. Sci. Appl. 7 (2014) 97–104, https://doi.org/10.2147/NSA.S70782.
[20] V.P. Ananikov, Organic-inorganic hybrid nanomaterials, Nanomaterials 9 (2019) 1197, https://doi.org/10.3390/nano9091197.
[21] C.A.H. Aguilar, A.B.P. Jiménez, A.R. Silva, N. Kaur, P. Thangarasu, J.M.V. Ramos, N. Singh, Organic-inorganic hybrid nanoparticles for bacterial inhibition: synthesis and characterization of doped and undoped ONPs with Ag/Au NPs, Molecules 20 (2015) 6002–6021, https://doi.org/10.3390/molecules20046002.
[22] A. Roy, M. Joshi, Enhancing antibacterial properties of polypropylene/Cu-loaded montmorillonite nanocomposite filaments through sheath-core morphology, Polym. Int. 67 (2018) 917–924, https://doi.org/10.1002/pi.5580.
[23] K. Delgado, R. Quijada, R. Palma, H. Palza, Polypropylene with embedded copper metal or copper oxide nanoparticles as a novel plastic antimicrobial agent, Lett. Appl. Microbiol. 53 (2011) 50–54, https://doi.org/10.1111/j.1472-765X.2011.03069.x.
[24] N. Tsao, T.-Y. Luh, C.-K. Chou, T.-Y. Chang, J.-J. Wu, C.-C. Liu, H.-Y. Lei, In vitro action of carboxyfullerene, J. Antimicrob. Chemother. 49 (2002) 641–649, https://doi.org/10.1093/jac/49.4.641.
[25] L.R. Arias, L. Yang, Inactivation of bacterial pathogens by carbon nanotubes in suspensions, Langmuir 25 (2009) 3003–3012, https://doi.org/10.1021/la802769m.
[26] A. Al-Jumaili, S. Alancherry, K. Bazaka, M.V. Jacob, Review on the antimicrobial properties of carbon nanostructures, Materials 10 (2017) 1066, https://doi.org/10.3390/ma10091066.
[27] C.D. Vecitis, K.R. Zodrow, S. Kang, M. Elimelech, Electronic-structure-dependent bacterial cytotoxicity of single-walled carbon nanotubes, ACS Nano 4 (2010) 5471–5479, https://doi.org/10.1021/nn101558x.
[28] T. Maisch, A new strategy to destroy antibiotic resistant microorganisms: antimicrobial photodynamic treatment, Mini Rev. Med. Chem. 9 (2009) 974–983, https://doi.org/10.2174/138955709788681582.
[29] P.T. Anastas, J.C. Warner, Green Chemistry: Theory and Practice, Oxford University Press, New York, Oxford, 1998.
[30] S. Küünal, P. Rauwel, E. Rauwel, Chapter 14—Plant extract mediated synthesis of nanoparticles, in: A. Barhoum, A.S.H. Makhlouf (Eds.), Emerging Applications of Nanoparticles and Architecture Nanostructures, Elsevier, 2018, pp. 411–446, https://doi.org/10.1016/B978-0-323-51254-1.00014-2pp.
[31] P. Rauwel, E. Rauwel, Emerging trends in nanoparticle synthesis using plant extracts for biomedical applications, Glob. J Nanomed. 1 (2017), 555562.
[32] J.A. Hernández-Díaz, J.J. Garza-García, A. Zamudio-Ojeda, J.M. León-Morales, J.C. López-Velázquez, S. García-Morales, Plant-mediated synthesis of nanoparticles and their antimicrobial activity against phytopathogens, J. Sci. Food Agric. (2020), https://doi.org/10.1002/jsfa.10767. *n/a*.

[33] S. Küünal, M. Visnapuu, O. Volubujeva, M. Soares Rosario, P. Rauwel, E. Rauwel, Optimisation of plant mediated synthesis of silver nanoparticles by common weed Plantago major and their antimicrobial properties, IOP Conf. Ser.: Mater. Sci. Eng. 613 (2019), https://doi.org/10.1088/1757-899x/613/1/012003, 012003.

[34] D.P. Linklater, V.A. Baulin, S. Juodkazis, E.P. Ivanova, Mechano-bactericidal mechanism of graphene nanomaterials, Interface Focus 8 (2018) 20170060, https://doi.org/10.1098/rsfs.2017.0060.

[35] Y. Wang, Y. Yang, Y. Shi, H. Song, C. Yu, Antibiotic-free antibacterial strategies enabled by nanomaterials: progress and perspectives, Adv. Mater. 32 (2020) 1904106, https://doi.org/10.1002/adma.201904106.

[36] M.A. Raza, Z. Kanwal, A. Rauf, A.N. Sabri, S. Riaz, S. Naseem, Size- and shape-dependent antibacterial studies of silver nanoparticles synthesized by wet chemical routes, Nanomaterials 6 (2016) 74, https://doi.org/10.3390/nano6040074.

[37] J.R. Morones, J.L. Elechiguerra, A. Camacho, K. Holt, J.B. Kouri, J.T. Ramírez, M.J. Yacaman, The bactericidal effect of silver nanoparticles, Nanotechnology 16 (2005) 2346–2353, https://doi.org/10.1088/0957-4484/16/10/059.

[38] A. Thill, O. Zeyons, O. Spalla, F. Chauvat, J. Rose, M. Auffan, A.M. Flank, Cytotoxicity of CeO2 nanoparticles for Escherichia coli. Physico-chemical insight of the cytotoxicity mechanism, Environ. Sci. Technol. 40 (2006) 6151–6156, https://doi.org/10.1021/es060999b.

[39] M. Arakha, M. Saleem, B.C. Mallick, S. Jha, The effects of interfacial potential on antimicrobial propensity of ZnO nanoparticle, Sci. Rep. 5 (2015) 9578, https://doi.org/10.1038/srep09578.

[40] L. Salvioni, E. Galbiati, V. Collico, G. Alessio, S. Avvakumova, F. Corsi, P. Tortora, D. Prosperi, M. Colombo, Negatively charged silver nanoparticles with potent antibacterial activity and reduced toxicity for pharmaceutical preparations, Int. J. Nanomedicine 12 (2017) 2517–2530, https://doi.org/10.2147/IJN.S127799.

[41] A. Raghunath, E. Perumal, Metal oxide nanoparticles as antimicrobial agents: a promise for the future, Int. J. Antimicrob. Agents 49 (2017) 137–152, https://doi.org/10.1016/j.ijantimicag.2016.11.011.

[42] R. Brayner, R. Ferrari-Iliou, N. Brivois, S. Djediat, M.F. Benedetti, F. Fiévet, Toxicological impact studies based on Escherichia coli Bacteria in ultrafine ZnO nanoparticles colloidal medium, Nano Lett. 6 (2006) 866–870, https://doi.org/10.1021/nl052326h.

[43] N. Jones, B. Ray, K.T. Ranjit, A.C. Manna, Antibacterial activity of ZnO nanoparticle suspensions on a broad spectrum of microorganisms, FEMS Microbiol. Lett. 279 (2008) 71–76, https://doi.org/10.1111/j.1574-6968.2007.01012.x %J FEMS Microbiology Letters.

[44] Code of Federal Regulations, Title 21—Food and Drugs, 2021, Availabe online: https://www.accessdata.fda.gov/scripts/cdrh/cfdocs/cfcfr/CFRSearch.cfm?fr=182.8991.

[45] E. Rauwel, A. Galeckas, P. Rauwel, M.F. Sunding, H. Fjellvåg, Precursor-dependent blue-green photoluminescence emission of ZnO nanoparticles, J. Phys. Chem. C 115 (2011) 25227–25233.

[46] P. Rauwel, M. Salumaa, A. Aasna, A. Galeckas, E. Rauwel, A review of the synthesis and photoluminescence properties of hybrid ZnO and carbon nanomaterials, J. Nanomater. 2016 (2016) 12, https://doi.org/10.1155/2016/5320625.

[47] E. Rauwel, A. Galeckas, M.R. Soares, P. Rauwel, Influence of the interface on the photoluminescence properties in ZnO carbon-based nanohybrids, J. Phys. Chem. C 121 (2017) 14879–14887, https://doi.org/10.1021/acs.jpcc.7b03070.

[48] A. Kołodziejczak-Radzimska, T. Jesionowski, Zinc oxide—from synthesis to application: a review, Materials 7 (2014) 2833–2881.

[49] P.J.P. Espitia, N.D.F.F. Soares, J.S.D.R. Coimbra, N.J. de Andrade, R.S. Cruz, E.A.A. Medeiros, Zinc oxide nanoparticles: synthesis, antimicrobial activity and food

packaging applications, Food Bioproc. Tech. 5 (2012) 1447–1464, https://doi.org/10.1007/s11947-012-0797-6.
[50] A. Hatamie, A. Khan, M. Golabi, A.P.F. Turner, V. Beni, W.C. Mak, A. Sadollahkhani, H. Alnoor, B. Zargar, S. Bano, et al., Zinc oxide nanostructure-modified textile and its application to biosensing, photocatalysis, and as antibacterial material, Langmuir 31 (2015) 10913–10921, https://doi.org/10.1021/acs.langmuir.5b02341.
[51] Z.L. Wang, Zinc oxide nanostructures: growth, properties and applications, J. Phys. Condens. Matter 16 (2004) R829–R858, https://doi.org/10.1088/0953-8984/16/25/r01.
[52] B. Lallo da Silva, M.P. Abuçafy, E. Berbel Manaia, J.A. Oshiro Junior, B.G. Chiari-Andréo, R.C.R. Pietro, L.A. Chiavacci, Relationship between structure and antimicrobial activity of zinc oxide nanoparticles: an overview, Int. J. Nanomedicine 14 (2019) 9395–9410, https://doi.org/10.2147/IJN.S216204.
[53] K.R. Raghupathi, R.T. Koodali, A.C. Manna, Size-dependent bacterial growth inhibition and mechanism of antibacterial activity of zinc oxide nanoparticles, Langmuir 27 (2011) 4020–4028, https://doi.org/10.1021/la104825u.
[54] J. Pasquet, Y. Chevalier, E. Couval, D. Bouvier, G. Noizet, C. Morlière, M.-A. Bolzinger, Antimicrobial activity of zinc oxide particles on five micro-organisms of the challenge tests related to their physicochemical properties, Int. J. Pharm. 460 (2014) 92–100, https://doi.org/10.1016/j.ijpharm.2013.10.031.
[55] V.C. Reyes, M. Li, E.M.V. Hoek, S. Mahendra, R. Damoiseaux, Genome-wide assessment in Escherichia coli reveals time-dependent nanotoxicity paradigms, ACS Nano 6 (2012) 9402–9415, https://doi.org/10.1021/nn302815w.
[56] N. Padmavathy, R. Vijayaraghavan, Enhanced bioactivity of ZnO nanoparticles-an antimicrobial study, Sci. Technol. Adv. Mater. 9 (2008) 035004, https://doi.org/10.1088/1468-6996/9/3/035004.
[57] A. Azam, M.M. Ahmed Oves, S. Khan Habib, A. Memic, Antimicrobial activity of metal oxide nanoparticles against gram-positive and gram-negative bacteria: a comparative study, Int. J. Nanomedicine 7 (2012) 6003–6009, https://doi.org/10.2147/IJN.S35347.
[58] Ş.S. Uğur, M. Sarıışık, A.H. Aktaş, M.Ç. Uçar, E. Erden, Modifying of cotton fabric surface with nano-ZnO multilayer films by layer-by-layer deposition method, Nanoscale Res. Lett. 5 (2010) 1204, https://doi.org/10.1007/s11671-010-9627-9.
[59] T.J. McCarthy, J.J. Zeelie, D.J. Krause, The antimicrobial action of zinc ion/antioxidant combinations, J. Clin. Pharm. Ther. 17 (1992) 51–54, https://doi.org/10.1111/j.1365-2710.1992.tb01265.x.
[60] P. Rauwel, S. Küünal, S. Ferdov, E. Rauwel, A review on the green synthesis of silver nanoparticles and their morphologies studied via TEM, Adv. Mater. Sci. Eng. 15 (2015) 682741–682749. Article ID 682749.
[61] J. Santhoshkumar, S.V. Kumar, S. Rajeshkumar, Synthesis of zinc oxide nanoparticles using plant leaf extract against urinary tract infection pathogen, Resour. Eff. Technol. 3 (2017) 459–465, https://doi.org/10.1016/j.reffit.2017.05.001.
[62] S.O. Ogunyemi, Y. Abdallah, M. Zhang, H. Fouad, X. Hong, E. Ibrahim, M.M.I. Masum, A. Hossain, J. Mo, B. Li, Green synthesis of zinc oxide nanoparticles using different plant extracts and their antibacterial activity against Xanthomonas oryzae pv. oryzae, Artif Cells Nanomed Biotechnol. 47 (2019) 341–352, https://doi.org/10.1080/21691401.2018.1557671.
[63] M. Naseer, U. Aslam, B. Khalid, B. Chen, Green route to synthesize zinc oxide nanoparticles using leaf extracts of Cassia fistula and Melia azadarach and their antibacterial potential, Sci. Rep. 10 (2020) 9055, https://doi.org/10.1038/s41598-020-65949-3.
[64] W. Ahmad, D. Kalra, Green synthesis, characterization and anti microbial activities of ZnO nanoparticles using Euphorbia hirta leaf extract, J. King Saud Univ.—Sci. 32 (2020) 2358–2364, https://doi.org/10.1016/j.jksus.2020.03.014.

[65] Z. Huang, X. Zheng, D. Yan, G. Yin, X. Liao, Y. Kang, Y. Yao, D. Huang, B. Hao, Toxicological effect of ZnO nanoparticles based on bacteria, Langmuir 24 (2008) 4140–4144, https://doi.org/10.1021/la7035949.

[66] M.A. Ansari, H.M. Khan, A.A. Khan, A. Sultan, A. Azam, Characterization of clinical strains of MSSA, MRSA and MRSE isolated from skin and soft tissue infections and the antibacterial activity of ZnO nanoparticles, World J. Microbiol. Biotechnol. 28 (2012) 1605–1613, https://doi.org/10.1007/s11274-011-0966-1.

[67] K. Feris, C. Otto, J. Tinker, D. Wingett, A. Punnoose, A. Thurber, M. Kongara, M. Sabetian, B. Quinn, C. Hanna, et al., Electrostatic interactions affect nanoparticle-mediated toxicity to gram-negative bacterium Pseudomonas aeruginosa PAO1, Langmuir 26 (2010) 4429–4436, https://doi.org/10.1021/la903491z.

[68] S. Vijayakumar, C. Krishnakumar, P. Arulmozhi, S. Mahadevan, N. Parameswari, Biosynthesis, characterization and antimicrobial activities of zinc oxide nanoparticles from leaf extract of Glycosmis pentaphylla (Retz.) DC, Microb. Pathog. 116 (2018) 44–48, https://doi.org/10.1016/j.micpath.2018.01.003.

[69] J. Jayabalan, G. Mani, N. Krishnan, J. Pernabas, J.M. Devadoss, H.T. Jang, Green biogenic synthesis of zinc oxide nanoparticles using pseudomonas putida culture and its in vitro antibacterial and anti-biofilm activity, Biocatal. Agric. Biotechnol. 21 (2019), https://doi.org/10.1016/j.bcab.2019.101327, 101327.

[70] H. Agarwal, A. Nakara, S. Menon, V. Shanmugam, Eco-friendly synthesis of zinc oxide nanoparticles using Cinnamomum Tamala leaf extract and its promising effect towards the antibacterial activity, J. Drug Delivery Sci. Technol. 53 (2019), https://doi.org/10.1016/j.jddst.2019.101212, 101212.

[71] P.C. Nagajyothi, T.V.M. Sreekanth, C.O. Tettey, Y.I. Jun, S.H. Mook, Characterization, antibacterial, antioxidant, and cytotoxic activities of ZnO nanoparticles using Coptidis Rhizoma, Bioorg. Med. Chem. Lett. 24 (2014) 4298–4303, https://doi.org/10.1016/j.bmcl.2014.07.023.

[72] O. Carp, C.L. Huisman, A. Reller, Photoinduced reactivity of titanium dioxide, Prog. Solid State Chem. 32 (2004) 33–177, https://doi.org/10.1016/j.progsolidstchem.2004.08.001.

[73] M.A. Fox, M.T. Dulay, Heterogeneous photocatalysis, Chem. Rev. 93 (1993) 341–357, https://doi.org/10.1021/cr00017a016.

[74] E.I. Cedillo-González, R. Riccò, M. Montorsi, M. Montorsi, P. Falcaro, C. Siligardi, Self-cleaning glass prepared from a commercial TiO2 nano-dispersion and its photocatalytic performance under common anthropogenic and atmospheric factors, Build. Environ. 71 (2014) 7–14, https://doi.org/10.1016/j.buildenv.2013.09.007.

[75] C. Maneerat, Y. Hayata, Antifungal activity of TiO2 photocatalysis against Penicillium expansum in vitro and in fruit tests, Int. J. Food Microbiol. 107 (2006) 99–103, https://doi.org/10.1016/j.ijfoodmicro.2005.08.018.

[76] A. Fujishima, T.N. Rao, D.A. Tryk, Titanium dioxide photocatalysis, J Photochem Photobiol C: Photochem Rev 1 (2000) 1–21, https://doi.org/10.1016/S1389-5567(00)00002-2.

[77] V. Krishna, S. Pumprueg, S.H. Lee, J. Zhao, W. Sigmund, B. Koopman, B.M. Moudgil, Photocatalytic disinfection with titanium dioxide coated multi-wall carbon nanotubes, Process Saf. Environ. Prot. 83 (2005) 393–397, https://doi.org/10.1205/psep.04387.

[78] C. Wei, W.Y. Lin, Z. Zainal, N.E. Williams, K. Zhu, A.P. Kruzic, R.L. Smith, K. Rajeshwar, Bactericidal activity of TiO2 photocatalyst in aqueous media: toward a solar-assisted water disinfection system, Environ. Sci. Technol. 28 (1994) 934–938, https://doi.org/10.1021/es00054a027.

[79] K. Gupta, R.P. Singh, A. Pandey, A. Pandey, Photocatalytic antibacterial performance of TiO_2 and Ag-doped TiO_2 against *S. aureus. P. aeruginosa* and *E. coli*, Beilstein J. Nanotechnol. 4 (2013) 345–351, https://doi.org/10.3762/bjnano.4.40.

[80] H.A. Foster, I.B. Ditta, S. Varghese, A. Steele, Photocatalytic disinfection using titanium dioxide: spectrum and mechanism of antimicrobial activity, Appl. Microbiol. Biotechnol. 90 (2011) 1847–1868, https://doi.org/10.1007/s00253-011-3213-7.
[81] K.P. Kühn, I.F. Chaberny, K. Massholder, M. Stickler, V.W. Benz, H.-G. Sonntag, L. Erdinger, Disinfection of surfaces by photocatalytic oxidation with titanium dioxide and UVA light, Chemosphere 53 (2003) 71–77, https://doi.org/10.1016/S0045-6535(03)00362-X.
[82] A.S. Roy, A. Parveen, A.R. Koppalkar, M. Prasad, Effect of Nano—titanium dioxide with different antibiotics against methicillin-resistant staphylococcus aureus, J. Biomater. Nanobiotechnol. 1 (2010) 37–41, https://doi.org/10.4236/jbnb.2010.11005.
[83] G. Carré, E. Hamon, S. Ennahar, M. Estner, M.-C. Lett, P. Horvatovich, J.-P. Gies, V. Keller, N. Keller, P. Andre, TiO_2 Photocatalysis damages lipids and proteins in Escherichia coli, J. Appl. Environ. Microbiol. 80 (2014) 2573–2581, https://doi.org/10.1128/AEM.03995-13.
[84] Q. Li, S. Mahendra, D.Y. Lyon, L. Brunet, M.V. Liga, D. Li, P.J.J. Alvarez, Antimicrobial nanomaterials for water disinfection and microbial control: potential applications and implications, Water Res. 42 (2008) 4591–4602, https://doi.org/10.1016/j.watres.2008.08.015.
[85] W. Xia, K. Grandfield, A. Hoess, A. Ballo, Y. Cai, H. Engqvist, Mesoporous titanium dioxide coating for metallic implants, J. Biomed. Mater. Res. 100B (2012) 82–93, https://doi.org/10.1002/jbm.b.31925.
[86] J.-Y. Choi, K.-H. Kim, K.-C. Choy, K.-T. Oh, K.-N. Kim, Photocatalytic antibacterial effect of TiO_2 film formed on Ti and TiAg exposed to Lactobacillus acidophilus, J. Biomed. Mater. Res. 80B (2007) 353–359, https://doi.org/10.1002/jbm.b.30604.
[87] H. Zhang, S. Yu, A. Tian, X. Xue, W. Lin, A. Ali, X. Bai, Improved antibacterial activity and biocompatibility on vancomycin-loaded TiO_2 nanotubes: in vivo and in vitro studies, Int. J. Nanomedicine 8 (2013) 4379–4389, https://doi.org/10.2147/IJN.S53221.
[88] W. Kangwansupamonkon, V. Lauruengtana, S. Surassmo, U. Ruktanonchai, Antibacterial effect of apatite-coated titanium dioxide for textiles applications, Nanomedicine 5 (2009) 240–249, https://doi.org/10.1016/j.nano.2008.09.004.
[89] A. Weir, P. Westerhoff, L. Fabricius, K. Hristovski, N. von Goetz, Titanium dioxide nanoparticles in food and personal care products, Environ. Sci. Technol. 46 (2012) 2242–2250, https://doi.org/10.1021/es204168d.
[90] L.K. Adams, D.Y. Lyon, P.J.J. Alvarez, Comparative eco-toxicity of nanoscale TiO2, SiO2, and ZnO water suspensions, Water Res. 40 (2006) 3527–3532, https://doi.org/10.1016/j.watres.2006.08.004.
[91] Y.H. Tsuang, J.S. Sun, Y.C. Huang, C.H. Lu, W.H. Chang, C.C. Wang, Studies of photokilling of bacteria using titanium dioxide nanoparticles, Artif. Organs 32 (2008) 167–174, https://doi.org/10.1111/j.1525-1594.2007.00530.x.
[92] K. Sundaresan, A. Sivakumar, C. Vigneswaran, T. Ramachandran, Influence of nano titanium dioxide finish, prepared by sol-gel technique, on the ultraviolet protection, antimicrobial, and self-cleaning characteristics of cotton fabrics, J. Ind. Text. 41 (2012) 259–277, https://doi.org/10.1177/1528083711414962.
[93] A. Simon-Deckers, S. Loo, M. Mayne-L'hermite, N. Herlin-Boime, N. Menguy, C. Reynaud, B. Gouget, M. Carrière, Size-, Composition- and shape-dependent toxicological impact of metal oxide nanoparticles and carbon nanotubes toward bacteria, Environ. Sci. Technol. 43 (2009) 8423–8429, https://doi.org/10.1021/es9016975.
[94] P.C. Maness, S. Smolinski, D.M. Blake, Z. Huang, E.J. Wolfrum, W.A. Jacoby, Bactericidal activity of photocatalytic TiO(2) reaction: toward an understanding of its killing mechanism, Appl. Environ. Microbiol. 65 (1999) 4094–4098, https://doi.org/10.1128/AEM.65.9.4094-4098.1999.

[95] A. Kumar, A.K. Pandey, S.S. Singh, R. Shanker, A. Dhawan, Cellular uptake and mutagenic potential of metal oxide nanoparticles in bacterial cells, Chemosphere 83 (2011) 1124–1132, https://doi.org/10.1016/j.chemosphere.2011.01.025.

[96] H.C. Lukaski, Magnesium, zinc, and chromium nutrition and athletic performance, Can. J. Appl. Physiol. 26 (2001) S13–S22, https://doi.org/10.1139/h2001-038.

[97] Y. He, S. Ingudam, S. Reed, A. Gehring, T.P. Strobaugh, P. Irwin, Study on the mechanism of antibacterial action of magnesium oxide nanoparticles against foodborne pathogens, J. Nanobiotechnol. 14 (2016) 54, https://doi.org/10.1186/s12951-016-0202-0.

[98] Z.-X. Tang, B.-F. Lv, MgO nanoparticles as antibacterial agent: preparation and activity, J. Braz. J. Chem. Eng. 31 (2014) 591–601, https://doi.org/10.1590/0104-6632.20140313s00002813.

[99] Y.H. Leung, A.M.C. Ng, X. Xu, Z. Shen, L.A. Gethings, M.T. Wong, C.M.N. Chan, M.Y. Guo, Y.H. Ng, A.B. Djurišić, et al., Mechanisms of antibacterial activity of MgO: non-ROS mediated toxicity of MgO nanoparticles towards Escherichia coli, Small 10 (2014) 1171–1183, https://doi.org/10.1002/smll.201302434.

[100] J. Sawai, H. Kojima, H. Igarashi, A. Hashimoto, S. Shoji, A. Takehara, T. Sawaki, T. Kokugan, M. Shimizu, Escherichia coli damage by ceramic powder slurries, J. Chem. Eng. Jpn 30 (1997) 1034–1039, https://doi.org/10.1252/jcej.30.1034.

[101] A. Sellik, T. Pollet, L. Ouvry, S. Briançon, H. Fessi, D.J. Hartmann, F.N.R. Renaud, Degradation of paraoxon (VX chemical agent simulant) and bacteria by magnesium oxide depends on the crystalline structure of magnesium oxide, Chem. Biol. Interact. 267 (2017) 67–73, https://doi.org/10.1016/j.cbi.2016.11.023.

[102] F. Al-Hazmi, F. Alnowaiser, A.A. Al-Ghamdi, A.A. Al-Ghamdi, M.M. Aly, R.M. Al-Tuwirqi, F. El-Tantawy, A new large—scale synthesis of magnesium oxide nanowires: structural and antibacterial properties, Superlattice. Microst. 52 (2012) 200–209, https://doi.org/10.1016/j.spmi.2012.04.013.

[103] T. Jin, Y. He, Antibacterial activities of magnesium oxide (MgO) nanoparticles against foodborne pathogens, J. Nanopart. Res. 13 (2011) 6877–6885, https://doi.org/10.1007/s11051-011-0595-5.

[104] M. Mirhosseini, Evaluation of antibacterial effect of magnesium oxide nanoparticles with nisin and heat in milk, Nanomed. J. 3 (2016) 135–142, https://doi.org/10.7508/nmj.2016.02.007.

[105] G.N. Moll, J. Clark, W.C. Chan, B.W. Bycroft, G.C. Roberts, W.N. Konings, A.J. Driessen, Role of transmembrane pH gradient and membrane binding in nisin pore formation, J. Bacteriol. 179 (1997) 135–140, https://doi.org/10.1128/jb.179.1.135-140.1997.

[106] K. Ramanujam, M. Sundrarajan, Antibacterial effects of biosynthesized MgO nanoparticles using ethanolic fruit extract of Emblica officinalis, J. Photochem. Photobiol. B Biol. 141 (2014) 296–300, https://doi.org/10.1016/j.jphotobiol.2014.09.011.

[107] J. Jeevanandam, Y.S. Chan, M.K. Danquah, Biosynthesis and characterization of MgO nanoparticles from plant extracts via induced molecular nucleation, New J. Chem. 41 (2017) 2800–2814, https://doi.org/10.1039/C6NJ03176E.

[108] E.R. Essien, V.N. Atasie, A.O. Okeafor, D.O. Nwude, Biogenic synthesis of magnesium oxide nanoparticles using Manihot esculenta (Crantz) leaf extract, Int. Nano Lett. 10 (2020) 43–48, https://doi.org/10.1007/s40089-019-00290-w.

[109] P.K. Stoimenov, R.L. Klinger, G.L. Marchin, K.J. Klabunde, Metal oxide nanoparticles as bactericidal agents, Langmuir 18 (2002) 6679–6686, https://doi.org/10.1021/la0202374.

[110] J. Lellouche, E. Kahana, S. Elias, A. Gedanken, E. Banin, Antibiofilm activity of nanosized magnesium fluoride, Biomaterials 30 (2009) 5969–5978, https://doi.org/10.1016/j.biomaterials.2009.07.037.

[111] R. Prasanth, S. Dinesh Kumar, A. Jayalakshmi, G. Singaravelu, K. Govindaraju, V. Ganesh Kumar, Green synthesis of magnesium oxide nanoparticles and their antibacterial activity, Indian J. Geo Mar. Sci. 48 (2019) 1210–1215.
[112] O. Akhavan, E. Ghaderi, Cu and CuO nanoparticles immobilized by silica thin films as antibacterial materials and photocatalysts, Surf. Coat. Technol. 205 (2010) 219–223, https://doi.org/10.1016/j.surfcoat.2010.06.036.
[113] K. Giannousi, A. Pantazaki, C. Dendrinou-Samara, Chapter 23—copper-based nanoparticles as antimicrobials, in: A. Ficai, A.M. Grumezescu (Eds.), Nanostructures for Antimicrobial Therapy, Elsevier, 2017, pp. 515–529, https://doi.org/10.1016/B978-0-323-46152-8.00023-8pp.
[114] G. Applerot, J. Lellouche, A. Lipovsky, Y. Nitzan, R. Lubart, A. Gedanken, E. Banin, Understanding the antibacterial mechanism of CuO nanoparticles: revealing the route of induced oxidative stress, Small 8 (2012) 3326–3337, https://doi.org/10.1002/smll.201200772.
[115] G. Ren, D. Hu, E.W.C. Cheng, M.A. Vargas-Reus, P. Reip, R.P. Allaker, Characterisation of copper oxide nanoparticles for antimicrobial applications, Int. J. Antimicrob. Agents 33 (2009) 587–590, https://doi.org/10.1016/j.ijantimicag.2008.12.004.
[116] S. Moniri Javadhesari, S. Alipour, S. Mohammadnejad, M.R. Akbarpour, Antibacterial activity of ultra-small copper oxide (II) nanoparticles synthesized by mechanochemical processing against *S. aureus* and *E. coli*, Mater. Sci. Eng. C 105 (2019) 110011, https://doi.org/10.1016/j.msec.2019.110011.
[117] V.V. Padil, M. Černík, Green synthesis of copper oxide nanoparticles using gum karaya as a biotemplate and their antibacterial application, Int. J. Nanomedicine 8 (2013) 889–898, https://doi.org/10.2147/IJN.S40599.
[118] A.F. Jaramillo, S.A. Riquelme, G. Sánchez-Sanhueza, C. Medina, F. Solís-Pomar, D. Rojas, C. Montalba, M.F. Melendrez, E. Pérez-Tijerina, Comparative study of the antimicrobial effect of nanocomposites and composite based on poly(butylene adipate-co-terephthalate) using Cu and Cu/Cu2O nanoparticles and CuSO4, Nanoscale Res. Lett. 14 (2019) 158, https://doi.org/10.1186/s11671-019-2987-x.
[119] S. Vasantharaj, S. Sathiyavimal, M. Saravanan, P. Senthilkumar, K. Gnanasekaran, M. Shanmugavel, E. Manikandan, A. Pugazhendhi, Synthesis of ecofriendly copper oxide nanoparticles for fabrication over textile fabrics: characterization of antibacterial activity and dye degradation potential, J. Photochem. Photobiol. B Biol. 191 (2019) 143–149, https://doi.org/10.1016/j.jphotobiol.2018.12.026.
[120] G. Borkow, S.S. Zhou, T. Page, J. Gabbay, A novel anti-influenza copper oxide containing respiratory face mask, PLoS One 5 (2010), https://doi.org/10.1371/journal.pone.0011295, e11295.
[121] P.P.N.V. Kumar, U. Shameem, P. Kollu, R.L. Kalyani, S.V.N. Pammi, Green synthesis of copper oxide nanoparticles using Aloe vera leaf extract and its antibacterial activity against fish bacterial pathogens, BioNanoScience 5 (2015) 135–139, https://doi.org/10.1007/s12668-015-0171-z.
[122] Y. Abboud, T. Saffaj, A. Chagraoui, A. El Bouari, K. Brouzi, O. Tanane, B. Ihssane, Biosynthesis, characterization and antimicrobial activity of copper oxide nanoparticles (CONPs) produced using brown alga extract (Bifurcaria bifurcata), Appl. Nanosci. 4 (2014) 571–576, https://doi.org/10.1007/s13204-013-0233-x.
[123] H. Gu, X. Chen, F. Chen, X. Zhou, Z. Parsaee, Ultrasound-assisted biosynthesis of CuO-NPs using brown alga Cystoseira trinodis: characterization, photocatalytic AOP, DPPH scavenging and antibacterial investigations, Ultrason. Sonochem. 41 (2018) 109–119, https://doi.org/10.1016/j.ultsonch.2017.09.006.
[124] S.V. Kumar, A.P. Bafana, P. Pawar, M. Faltane, A. Rahman, S.A. Dahoumane, A. Kucknoor, C.S. Jeffryes, Optimized production of antibacterial copper oxide nanoparticles in a microwave-assisted synthesis reaction using response surface methodology,

Colloids Surf. A Physicochem. Eng. Asp. 573 (2019) 170–178, https://doi.org/10.1016/j.colsurfa.2019.04.063.
[125] Rajgovind, G. Sharma, K. Deepak Gupta, N.D. Jasuja, C. Suresh Joshi, Pterocarpus marsupium derived phyto-synthesis of copper oxide nanoparticles and their antimicrobial activities, J. Microb. Biochem. Technol. 7 (2015) 140–144, https://doi.org/10.4172/1948-5948.1000195.
[126] P. Pandey, S. Merwyn, G.S. Agarwal, B.K. Tripathi, S.C. Pant, Electrochemical synthesis of multi-armed CuO nanoparticles and their remarkable bactericidal potential against waterborne bacteria, J. Nanopart. Res. 14 (2012) 709, https://doi.org/10.1007/s11051-011-0709-0.
[127] A. Dhall, W. Self, Cerium oxide nanoparticles: a brief review of their synthesis methods and biomedical applications, Antioxidants 7 (2018) 97, https://doi.org/10.3390/antiox7080097.
[128] C. Xu, X. Qu, Cerium oxide nanoparticle: a remarkably versatile rare earth nanomaterial for biological applications, NPG Asia Mater. 6 (2014) e90, https://doi.org/10.1038/am.2013.88.
[129] S. Rajeshkumar, P. Naik, Synthesis and biomedical applications of cerium oxide nanoparticles—a review, Biotechnol. Rep. 17 (2018) 1–5, https://doi.org/10.1016/j.btre.2017.11.008.
[130] M. Nadeem, R. Khan, K. Afridi, A. Nadhman, S. Ullah, S. Faisal, Z.U. Mabood, C. Hano, B.H. Abbasi, Green synthesis of cerium oxide nanoparticles (CeO2 NPs) and their antimicrobial applications: a review, Int. J. Nanomedicine 15 (2020) 5951–5961, https://doi.org/10.2147/IJN.S255784.
[131] S.K. Kannan, M. Sundrarajan, A green approach for the synthesis of a cerium oxide nanoparticle: characterization and antibacterial activity, Int. J. Nanosci. 13 (2014) 1450018, https://doi.org/10.1142/s0219581x14500185.
[132] J. Malleshappa, H. Nagabhushana, S.C. Sharma, Y.S. Vidya, K.S. Anantharaju, S.C. Prashantha, B. Daruka Prasad, H. Raja Naika, K. Lingaraju, B.S. Surendra, Leucas aspera mediated multifunctional CeO2 nanoparticles: structural, photoluminescent, photocatalytic and antibacterial properties, Spectrochim. Acta A Mol. Biomol. Spectrosc. 149 (2015) 452–462, https://doi.org/10.1016/j.saa.2015.04.073.
[133] K. Gopinath, V. Karthika, C. Sundaravadivelan, S. Gowri, A. Arumugam, Mycogenesis of cerium oxide nanoparticles using Aspergillus Niger culture filtrate and their applications for antibacterial and larvicidal activities, J. Nanostruct. Chem. 5 (2015) 295–303, https://doi.org/10.1007/s40097-015-0161-2.
[134] A. Arumugam, C. Karthikeyan, A.S. Haja Hameed, K. Gopinath, S. Gowri, V. Karthika, Synthesis of cerium oxide nanoparticles using Gloriosa superba L. leaf extract and their structural, optical and antibacterial properties, Mater. Sci. Eng. C 49 (2015) 408–415, https://doi.org/10.1016/j.msec.2015.01.042.
[135] D.A. Pelletier, et al., Effects of engineered cerium oxide nanoparticles on bacterial growth and viability, Appl. Environ. Microbiol. 76 (2010) 7981–7989, https://doi.org/10.1128/AEM.00650-10.
[136] R. Dinali, A. Ebrahiminezhad, M. Manley-Harris, Y. Ghasemi, A. Berenjian, Iron oxide nanoparticles in modern microbiology and biotechnology, Crit. Rev. Microbiol. 43 (2017) 493–507, https://doi.org/10.1080/1040841X.2016.1267708.
[137] S. Majidi, F. Zeinali Sehrig, S.M. Farkhani, M. Soleymani Goloujeh, A. Akbarzadeh, Current methods for synthesis of magnetic nanoparticles, Artif. Cells Nanomed. Biotechnol. 44 (2016) 722–734, https://doi.org/10.3109/21691401.2014.982802.
[138] N.G. Durmus, E.N. Taylor, K.M. Kummer, T.J. Webster, Enhanced efficacy of superparamagnetic iron oxide nanoparticles against antibiotic-resistant biofilms in the presence of metabolites, Adv. Mater. 25 (2013) 5706–5713, https://doi.org/10.1002/adma.201302627.

[139] N. Tran, A. Mir, D. Mallik, A. Sinha, S. Nayar, T. Webster, Bactericidal effect of iron oxide nanoparticles on Staphylococcus aureus, Int. J. Nanomedicine 5 (2010) 277–283, https://doi.org/10.2147/IJN.S9220.

[140] B.M. Geilich, I. Gelfat, S. Sridhar, A.L. van de Ven, T.J. Webster, Superparamagnetic iron oxide-encapsulating polymersome nanocarriers for biofilm eradication, Biomaterials 119 (2017) 78–85, https://doi.org/10.1016/j.biomaterials.2016.12.011.

[141] S. Vasantharaj, S. Sathiyavimal, P. Senthilkumar, F. LewisOscar, A. Pugazhendhi, Biosynthesis of iron oxide nanoparticles using leaf extract of Ruellia tuberosa: antimicrobial properties and their applications in photocatalytic degradation, J. Photochem. Photobiol. B Biol. 192 (2019) 74–82, https://doi.org/10.1016/j.jphotobiol.2018.12.025.

[142] S. Arokiyaraj, M. Saravanan, N.K. Udaya Prakash, M. Valan Arasu, B. Vijayakumar, S. Vincent, Enhanced antibacterial activity of iron oxide magnetic nanoparticles treated with Argemone mexicana L. leaf extract: an in vitro study, Mater. Res. Bull. 48 (2013) 3323–3327, https://doi.org/10.1016/j.materresbull.2013.05.059.

[143] E.N. Taylor, T.J. Webster, The use of superparamagnetic nanoparticles for prosthetic biofilm prevention, Int. J. Nanomedicine 4 (2009) 145–152, https://doi.org/10.2147/IJN.S5976.

[144] E.L. Chang, C. Simmers, D.A. Knight, Cobalt complexes as antiviral and antibacterial agents, Pharmaceuticals 3 (2010) 1700–1728, https://doi.org/10.3390/ph3061711.

[145] D. Cavallo, A. Ciervo, A.M. Fresegna, R. Maiello, P. Tassone, G. Buresti, S. Casciardi, S. Iavicoli, C.L. Ursini, Investigation on cobalt-oxide nanoparticles cyto-genotoxicity and inflammatory response in two types of respiratory cells, J. Appl. Toxicol. 35 (2015) 1102–1113, https://doi.org/10.1002/jat.3133.

[146] E. Alsharaeh, Y. Mussa, F. Ahmed, Y. Aldawsari, M. Al-Hindawi, G.K. Sing, Novel route for the preparation of cobalt oxide nanoparticles/reduced graphene oxide nanocomposites and their antibacterial activities, Ceram. Int. 42 (2016) 3407–3410, https://doi.org/10.1016/j.ceramint.2015.10.135.

[147] L. Syam Sundar, N.A. Anjum, M.C. Ferro, E. Pereira, M.K. Singh, A.C.M. Sousa, Biocompatibility and biotoxicity of in-situ synthesized carboxylated nanodiamond-cobalt oxide nanocomposite, J. Mater. Sci. Technol. 33 (2017) 879–888, https://doi.org/10.1016/j.jmst.2017.03.016.

[148] S. Khan, A.A. Ansari, A.A. Khan, R. Ahmad, O. Al-Obaid, W. Al-Kattan, In vitro evaluation of anticancer and antibacterial activities of cobalt oxide nanoparticles, JBIC, J. Biol. Inorg. Chem. 20 (2015) 1319–1326, https://doi.org/10.1007/s00775-015-1310-2.

[149] B.A. Omran, H.N. Nassar, S.A. Younis, R.A. El-Salamony, N.A. Fatthallah, A. Hamdy, E.H. El-Shatoury, N.S. El-Gendy, Novel mycosynthesis of cobalt oxide nanoparticles using Aspergillus brasiliensis ATCC 16404—optimization, characterization and antimicrobial activity, J. Appl. Microbiol. 128 (2020) 438–457, https://doi.org/10.1111/jam.14498.

[150] A.T. Khalil, M. Ovais, I. Ullah, M. Ali, Z.K. Shinwari, M. Maaza, Physical properties, biological applications and biocompatibility studies on biosynthesized single phase cobalt oxide (Co3O4) nanoparticles via Sageretia thea (Osbeck.), Arab. J. Chem. 13 (2020) 606–619, https://doi.org/10.1016/j.arabjc.2017.07.004.

[151] M. Hafeez, R. Shaheen, B. Akram, A. Zain ul, S. Haq, S. Mahsud, S. Ali, R.T. Khan, Green synthesis of cobalt oxide nanoparticles for potential biological applications, Mater. Res. Express 7 (2020), https://doi.org/10.1088/2053-1591/ab70dd, 025019.

[152] E. Bossi, D. Zanella, R. Gornati, G. Bernardini, Cobalt oxide nanoparticles can enter inside the cells by crossing plasma membranes, Sci. Rep. 6 (2016) 22254, https://doi.org/10.1038/srep22254.

[153] M. Sivachidambaram, J.J. Vijaya, K. Kaviyarasu, L.J. Kennedy, H.A. Al-Lohedan, R. Jothi Ramalingam, A novel synthesis protocol for Co3O4 nanocatalysts and their catalytic applications, RSC Adv. 7 (2017) 38861–38870, https://doi.org/10.1039/C7RA06996K.

[154] S. Dubey, J. Kumar, A. Kumar, Y.C. Sharma, Facile and green synthesis of highly dispersed cobalt oxide (Co3O4) nano powder: characterization and screening of its ecotoxicity, Adv. Powder Technol. 29 (2018) 2583–2590, https://doi.org/10.1016/j.apt.2018.03.009.
[155] S. Iravani, Green synthesis of metal nanoparticles using plants, Green Chem. 13 (2011) 2638–2650, https://doi.org/10.1039/C1GC15386B.
[156] S. Chernousova, M. Epple, Silver as antibacterial agent: ion, nanoparticle, and metal, Angew. Chem. Int. Ed. 52 (2013) 1636–1653, https://doi.org/10.1002/anie.201205923.
[157] T.C. Dakal, A. Kumar, R.S. Majumdar, V. Yadav, Mechanistic basis of antimicrobial actions of silver nanoparticles, Front. Microbiol. 2016 (1831) 7, https://doi.org/10.3389/fmicb.2016.01831.
[158] S. Küünal, S. Kutti, M. Guha, P. Rauwel, D. Wragg, G. Nurk, E. Rauwel, Silver nanoparticles study for application in green housing, ECS Trans. 64 (2015) 15–24, https://doi.org/10.1149/06447.0015ecst.
[159] P. Orlowski, M. Zmigrodzka, E. Tomaszewska, K. Ranoszek-Soliwoda, M. Czupryn, M. Antos-Bielska, J. Szemraj, G. Celichowski, J. Grobelny, M. Krzyzowska, Tannic acid-modified silver nanoparticles for wound healing: the importance of size, Int. J. Nanomedicine 13 (2018) 991–1007, https://doi.org/10.2147/IJN.S154797.
[160] P. Rauwel, E. Rauwel, S. Ferdov, M.P. Singh, Silver nanoparticles: synthesis, properties, and applications, Adv. Mater. Sci. Eng. 2015 (2015) 2, https://doi.org/10.1155/2015/624394.
[161] A.-C. Burdușel, O. Gherasim, A.M. Grumezescu, L. Mogoantă, A. Ficai, E. Andronescu, Biomedical applications of silver nanoparticles: an up-to-date overview, Nanomaterials 8 (2018) 681, https://doi.org/10.3390/nano8090681.
[162] A. Vassallo, M.F. Silletti, I. Faraone, L. Milella, Nanoparticulate antibiotic systems as antibacterial agents and antibiotic delivery platforms to fight infections, J. Nanomater. 2020 (2020) 6905631, https://doi.org/10.1155/2020/6905631.
[163] S. Srikar, D. Giri, D. Pal, P. Mishra, S. Upadhyay, Green synthesis of silver nanoparticles: a review, Green Sustain. Chem. 6 (2016) 34–56, https://doi.org/10.4236/gsc.2016.61004.
[164] W.-R. Li, X.-B. Xie, Q.-S. Shi, H.-Y. Zeng, Y.-S. Ou-Yang, Y.-B. Chen, Antibacterial activity and mechanism of silver nanoparticles on Escherichia coli, Appl. Microbiol. Biotechnol. 85 (2010) 1115–1122, https://doi.org/10.1007/s00253-009-2159-5.
[165] K.J. Kim, W.S. Sung, S.K. Moon, J.S. Choi, J.G. Kim, D.G. Lee, Antifungal effect of silver nanoparticles on dermatophytes, J. Microbiol. Biotechnol. 18 (2008).
[166] A. Nagy, A. Harrison, S. Sabbani, J.R.S. Munson, P. Dutta, W.J. Waldman, Silver nanoparticles embedded in zeolite membranes: release of silver ions and mechanism of antibacterial action, Int. J. Nanomedicine 6 (2011) 1833–1852, https://doi.org/10.2147/IJN.S24019.
[167] Y. Long, L. Hu, X. Yan, X. Zhao, Q. Zhou, Y. Cai, G. Jiang, Surface ligand controls silver ion release of nanosilver and its antibacterial activity against Escherichia coli, Int. J. Nanomedicine 12 (2017) 3193, https://doi.org/10.2147/IJN.S132327.
[168] H.L. Su, C.C. Chou, D.J. Hung, S.H. Lin, I.C. Pao, J.H. Lin, F.L. Huang, R.X. Dong, J.J. Lin, The disruption of bacterial membrane integrity through ROS generation induced by nanohybrids of silver and clay, Biomaterials 30 (2009) 5979–5987, https://doi.org/10.1016/j.biomaterials.2009.07.030.
[169] P. Jena, M. Bhattacharya, G. Bhattacharjee, B. Satpati, P. Mukherjee, D. Senapati, R. Srinivasan, Bimetallic gold-silver nanoparticles mediate bacterial killing by disrupting the actin cytoskeleton MreB, Nanoscale 12 (2020) 3731–3749, https://doi.org/10.1039/C9NR10700B.
[170] N. Zafar, S. Shamaila, J. Nazir, R. Sharif, M. Shahid Rafique, J. Ul-Hasan, S. Ammara, H. Khalid, Antibacterial action of chemically synthesized and laser generated silver nanoparticles against human pathogenic bacteria, J. Mater. Sci. Technol. 32 (2016) 721–728, https://doi.org/10.1016/j.jmst.2016.05.009.

[171] S. Arokiyaraj, M. Valan Arasu, S. Vincent, N.U. Prakash, S.H. Choi, Y. Oh, K. Choi, K.H. Kim, Rapid green synthesis of silver nanoparticles from Chrysanthemum indicum L and its antibacterial and cytotoxic effects: an in vitro study, Int. J. Nanomedicine 9 (2014) 379–388, https://doi.org/10.2147/IJN.S53546.

[172] S. Liao, Y. Zhang, X. Pan, F. Zhu, C. Jiang, Q. Liu, Z. Cheng, G. Dai, G. Wu, L. Wang, et al., Antibacterial activity and mechanism of silver nanoparticles against multidrug-resistant Pseudomonas aeruginosa, Int. J. Nanomedicine 14 (2019) 1469–1487, https://doi.org/10.2147/IJN.S191340.

[173] C. Tao, Antimicrobial activity and toxicity of gold nanoparticles: research progress, challenges and prospects, Lett. Appl. Microbiol. 67 (2018) 537–543, https://doi.org/10.1111/lam.13082.

[174] M. Nadeem, B.H. Abbasi, M. Younas, W. Ahmad, T. Khan, A review of the green syntheses and anti-microbial applications of gold nanoparticles, Green Chem. Lett. Rev. 10 (2017) 216–227, https://doi.org/10.1080/17518253.2017.1349192.

[175] A. Nirmala Grace, K. Pandian, Antibacterial efficacy of aminoglycosidic antibiotics protected gold nanoparticles—a brief study, Colloids Surf. A Physicochem. Eng. Asp. 297 (2007) 63–70, https://doi.org/10.1016/j.colsurfa.2006.10.024.

[176] M. Chamundeeswari, S.S.L. Sobhana, J.P. Jacob, M.G. Kumar, M.P. Devi, T.P. Sastry, A.-B. Mandal, Preparation, characterization and evaluation of a biopolymeric gold nanocomposite with antimicrobial activity, Biotechnol. Appl. Biochem. 55 (2010) 29–35, https://doi.org/10.1042/ba20090198.

[177] J.C. Castillo-Martínez, G.A. Martínez-Castañón, F. Martínez-Gutierrez, N.V. Zavala-Alonso, N. Patiño-Marín, N. Niño-Martinez, V. Zaragoza-Magaña, C. Cabral-Romero, Antibacterial and antibiofilm activities of the photothermal therapy using gold nanorods against seven different bacterial strains, J. Nanomater. 2015 (2015), https://doi.org/10.1155/2015/783671, 783671.

[178] Y. Cui, Y. Zhao, Y. Tian, W. Zhang, X. Lü, X. Jiang, The molecular mechanism of action of bactericidal gold nanoparticles on Escherichia coli, Biomaterials 33 (2012) 2327–2333, https://doi.org/10.1016/j.biomaterials.2011.11.057.

[179] S.K. Das, A.R. Das, A.K. Guha, Gold nanoparticles: microbial synthesis and application in water hygiene management, Langmuir 25 (2009) 8192–8199, https://doi.org/10.1021/la900585p.

[180] A. Mishra, S.K. Tripathy, S.-I. Yun, Bio-synthesis of Gold and silver nanoparticles from Candida guilliermondii and their antimicrobial effect against pathogenic bacteria, J. Nanosci. Nanotechnol. 11 (2011) 243–248, https://doi.org/10.1166/jnn.2011.3265.

[181] D. MubarakAli, N. Thajuddin, K. Jeganathan, M. Gunasekaran, Plant extract mediated synthesis of silver and gold nanoparticles and its antibacterial activity against clinically isolated pathogens, Colloids Surf. B Biointerfaces 85 (2011) 360–365, https://doi.org/10.1016/j.colsurfb.2011.03.009.

[182] R. Geethalakshmi, D.V.L. Sarada, Characterization and antimicrobial activity of gold and silver nanoparticles synthesized using saponin isolated from Trianthema decandra L, Ind. Crop Prod. 51 (2013) 107–115, https://doi.org/10.1016/j.indcrop.2013.08.055.

[183] M.R. Bindhu, P. Vijaya Rekha, T. Umamaheswari, M. Umadevi, Antibacterial activities of Hibiscus cannabinus stem-assisted silver and gold nanoparticles, Mater. Lett. 131 (2014) 194–197, https://doi.org/10.1016/j.matlet.2014.05.172.

[184] M. Nagalingam, V.N. Kalpana, V.D. Rajeswari, A. Panneerselvam, Biosynthesis, characterization, and evaluation of bioactivities of leaf extract-mediated biocompatible gold nanoparticles from Alternanthera bettzickiana, Biotechnol. Rep. 19 (2018), https://doi.org/10.1016/j.btre.2018.e00268, e00268.

[185] J. Li, Q. Li, X. Ma, B. Tian, T. Li, J. Yu, S. Dai, Y. Weng, Y. Hua, Biosynthesis of gold nanoparticles by the extreme bacterium Deinococcus radiodurans and an evaluation of their antibacterial properties, Int. J. Nanomedicine 11 (2016) 5931–5944, https://doi.org/10.2147/IJN.S119618.

[186] F. Arockiya Aarthi Rajathi, C. Parthiban, V. Ganesh Kumar, P. Anantharaman, Biosynthesis of antibacterial gold nanoparticles using brown alga, Stoechospermum marginatum (kützing), Spectrochim. Acta A Mol. Biomol. Spectrosc. 99 (2012) 166–173, https://doi.org/10.1016/j.saa.2012.08.081.
[187] M.B. Gawande, A. Goswami, F.-X. Felpin, T. Asefa, X. Huang, R. Silva, X. Zou, R. Zboril, R.S. Varma, Cu and cu-based nanoparticles: synthesis and applications in catalysis, Chem. Rev. 116 (2016) 3722–3811, https://doi.org/10.1021/acs.chemrev.5b00482.
[188] L.P. Arendsen, R. Thakar, A.H. Sultan, The use of copper as an antimicrobial agent in health care, including obstetrics and gynecology, J. Clin. Microbiol. Rev. 32 (2019), https://doi.org/10.1128/CMR.00125-18, e00125–18.
[189] S.W.J. Gould, M.D. Fielder, A.F. Kelly, M. Morgan, J. Kenny, D.P. Naughton, The antimicrobial properties of copper surfaces against a range of important nosocomial pathogens, Ann. Microbiol. 59 (2009) 151–156, https://doi.org/10.1007/BF03175613.
[190] D. Longano, N. Ditaranto, L. Sabbatini, L. Torsi, N. Cioffi, Synthesis and antimicrobial activity of copper nanomaterials, in: N. Cioffi, M. Rai (Eds.), Nano-Antimicrobials: Progress and Prospects, Springer Berlin Heidelberg, Berlin, Heidelberg, 2012, pp. 85–117, https://doi.org/10.1007/978-3-642-24428-5_3pp.
[191] M.I. Din, F. Arshad, Z. Hussain, M. Mukhtar, Green adeptness in the synthesis and stabilization of copper nanoparticles: catalytic, antibacterial, cytotoxicity, and antioxidant activities, Nanoscale Res. Lett. 12 (2017) 638, https://doi.org/10.1186/s11671-017-2399-8.
[192] S.-M. Hasheminya, J. Dehghannya, Green synthesis and characterization of copper nanoparticles using Eryngium caucasicum Trautv aqueous extracts and its antioxidant and antimicrobial properties, Part. Sci. Technol. (2019), https://doi.org/10.1080/02726351.2019.1658664, 1-8, doi:10.1080/02726351.2019.1658664.
[193] Q. Lv, B. Zhang, X. Xing, Y. Zhao, R. Cai, W. Wang, Q. Gu, Biosynthesis of copper nanoparticles using Shewanella loihica PV-4 with antibacterial activity: novel approach and mechanisms investigation, J. Hazard. Mater. 347 (2018) 141–149, https://doi.org/10.1016/j.jhazmat.2017.12.070.
[194] M. Usman, M. El Zowalaty, K. Shameli, N. Zainuddin, M. Salama, N.A. Ibrahim, Synthesis, characterization, and antimicrobial properties of copper nanoparticles, Int. J. Nanomedicine 8 (2013) 4467–4479, https://doi.org/10.2147/IJN.S50837.
[195] N. Cioffi, L. Torsi, N. Ditaranto, G. Tantillo, L. Ghibelli, L. Sabbatini, T. Bleve-Zacheo, M. D'Alessio, P.G. Zambonin, E. Traversa, Copper nanoparticle/polymer composites with antifungal and bacteriostatic properties, Chem. Mater. 17 (2005) 5255–5262, https://doi.org/10.1021/cm0505244.
[196] K.-Y. Yoon, J. Hoon Byeon, J.-H. Park, J. Hwang, Susceptibility constants of Escherichia coli and Bacillus subtilis to silver and copper nanoparticles, Sci. Total Environ. 373 (2007) 572–575, https://doi.org/10.1016/j.scitotenv.2006.11.007.
[197] Y. Kaneko, M. Thoendel, O. Olakanmi, B.E. Britigan, P.K. Singh, The transition metal gallium disrupts Pseudomonas aeruginosa iron metabolism and has antimicrobial and antibiofilm activity, J. Clin. Invest. 117 (2007) 877–888, https://doi.org/10.1172/JCI30783.
[198] K. Richter, N. Thomas, J. Claeys, J. McGuane, C.A. Prestidge, T. Coenye, P.-J. Wormald, S. Vreugde, A topical hydrogel with deferiprone and gallium-protoporphyrin targets bacterial Iron metabolism and has antibiofilm activity, J Antimicrob. Agents Chemother. 61 (2017), https://doi.org/10.1128/AAC.00481-17, e00481–17.
[199] P. Narayanasamy, B.L. Switzer, B.E. Britigan, Prolonged-acting, multi-targeting gallium nanoparticles potently inhibit growth of both HIV and mycobacteria in co-infected human macrophages, Sci. Rep. 5 (2015) 8824, https://doi.org/10.1038/srep08824.
[200] S.-R. Choi, B.E. Britigan, D.M. Moran, P. Narayanasamy, Gallium nanoparticles facilitate phagosome maturation and inhibit growth of virulent mycobacterium tuberculosis in macrophages, PLoS One 12 (2017), https://doi.org/10.1371/journal.pone.0177987, e0177987.

CHAPTER

7

Surface functionalization of nanoparticles: Structure determines function

Gabriela Navarro-Tovar[a,b,*], *Daniela Salado-Leza*[b,*], *Clara Carreón-Álvarez*[c,*], *Brenda Jeanneth Acosta-Ruelas*[d], *and José Luis Rodríguez-López*[e,*]

[a]CONACYT, Centro de Investigación en Ciencias de la Salud y Biomedicina, Universidad Autónoma de San Luis Potosí, San Luis Potosí, Mexico, [b]CONACYT, Facultad de Ciencias Químicas, Universidad Autónoma de San Luis Potos, San Luis Potosí, Mexico, [c]Departamento de Ciencias Naturales y Exactas, Centro Universitario de los Valles, Universidad de Guadalajara, Ameca, Jalisco, Mexico, [d]CONACYT, Coordinación para la Innovación y Aplicación de la Ciencia y la Tecnología, Universidad Autónoma de San Luis Potosí, San Luis Potosí, Mexico, [e]Advanced Materials Department, Instituto Potosino de Investigación Científica y Tecnológica, A.C, 78216 San Luis Potosí, Mexico

1 General aspects

Nowadays, the term "nanoparticles" (NPs) is widely used in many diverse research fields, such as materials science, physics, chemistry, biomedical applications, and even in social sciences. By reviewing the data from Scopus (07-2020), scientific publications devoted to NPs have increased in more than 2700% since the beginning of the XXI century.

[*]Contributed equally to the manuscript.

Antimicrobial Activity of Nanoparticles
https://doi.org/10.1016/B978-0-12-821637-8.00004-3

Copyright © 2023 Elsevier Inc. All rights reserved.

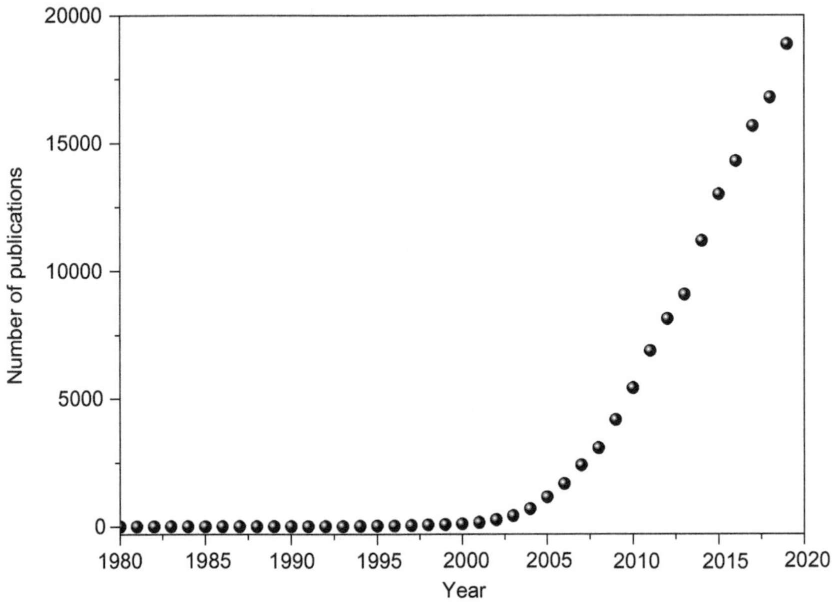

FIG. 1 Publications devoted to nanoparticles for biomedical applications. *Data were collected from Scopus (07-2020) under the search of words "nanoparticles" and "biomedical."*

Therefore, after this birth flight review, two questions are highlighted: why did NPs become so attractive? And, what is the reason behind the exponential rise in the development of NPs for biomedical applications? In the last 20 years, we pass from less than 100 to almost 19,000 publications devoted to the study of NPs for biomedical applications (see Fig. 1).

Nanoparticles are structures that consist in aggregated atoms, even scanty of them, consolidated in a well-defined shape and size, which should not be larger than few hundreds of nanometers in any direction [1]. The performance of NPs, indistinctly of their application, is strongly linked to their size. The NPs' high surface/volume ratio enables the exposition of high number of atoms with low coordination; thus, smaller the nanoparticle, more unusual features from those at the bulk state are found [2]. However, the definition of nanoparticles is quite complex and extended when structural characteristics are involved. For example, nanoparticles can exhibit single or complex compositions, different chemical nature (metallic, polymers, metallic oxides, biomolecules, carbon-based, semiconductor-based, a combination of many, etc.), structure (be solid, porous, hard, or soft), and habits or shape (spheres, rods, platonic, Archimedean, Kepler-Poison solids, etc.), as well as evidence of amorphous or crystalline structures [3–5]. It is also well known that the NPs' habit strongly influences many physicochemical phenomena. In some applications, NPs' shape has been adapted as a traditional sphere, or being faceted, for convenience. However, the concept of NPs has evolved to include different dimensions (0D, 1D, 2D, or 3D), such as nanodiamonds, nanotubes, nanocapsules, nanospheres, nanostars, nanorods, nanodisks, nanopyramids, and certainly, many more nanostructures that are characterized with nanodomains at their surface similar to that found on the nanospheres [6]. Undisputed development in nanoparticle research, in both synthesis (with focus on control of size and shape) and characterization, is accompanied by their bold and efficient incorporation in many scientific and technological advances.

According to Jana et al. [7], nanoparticles have been broadly used in biomedical applications due to the similarity in scales with biomolecules and biological events. The results have been promising for the treatment of a wide range of non-communicable diseases such as neurodegenerative [8], cancer [9,10], communicable diseases [11–13], early diagnosis [14], therapy [15], photothermal therapy [16,17], drug delivery [18,19], imaging probes [20] (cell and tissues), contrast agents [21,22], immune assays [23], among others. However, the initial scenario for the open-ended use of NPs in this field was kept cautious. Many warnings were spread worldwide related to the exposure to NPs, claiming possible unknown risks since the response of biological systems to novel materials of sizes less than 100nm [24–27]. Due to researchers' concern, over the safe use of NPs has been growing as well. Therefore, among current strategies for reducing toxicity, hazards and to control the biodistribution of NPs in the human being organism, the study of NPs functionalization for biomedical applications has been widely proposed with emphasis in understanding the interactions between biomolecular attachments and the NPs´ surface [28].

Functionalization is widely described as the act to rationally attach different species (such as ions, molecules, polymers, biomolecules, metals, etc.) on the surface of NPs. Specifically, for biomedical applications, Raliya et al. [29] claimed that the principle to achieve a rational, designed surface functionalization is to imitate the cellular surface compatibility without comprising the ability of NPs to induce a therapeutic effect.

Therefore, functionalization of NPs' surface has been conducted via chemical junction or physical deposition of those specific species, by means of well-established fundamental laws and principles of physics, chemistry, and biology, among other crossed research fields. It is important to highlight that functionalized NPs result in highly hierarchically nanostructured functional systems, characterized by superior performance than those non-functionalized, which if the projected biomedical applications succeed, there would be not only toxicity effects surpassed, but also a cascade of benefits come depending on the particular system and the application [30]. Following the above ideas, we can say that physical functionalization, also interpreted as decoration, is generally achieved after the NP synthesis, and generates a weak nanoparticle-stabilized agent interaction that could be unstable, or even broken, under harsh conditions [31]. On the other hand, chemical functionalization results in covalent bonding, which typically means stable interaction between NPs' surface and the molecular attachment agents. Table 1 summarizes the more recent reports devoted to nanoparticles' functionalization focused on biomedical applications.

For the past three decades, there was being a notorious and increasingly research in the functionalization of nanoparticles; i.e., in the understanding of the ligands and the hierarchically nanostructured system properties; in the use of a broad spectrum of chemical ligands; and an incredible number of applications regarding biological problems. We even now talk about designed bio-nanostructured materials for nanomedicine, where materials research scientists, biologist, chemists, physicist, medical researchers, etc., continue in the objective of the improvement of the NP properties via their surface modification.

Nowadays, the main characteristics of modern nanostructured functionalized materials are those that show: (a) to be by design, following the laws and principles of main sciences as mentioned above (i.e., physics, chemistry; biology; biochemistry; inclusive mathematics; and now lately, we can count also artificial intelligence algorithms); (b) to replicate the behavior of the nature systems; (c) to be nanostructures with a high level of hierarchy; (d) to present better and even unexpected properties and performances; and finally; (e) to be applied in relevant technological and social problems such as thermomagnetic and photothermal therapies in cancer research, imaging, drug delivering, diagnostics, and the more conceptually advanced field of theranostics, among others. Thus and clearly, biofunctionalized nanoparticles are shown to be a light in the born of a new era in nanomedicine. Therefore, the most recent and novel methods developed for the functionalization of NPs directed to biomedical applications require an extensive presentation. This chapter aims to describe the most recent functionalization techniques reported for inorganic NPs with potential applications in nanomedicine, such as main functionalization routes using ligands exchange, phase transfer, bioconjugation, and orthogonal chemistry methods.

2 Methods to functionalize and characterize nanoparticles for biomedical applications

2.1 Ligand exchange to improve nanoparticles' physiological stability

The ligand exchange (LE) method consists of replacing one or more ligands onto the nanomaterial surface by ligands with more affinity to the target, and thus, the physicochemical and biological properties are

TABLE 1 Most recent advances in nanoparticles' functionalization for biomedical applications.

Nature of nanoparticles	Material of nanoparticles	Functionalization agent	Functionalization type	Application	Ref.
Oxides	Fe_2O_3	Dihexadecyl phosphate (DHP)	Chemical	Bioactivity on SW1353 and TCam-2	[32]
		Folic acid	Chemical	Cancer theranostic agent	[33]
	ZnO	*D. tortuosa* (capping agent)	Chemical	Cancer treatment	[34]
		Lipid	Chemical	Toxicity	[35]
Metals	Ag				
	Au	HS-FEG-NH_2	Chemical	Antitumor agent	[36]
		Sophorolipid	Physical	Antimicrobial agent against cholerae	[37]
	Pt	Corona protein	Chemical	Drug delivery	[38]
Hybrid	Ba-TiO_2	Corona protein	Chemical	Breast cancer treatment	[39]
	Au@SiO2	Fluorescein isothiocyanate isomer	Chemical	Prostate cancer	[40]
Non-metals	Se	Lysozyme	Physical	Resistance of *Escherichia coli* and *Staphylococcus aureus*	[41]
Polymers	Polylactic-*co*-glycolic acid	Albumin and acidity-triggered rational membrane	Chemical	Chemotherapeutic drug delivery systems	[42]
		Chitosan/doxycycline	Chemical	Antibacterial	[43]

Continued

TABLE 1 Most recent advances in nanoparticles' functionalization for biomedical applications—cont'd

Nature of nanoparticles	Material of nanoparticles	Functionalization agent	Functionalization type	Application	Ref.
Biomolecules	Alginate/ chitosan	Lovastatin	Chemical	Drug delivery	[44]
	mRNA/Lipid	RSV fusion protein	Chemical	Vaccines	[45]
	Albumin	Paclitaxel	Chemical	Breast cancer	[46]
	Lipid	Melittin	Physical	Vaccine	[47]

modified. This methodology is defined by nanoparticle parameters such as size, shape, crystalline structure, surface charge, and oxidation state [48]. Furthermore, it must consider the chemical structure of ligands, functional groups (most importantly head group), chain length, pKa, the extent of branching (sterics), and chirality [49]. Ligands can be used to tailor the surface properties of nanomaterials such as polarity, surface charge, as well as biocompatibility and/or selective binding sites for biomolecules in living organisms [50–56]. Several mechanisms of LE have been proposed, and novel techniques are currently applied to confirm or propose other mechanisms. Terraschke et al. [57] and Dinkel et al. [58] compiled the most recent information regarding LE processes.

From an inorganic chemistry point of view, two general mechanisms are proposed, i.e., associative pathway and dissociative mechanism. The associative mechanism is one step where the incoming ligand interacts onto the nanoparticle surface and the outcoming ligand desorbs in a concerted manner. Moreover, the dissociative mechanism is described as a two-step process where the outcoming ligand desorbs and later the incoming ligand coordinates onto the surface [59] (Fig. 2).

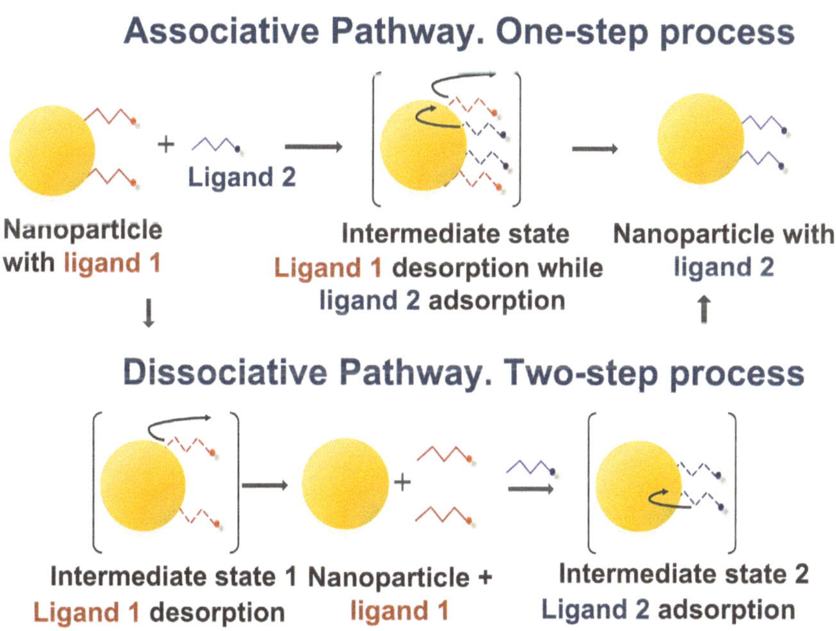

FIG. 2 Two general mechanisms of ligand exchange. Associative pathway in one step is described as a substitution nucleophilic (bi-molecular) or S_N2 reaction and dissociative pathway in two steps as substitution nucleophilic (unimolecular) or S_N1 reaction. *Adapted from A. Caragheorgheopol, V. Chechik, Mechanistic aspects of ligand exchange in Au nanoparticles, Phys. Chem. Chem. Phys. 10 (33) (2008): 5029–5041. https://doi.org/10.1039/b805551c.*

Santhosh et al. [54] reported the improvement in stability and biocompatibility of gold nanorods (GNRs) when a reported toxic cetyltrimethylammonium bromide (CTAB) was removed by ligand exchange and replaced by 1,2-dimyristoyl-sn-glycerol-3-phosphocholine (DMPC), a phospholipid molecule. CTAB molecules are weakly bound on the GNRs by electrostatic forces, and desorption was favored by ultrasonication in a solution containing both CTAB-GNRs and DMPC liposomes. The kinetic data obtained by surface-enhanced Raman scattering (633-nm laser) suggested an extremely slow ligand exchange process (12 h for effective replacement). The authors evaluated stability by dynamic light scattering analysis (DLS) proposing that GNPs with 22 mM of DMPC exhibited the highest stability since the hydrodynamic diameter (49 ± 2.7 nm) was very close to the one obtained by transmission electron microscopy (TEM) images (45 ± 1.6 nm) due to no particle aggregation. The biocompatibility study showed that cell viability (MCF-7, 3T3, and HMEC-1 cell lines) decreased significantly as the concentration of CTAB-GNRs increased (10 to 100 μM). In contrast, cell cultures exposed to DMPC-GNRs remained with high cell viability percent (>78% at 100 μM DMPC-GNRs). Thus, the ligand exchange process to replace toxic surfactants, such as CTAB, may offer biocompatible nanomaterials for biomedical applications.

Wei Kong et al. [51] proposed an exchange ligand process to enhance the biocompatibility of 25-nm spherical lanthanide-doped upconversion NPs ($NaYF_4$: Yb/Er@ $NaYF_4$). The authors proposed a two-step ligand exchange to remove the weakly bound oleate ligands (oleate acid and 1-octadecene) on the nanoparticle surface in acid pH. The ligand-free NPs in a basic pH aqueous solution were transferred to a diethylene glycol solution and the later evaporation of water (105–160°C) incorporated the new ligand such as poly(acrylic acid), poly(ethylenimine), cysteine, glycine, citric acid, biotin, and others, which could form chemical bonds with the lanthanide ions. NPs were characterized by TEM, IR spectroscopy, H^1 NMR, and TGA. Additionally, biocompatible evaluations conducted in human lung cancer cells (A549 cells) (4, 20, 100, and 500 μg/mL) showed ~88.9% of cell viability in cultures exposed to free ligand NPs and noticeable variation in cell viability in treatments with NPs coated with different types of ligands, being consisted with toxicity exhibited by free ligands. The design of an energy-transfer-based biosensor using biotinylated upconversion NPs to detect Cy3-conjugated streptavidin was successfully validated by emission spectra, suggesting that ligand-exchange process suggested by Wei Kong et al. could provide biocompatible NPs for the design of biosensor platforms.

Munjal et al. [53] synthesized by hydrothermal method a 4- to 7-nm nanohybrid $CoFe_2O_4$. The initial NPs were stabilized with oleic acid (OA) molecules and kept in non-polar solvents. The authors proposed the use of a rapid mechanochemical ligand exchange process to transform hydrophobic and unstable OA-capped $CoFe_2O_4$ (OA-CFO) to water-soluble and more stable NPs. The OA-CFO NPs were incorporated in a toluene solution with citric acid (CA) and dimethyl sulfoxide and stirred for 30h. The resulting NPs were collected and washed in ethanol, which was evaporated (60°C). The new CA-capped $CoFe_2O_4$ (CA-CFO) NPs exhibited a surface charge of −22.3mV and a hydrodynamic diameter of ~12nm at pH 7. Both CFO possessed the same particle size and similar saturation magnetization (M_s) and coercivity (H_c), indicating that NPs are near the superparamagnetic limit: ~ 47emu/gm and ~11, for OA-CFO; and ~42emu/gm and ~13, for CA-CFO. An L929 mouse fibroblast cell culture was exposed to 100 to 1000μg/mL CA-CFO for 24h, demonstrating high cell viability (> 90%) in all concentrations. Later, Munjal and Khare [60] reported that the same CA-CFO were stable over 120 days in aqueous solution (surface charge between −22.9 and −25.1mV) and showed high biocompatibility with cell viability >88% in L929 cells and 77% in glial G-1 cell cultures exposed to 50–1000μg/mL of CA-CFO for 24h. The potential use of this biocompatible nanohybrid was conducted in a magnetic hyperthermia in vitro model, exposing G-1 cells to a magnetic field (H=95.6 kA/m and frequency=329kHz) alone or magnetic field with CA-CFO. The cell viability when only magnetic field was applied was ~80%, while cell viability in CA-$CoFe_2O_4$ + magnetic field treatment was 21%–25%. Overall, results reported by Munjal demonstrated that ligand exchange processes, such as a rapid mechanochemical method, can enhance biocompatibility and stability of complex nanosystems for novel hyperthermia therapies.

Initial quantum dots (QD) were capped with trioctylphosphine (TOP) and trioctylphosphine oxide (TOPO). The ligand exchange process consisted of stirring at room temperature CdSe QD in chloroform with thiolated poly(ethylene glycol) (PEG-SH) polymer (30mg PEG-SH per 1nmol of CdSe). Despite that only 50%–85% of TOP/TOPO molecules were removed and few of them were replaced by PEG-SH, the ligand exchange was sufficient to disperse the QD in polar solvents such as chloroform and water.

Zeng et al. [56] prepared a water-soluble Eu^{3+}-doped luminescent hydroxyapatite (HAp) nanorods by replacing the original hydrophobic oleylamine coating for a biocompatible polymer. The amino groups in HAp nanorods were prepared using adenosine mnophosphate (AMP) as the ligand. Then, Br-containing initiators (HAp—Br) were incorporated onto the modified Hap-AMP in an amidation reaction. The polymer-functionalized HAp nanorods were finally obtained by metal-free ATRP method using poly(ethylene glycol) methacrylate (PEGMA) as a monomer and 10-phenylphenothiazine (PTH) as an organic photocatalyst. The resulting nanohybrids, with a diameter of 15nm and lengths between 50 and 80nm, were characterized by H^1 NMR, IR spectroscopy, XPS, and TGA. Furthermore, the emission spectrum from 420 to 750 nm and excitation wavelength set at 405 nm determined two peaks corresponding to red luminescence at 591 and 615 nm for Hap-PEGMA, demonstrating that the modification conducted in the Eu^{3+} HAp nanorods did not affect the luminescent properties of the nanomaterial. Moreover, biocompatibility properties were reported in an in vitro study with A549 cells showing 95% of cell viability.

The same research group prepared a HAp-based polymer composite using PEGMA (60 nm of length and width of 15–20 nm); the ligand exchange and polymer formation were similar to the previously reported process; however, a diacrylolyl fluorescein molecule (AcFI) was introduced in the PEGMA chain instead of Eu^{3+}-doped HAp. The nanocomposites (HAp-PEGMA-AcFI) exhibited visible fluorescence and dispersion in water. The biological evaluations demonstrated high biocompatibility in L929 cell cultures (~95%–100% of cell viability) after treatments with 80–160 μg/mL of HAp-PEGMA-AcFI. Moreover, L929 cell incubated with the nanocomposite for 2 h showed that particles were internalized into the cells through non-specific endocytosis and mainly distributed with strong green fluorescence in the cytoplasm [50]. Thus, the ligand exchange process in HAp-based material enhances biocompatibility and does not affect the biological application.

Ligand exchange could be also applied when initial ligands are strongly bond to the surface of NPs, such as CTAB onto GNRs (60 mV). In this sense, Lee et al. [52] proposed a method to achieve the complete replacement of CTAB and citrate onto the GNRs surface by a phospholipid bilayer of 1,2-dipalmitoyl-sn-glycerol-3-phosphothioethanol (DPPTE). The process was performed in a water/chloroform solution using PEG-SH. The complete replacement was reached by adding DPPTE in a two-phase solution (water/chloroform). The GNRs (~ 10 nm × 50 nm; ~ 5 mV) obtained were re-dispersed in water and characterization by UV–Vis, Raman, and TEM showed a complete ligand exchange. An evaluation in PBS buffer and NaCl solutions evidenced higher stability for DPPTE-GNRs and biological evaluation in *Homo sapiens* bone marrow neuroblastoma SH-SY5Y cells exhibited higher biocompatibility in DPPTE-GNRs treatments than in CTAB-GNRs.

The classical functionalization methods to favor nanomaterial interactions with cargo molecules are defined by both the chemistry of the nanomaterial and the cargo nature. The homogeneous coverage onto the nanomaterial surface remains a challenge for several materials, particularly in the development of biosensors. Different approaches have been described through this chapter for the functionalization of nanomaterials by non-covalent and covalent interaction with drugs and biomolecules. A single ligand can approach the surface nanoparticle interacting with different functional groups by van der Waals forces, stacking interactions, hydrogen bond, or covalent bonds. Thus, rarely, a single interaction occurs in functionalized nanomaterials. Kyaw et al. [61] reported four types of interfaces between (3-amino)-propyl-triethoxysilane (APTES) (Type I with Si-OH, Type II with two Si-OH, Type III with NH_2, and Type IV with three Si-OH) and glass surfaces, resulting in heterogeneous sites where gold NPs (AuNPs) later interact on the evaluated system. Atomic force microscopy (AFM) showed a partial coverage of AuNPs where Type III interaction was confirmed, while Type IV was the interaction allowing homogeneous coverage of AuNPs, by electrostatic forces with NH_2 groups exposed. Thus, the functionalization with APTES exhibits different interaction with nanomaterial surface, affecting the development of biosensors, nano-vehicles for drug delivery, and other applications.

Moreover, Bekdemir et al. [62] determined by thermodynamic studies that the nanoparticle/protein (AuNPs/albumin) ratio is mainly dependent on two parameters, the particle size (3–6 nm) and the patchy character of surface ligands, such as 1-octanethiol and 11-mercaptoundecane sulfonate. In this sense, the understanding of nanoparticle surface ligands and biomolecule interactions could improve the design of more sophisticated nanosystems.

In recent years, alternatives to improve specificity regarding nanomaterial functionalization have gained attention. In this sense, "click chemistry" is a potential candidate in the new proposed nanoplatforms. Click chemistry is a concept derived from orthogonality, which corresponds to organic chemistry reactions with high reactivity and binding selectivity. Orthogonal reactions occur between two molecules interacting as specific partners (or building blocks) in a reaction mechanism where secondary reactions are not favorable, and therefore, purification steps are not required [63]. Among orthogonal reactions, the "click chemistry" reactions and its variants stand for mechanisms where a single product has a high yield in aqueous solution and selectivity [64]. The first reaction considered under this term was azide-alkyne cycloaddition using copper as a catalyst [65]. However, the incorporation of copper comprises the biological applications due to its toxicity, and free cooper click chemistry reactions have been proposed as an alternative [66]. Two main click chemistry reactions can be distinguished in the literature, copper-catalyzed alkyne-azide cycloaddition or CuAAC, and strain promoted alkyne-azide cycloaddition or SPAAC (Fig. 3). The application of click chemistry in nanomaterials for medical applications has been directed to attach biological ligands onto nanoparticle surface with high specificity (orthogonality) [67–69], for control drug release [70–72], imaging [73,74], biosensor, and theranostic platforms [75–77].

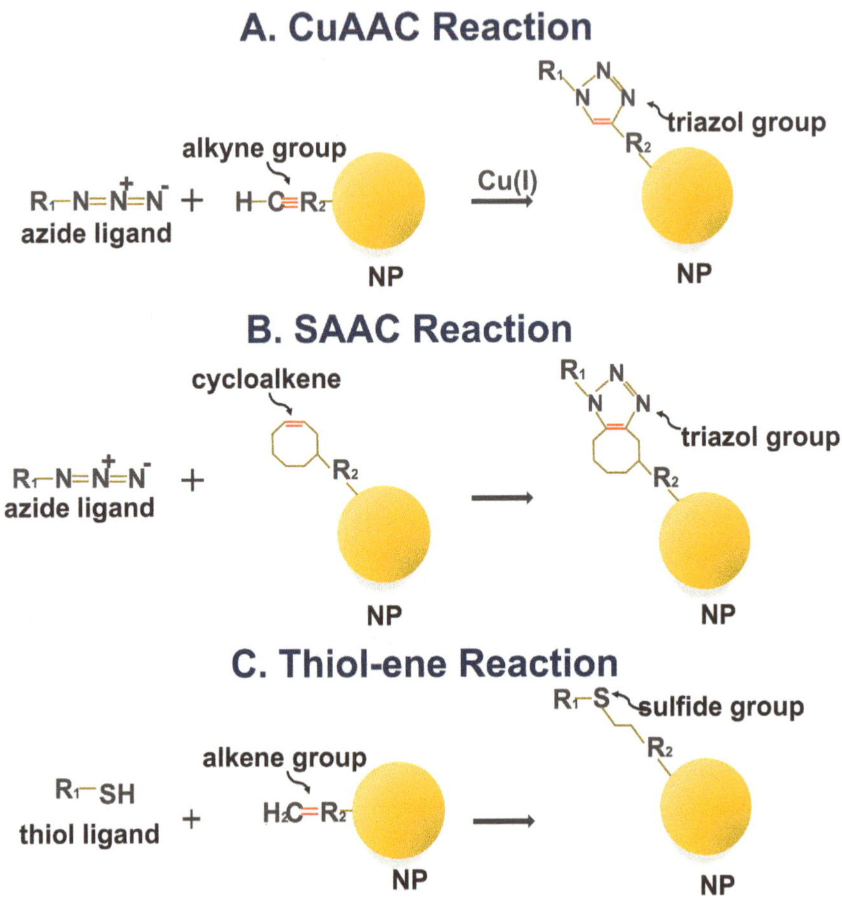

FIG. 3 Click chemistry reactions onto nanoparticle surfaces. (A) CuAAC, (B) SPAAC, and (C) Thiol-ene coupling.

AuNPs are the most reported metallic NPs for click chemistry using azide derivatives. For instance, Elliott et al. [78] reported the synthesis of 3.5-nm AuNPs with mixed ligands in a molar ratio of 1:19: an azide derivative, S-(2-(2-(2-azidoethoxy)ethoxy)ethyl)sulfothioate (N_3-EG_3-BS), and 2-[2-(2-mercaptoethoxy)ethoxy)ethanol (EG_3-BS), using a single-step flow-based synthesis of azide-terminated mixed monolayer AuNPs in aqueous medium at pH5. The nanoparticle stability was monitored by ^{1}H NMR and demonstrated that stocks of N_3-EG_3-AuNPs-EG_3 1mg/mL do not exhibit changes within 3-month period. UV–Vis spectrophotometry showed that the azide group remained with no changes over the course of 17months, and samples remain well dispersed in water. To verify the

N_3-EG_3-AuNPs-EG_3 reactivity, SPAAC reaction (copper-free click chemistry) and CuAAC reaction were conducted by Elliot et al., and the formation of the 7-triazole group onto the NPs was confirmed by ^{1}H NMR. Thus, the incorporation of azide derivatives in controlled conditions may add stability in aqueous solutions and provide a versatile nanoparticle surface to incorporate hybrid alkyne biomolecules in the search of highly specific biosensors.

Finetti et al. [67] proposed a hybrid nanoparticle with AuNP core and silica shell, which was stabilized and functionalized with a copolymer synthesized from N-acryloyloxy-succinimide (NAS), *N,N*-dimethylacrylamide (DMA), and γ-methacryloxypropyltrimethoxy silane (MAPS). Alkyne groups onto the copolymer surface were used in a copper-catalyzed azide/alkyne cycloaddition to bind azide-modified anti-mouse IgG antibodies. Finetti et al. used an interferometric reflectance imaging system (IRIS) to evaluate the hybrid nanosystem. The immobilized target (anti-CD63) (onto glass surface) was then recognized by the AuNP-labeled anti-mouse polyclonal IgG, an indication of the proper functionalization as well as the stability (no aggregation). Comparative studies in IRIS, with no clicked antibody incorporation onto AuNPs, showed that particles are detected with a covalent union as click chemistry does.

Another example of high selectivity to biomolecules, such as DNA and proteins, by thiol-ene reaction was reported by Liu et al. [69]. Hydrophobic lanthanide-doped upconversion NPs (UCNPs) were conjugated with either PEG-HS or DNA-HS, and cysteine-containing horseradish peroxidase, using triethylamine as a catalyst to promote the thiol-ene click reaction. Products were characterized by ^{13}C and ^{1}H NMR, and IR spectroscopy. HS-PEG-UCNPs (3.2 mV), and UCNP-S-DNA (−25.9 mV) NPs, showed monodispersion by TEM images. Moreover, when UCNPs were linked to FITC-modified HS-PEG3400 demonstrated fluorescence signals at 525 nm, indicating a covalent thiol-ene conjugation onto the particles. Lui et al. explored the targeting of protein PTK7, which is highly expressed on leukemic lymphoid CEM cells by using UCNPs-HS-aptamer (Sgc8). The fluorescence-labeled aptamer was followed in confocal laser scanning microscopy and flow cytometry, demonstrating the selective union of UCNPs-HS-aptamer (Sgc8) and protein PTK7 in the CEM cell culture.

Click chemistry could be used for improvements in the supramolecular targeting for potential theragnostic or imaging devices. For instance, Zhang et al. [77] propose an interesting biosensor consisting of a cation-exchange reaction of CuS NPs combined with the click chemistry-functionalized AuNPs using tris[(1-benzyl-1H-1,2,3-triazol4-yl)methyl] amine (TBTA) as a catalyst. The authors evaluated by UV–vis spectrophotometry the system by detecting single-nucleotide polymorphisms (SNPs) and DNA methyltransferase (MTase), and by validating the analytical performance by quantifying DNA MTase at various concentrations, demonstrating that this nanosystem based in cation-exchange reaction combined with click chemistry could be suitable for detection in biological samples.

Croissant et al. [79] developed ethynylene-based periodic mesoporous organosilica NPs (PMO NPs) for the potential application as imaging platforms and as drug delivery vehicles. The nanosystem consisted of a tetra-propargylated two-photon fluorophore onto the PMO as a precursor to performing click reaction (CuAAC) with (3-azidopropyl)triethoxysilane, resulting in spherical NPs with a diameter of ~80nm. The PMO NPs achieved a maximum drug release of 14.2wt%, when doxorubicin was used as a drug model in an aqueous medium with pH5.5 (lysosomal compartment). The authors reported in previous work the importance of loading ethenylene-bridged onto the NPs to enhance the drug diffusion throughout the porous cavities. Moreover, Croissant et al. proposed the PMO NPs as a potential platform for two-photon photodynamic therapy (TPE-PDT), which has an improved penetration depth in tissues and the option to selectively address confined volumes, compared with reduced scattering. The in vitro model was human breast cancer cells (MCF-7) culture where PMO NPs demonstrated endocytosis internalization and biocompatibility. The TPE-PDT study (confocal laser scans 1.57s in four different areas) determined that 48h later, 40% of cells were not viable, compared with the PMO NPs system, which only possessed the alkyne precursor (no significant reduction in cell viability), suggesting that PMO NPs combined with potent fluorophores and photosensitizers assembled by click chemistry are potential theranostic devices [75].

Bharathiraja et al. [80] used a ligand-exchange process to cover 18–26nm palladium NPs with water-soluble chitosan oligosaccharides (COS). Then, maleic anhydride was incorporated onto the surface of Pd@COS via the ring-opening reaction; as a result, "ene" groups were produced in the COS polymer, which was finally used to perform "thiol-ene" click chemistry with a peptide RGD (a motif that can bind $\alpha_v\beta_3$) to obtain Pd@COS-RGD. The authors reported high stability of the NPs (up to 6months) and biocompatibility when MDA-MB-231 cells were exposed to 0–50ppm of Pd@COS-RGD. This developed nanosystem demonstrated promising results as a theranostic platform using a tumor-bearing mice model following the tumor progression for up to 20days after treatments with Pd@COS-RGD, Pd@COS-RGD+laser (808nm at $2\,W\,cm^{-2}$), PBS+laser. Pd@COS-RGD did not cause adverse effects (loss of weight) of death and combined with laser achieved a significant reduction of tumors since day 6. Furthermore, photoacoustic imaging showed that Pd@COS-RGD accumulated in the tumor region and is acting as a good contrast agent. Overall, the nanohybrid material obtained could be a potential theranostic agent in cancer.

Moreover, van der Meer et al. [74] reported the obtention of ultrasmall AuNPs (~2 nm) decorated with azide-peptide K(N_3)CD (tripeptide lysine (N_3)-cysteine-asparagine with 117 peptide molecules per each nanoparticle) to promote azide-alkyle cycloaddition. AuNPs were synthesized by a modified one-phase Brust synthesis and AuNPs-K(N_3)CDs were characterized by conventional analysis, including ^{1}H NMR spectroscopy, to determine the peptide concentration. The azide-alkyl cycloaddition was performed with copper as a catalyst and using separately, propargyl alcohol, and fluorescein alkyne. Both reactions led to tetrazolium groups monitored by 2D total correlation spectroscopy TOCSY ^{1}H NMR. Additionally, van der Meer et al., evaluated the uptake in human cervix carcinoma cells (HeLa) using the fluorescein-clicked AuNPs (125 μg/mL per well with 200 μL), demonstrating that the nanosystem penetrated the nuclear membrane.

On the other hand, the functionalization of nanomaterials with click chemistry can result in nanodrug delivery vehicles, which respond to specific conditions to release bioactive molecules or drugs. Yan et al. [71] reported the use of thiol-functionalized mesoporous silica NPs (MSN-SH)-type MCM-41 as a vehicle for a silyl ether camptothecin prodrug; the thiol-alkene click chemistry achieved a covalent bond between nanoparticle surface and camptothecin (~293 nm and −22 mV). The cytotoxic effect of MSN-camptothecin in HeLa cells culture was similar to the one exhibited by the soluble drug (40%). The incorporation of silyl ether linkers makes MSN-camptothecin nanosystem pH-stimuli-responsive, about a 100% of loaded drug (loaded amount calculated ~12%wt) was released at pH 4, while at pH 6.8 the release profile is <20% in about 2 h. No drug release was achieved at pH 7.4, suggesting that the anticancer drug camptothecin-effective release could be in the tumor regions where pH is acidic, and avoid adverse effects on healthy tissues.

Similarly, Wei-Chen et al. [70] developed bipyramidal metal–organic frame NPs (NMOFs) with nucleic acid tethers (250–300 nm). Functionalization was conducted with click chemistry onto the nanoparticle surface. In this work, the functionalized DNA-NMOFs were loaded with doxorubicin, which was locked in the nanomaterial by acid pH-responsive DNA duplex-capping units. The drug release was monitored by UV–vis spectrophotometry at pH 7.4 and 5.0, and drug release profile showed an enhanced liberation at pH 5.0 (52.8 μmol doxorubicin per g of nanoparticle), suggesting that the use of nucleic acid-functionalized NMOFs by click chemistry could also be applied in stimulus-dependent drug release.

A relevant application of click chemistry relies on the concept of bioorthogonality, which is the design of a chemical reaction between two components in a biological medium. The concept was introduced by Bertozzi et al. [81,82]. The reaction must occur even in complex biological environments (e.g., whole organism); thus, the selection of functional chemical groups such as nucleophiles, electrophiles, reductants, and oxidants requires organic chemistry, materials, and analytical chemistry tools [83]. The potential use of bio-orthogonality in nanomaterials could provide successful platforms for drug delivery, biosensors, and imaging with high selectivity and sensitivity. For instance, Fratila et al. [68] synthesized 6-nm spherical magnetic NPs (MNPs) stabilized with oleic acid ligands and functionalized with two cyclooctynylamine derivatives (~27% surface coverage). The MNPs were immobilized onto silicon surface to promote azide alkyne cycloaddition by SPAAC reaction with two different azide derivatives: polyethylene glycol or a glucose moiety. Moreover, the authors reported an excellent colloidal stability in both physiological media and cell culture medium (up to 15 months, monitoring by NMR). The system is a potential platform for bio-orthogonal click chemistry applications.

Lee et al. [73] developed an interesting probe to track stem cells by applying metabolic glycol-engineering using tetra-acetylated *N*-azidoacetyl-D-mannosamine (Ac4ManNAz) to express chemical receptors containing azide groups on the surface of adipose tissue-derived human mesenchymal stem cells. The NPs to track cells were iron oxide NPs (10 nm) and AuNPs conjugated (3 nm) or encapsulated to bicyclo[6.1.0] nonyne-modified glycol chitosan NPs (BCN-CNPs) with a final hydrodynamic diameter size from ~314 to 340 nm; thus, the nanosystem had an alkyne group onto the surface to perform click chemistry onto the steam cells. The Cy5.5 fluorescent molecule was incorporated into the nanosystem to label cells with particles by the optical phantom study. The success of the in vitro evaluations foresees the potential use of the nanosystem to elucidate the in vivo fates of transplanted steam cells.

An example of bio-orthogonal chemistry in the search for selective drug delivery vehicles is reported by Yoo et al. [72]. The authors modified the antibody trastuzumab (Trb) with trans-cyclooctene (TCO). The antibody trastuzumab binds to human epidermal growth factor receptor 2 (HER2) receptor overexpressed some in breast cancer. Then, tetrazine-polymeric nanoparticles loaded with the chemotherapeutic drug SN38 (SN38-Tz-NP) (encapsulation efficiency of 82%) were obtained. NIH3T6.7 tumor cells were firstly treated with 40 μg/mL TCO-Trb and later with SN38-Tz-NP. The cell viability by MTT assay showed that 0.5/0.5 SN38-Tz-NP could low the cell viability in 40–50% when cells were previously treated with TCO-Trb. Real-time in vivo images of tumor-bearing mice showed that the administration of Tz-NPs containing DiI dye reached the tumor region 12 h post-injection when mice were previously treated with TCO-Trb (1.5 mg/kg) via intravenously.

In the search for low cost and portable devices for the detection of pathogenic bacteria, Mou et al. [76] reported a potential platform for the detection of *Escherichia coli* by applying the concept of bacteria-instructed click chemistry. The authors propose the use of 14 nm AuNPs functionalized with azide and alkyne groups (by a ligand-exchange process) onto the point of care, a potential place contaminated with pathogenic bacteria, where organisms can reduce Cu^{2+} to Cu^{+}. The produced Cu^{+} plays the role of catalyst to trigger the click chemistry reaction between AuNPs. The colorimetric assay is suggested to be integrated into a smartphone app as a portable device to detect contaminated point of care. The app smartphone platform is based on red-green-blue (RGB) pixel values recording and quantification ability to analyze the color differences of the bacteria-instructed click chemistry. It can detect *E. coli* within 1 h with a sensitive of 40 colony-forming units/mL.

2.2 Polarity and surface charge modification to enhance drug loading and cell adhesion

NPs' polarity and surface charge are key parameters that may determine properties such as wettability, biochemical affinity, loading capacity, cell adhesion, and intracellular delivery. The surface of most of the technological anthropogenic nanostructures is functionalized with organic or inorganic ligands, in situ or post-synthesis, to enrich their physicochemical characteristics and application.

For biomedical applications, the nanostructures synthesized in the presence of surfactants are commonly found dispersed in toxic organic non-polar solvents, restricting their direct use in biological systems, adding costly, energetic, and time-consuming purification steps. Indeed, the remained greatest challenge in nanomedicine is the development and implementation of low cost, scalable, eco-friendly, efficient, and straightforward production routes suitable for biomedical research and clinic translation [84].

Currently, inorganic potential medical nano-agents are mainly treated ex situ to acquire biologically relevant surface competences by performing, for instance, (i) phase transfer and (ii) ligand exchange (as previously mentioned in Section 2.1) protocols to make them hydrophilic and thus soluble into polar biocompatible solutions. These protocols consist of using synthetic or natural agents to transfer hydrophobic NPs into an aqueous phase or to replace the original hydrophobic surfactant, used to control the nanostructure size and shape during synthesis, by functional hydrophilic molecules, which are frequently based on amphiphilic polymeric compounds of variable chain length. Particularly, polymers may contain anchoring groups (-COOH, $-NH_2$, -SH, etc.) for a successful surface modification, as well as a complementary ending group conferring solubility, colloidal stability in aqueous media, and reactive species for further conjugation. These post-synthesis methods are not exclusive and other in situ approaches can be adapted to the synthesis routes.

Metal and metal oxide NPs are very attractive inorganic nanostructures with enhanced physicochemical properties under, e.g., optic and magnetic stimuli, whose biological impact depends on their behavior in biomimetic environments. In this section, we present some current strategies used to modify the polarity and surface charge of inorganic core NPs to enhance drug-loading capacity, cell adhesion, and treatment strategies.

For instance, when NPs are embedded in physiological fluids, specific proteins bound their surface to form a layer known as the protein corona, which may critically affect their bio-recognition and circulation lifetime, interfering thus with the NPs' designed therapeutic function. Inspired in this NP-protein interface, proteins can be used as phase-transfer agents to modify NPs' polarity for biomedical applications (Fig. 4).

In this perspective, the bovine serum albumin (BSA) is a well-known protein used to transfer hydrophobic NPs to aqueous suspensions. Albumin is the most abundant protein in serum and has the ability to coat inorganic NPs forming a hard corona, which prevents the adsorption of minor serum proteins that should affect the delivery of active molecules via passive targeting. Sabri et al. [85] used a modified desolvation method to render lanthanide-based upconversion $NaGdF_4$:Yb^{3+}/Er^{3+} nanoparticles (Ln-UCNPs) hydrophilic and biocompatible for photodynamic therapy. The Ln-UCNPs were initially prepared by thermal decomposition. They were soluble in non-polar organic solvents due to the presence of oleate ligands at their surface. The NPs' polarity, biocompatibility, and solubility were then modified following an ethanol-based desolvation procedure

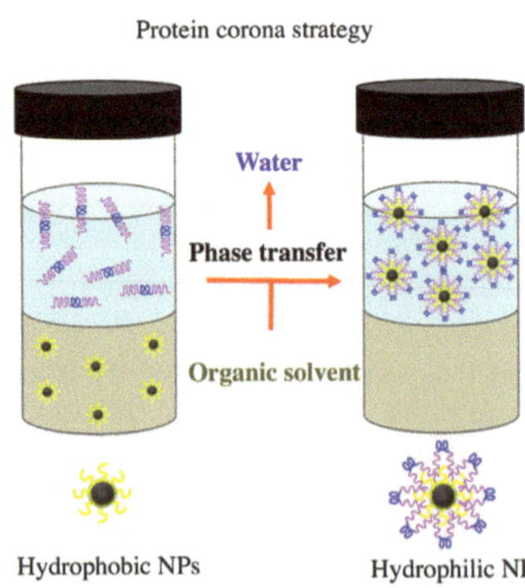

FIG. 4 Schematic representation of the phase transfer methodology inspired by the NP-protein corona interaction.

where the BSA shell thickness and water dispersion strongly depended on the use of a cross-linking agent, glutaraldehyde. Ethanol promoted the unfolding and polarization of BSA to coat the hydrophobic oleate-capped Ln-UCNPs. Indeed, far-UV circular dichroism (CD) revealed that the protein helical content decreased by about 80% when adsorbed to NPs. Moreover, glutaraldehyde allowed coating individual NPs with a 2nm thickness layer. Thermogravimetric analysis showed a 15% weight loss, which suggested that one single NP had been coated with 112 BSA molecules. BSA-functionalized Ln-UCNPs were covalently conjugated with Rose Bengal (RB), a commonly used photosensitizer capable of converting triplet oxygen to singlet oxygen, a very reactive and cytotoxic species. The authors showed by confocal fluorescence microscopy that RB-conjugated upconverting NPs are internalized by A549 cells and localized the cytosol, contrary to BSA-functionalized NPs that did not enter A549 cells. These in vitro findings, together with a laser (980nm)-activated cytotoxicity, positioned the RB-BSA-Ln-UCNPs as promising targeting nano-agents for improving the performance of molecular phototherapy.

Huang et al. [86] reported a simple phase transfer and surface bioengineering method, which may be applied to a wide variety of nanostructures, using β-sheet-rich silk fibroin, an FDA-approved natural protein. The enriched β-sheet structure confers remarkable physicochemical properties to the protein, which enhance its interaction with the hydrophobic surfactant molecules at the NPs surface and thus, the NP encapsulation within a hydrophilic stable corona. In this work, the authors used OA-capped superparamagnetic iron oxide nanoparticles (SPIONs) and Ln-UCNPs as models. The phase transfer was performed at atmospheric conditions under stirring mixing together the NPs suspended in chloroform and an aqueous solution containing the protein (49% β-sheet). Briefly, the resulting bioengineered SPIONs were homogeneously dispersed in water with a core-shell hybridized structure without compromising the intrinsic superparamagnetic core property. For instance, the colloidal stability was evaluated by DLS and zeta potential. The corona presence was confirmed by doubling the hydrodynamic diameter of NPs after phase transfer. Moreover, the NPs shell containing amphiphilic silk fibroin molecules, left reactive functional groups at the surface to assure a further bioconjugation and thus, the development of multi-modal nano-agents. Additionally, the authors showed the capability of those bioengineered NPs to load and release curcumin (CUR), a natural and hydrophobic antineoplastic drug. The encapsulated CUR induced more toxicity (MTT assay) to MCF-7 cells than free CUR, showing the potential use of silk fibroin-coated NPs as drug delivery systems.

Hydrophobic NPs may also be coated with amphiphilic neutral or charged polymers that anchor their surface-exposing hydrophilic tails outward for further targeting conjugation. The polymer choice impacts the stability and hydrodynamic size of NPs, which strongly depends on chain lengths and molecular weights. If NPs are poorly coated with those polymers and hydrophobic groups stay exposed, non-specific interactions with biological systems may appear. A better NP performance is expected performing a robust coating procedure via cross-linking. This extensive surface coverage minimizes non-specific interactions, confers biocompatibility and hydro-solubility, and allows simple conjugation.

Debnath et al. [87] reported a one-pot coating and functionalization strategy using an amphiphilic poly (amino acid), the polyaspartic acid, which has been modified as a functional polymer to develop therapeutic nanocarriers. The authors showed the versatility of this polymer to overcome the hydrophobicity of a wide variety of NPs, e.g., γ-Fe_2O_3, Ag, and CdSe@ZnS. A polyaspartimide (PSI) structure was developed with succinimide, octadecylamine, and PEG-diamine blocks with specific tasks in the polarity modification procedure. The octadecyl group anchored the hydrophobic NP surface, while polar succinimide and PEG molecules stay exposed outward. Indeed, it is the succinimide, the reactive specie for bioconjugation or further thiol functionalization. The bifunctional PEG allows performing conventional conjugation chemistry; at the time, it minimizes unspecific interactions. In brief, this one-pot approach consists in a phase transfer via in situ polymer coating. The aqueous phase contains the desired biomolecule or chemical agent, and NPs, in the presence of the functional PSI, are suspended in chloroform. Under warm conditions, the chloroform evaporation is favored, resulting in a quick phase transfer and coating. Then, basic pH may be used to catalyze ring-opening reactions for succinimide conjugation. An optimized NPs/polymer ratio is needed to avoid by-products, such as empty micelles, or multiple NP encapsulation. The methodology developed by Debnath et al. [87] allowed obtaining hybrid water-soluble NPs functionalized with primary amine, thiol, PEG, arginine, spermine, and glucosamine, as confirmed by standard fluorescent assays. Notably, these hybrid NPs showed <10 nm hydrodynamic diameters, suggesting a monolayer polymer capping, and displayed negative surface charge at pH of 4.5, 7.4, and 10.0. The surface charge was strongly impacted by the shell-exposed groups (e.g., –COOH), solution pH, and ligand protonation state, independently from the core nature. The cationic behavior of protonated species played an important role in promoting NPs-cell electrostatic interactions and thus, cell labeling. For instance, the tested CdSe@ZnS QD coated with PEG (charge of -33 mV, pH 7.4) did not label mammalian cells (CHO model), while spermine and arginine functionalized QD-labeled (charge of -16 mV and -9 mV, respectively, at pH 7.4) CHO cells, indicating a preferential cell interaction with protonated NPs.

Hybrid NPs having magnetic properties (e.g., Fe_2O_3-based nanostructures) are being widely studied as they can act as drug delivery systems,

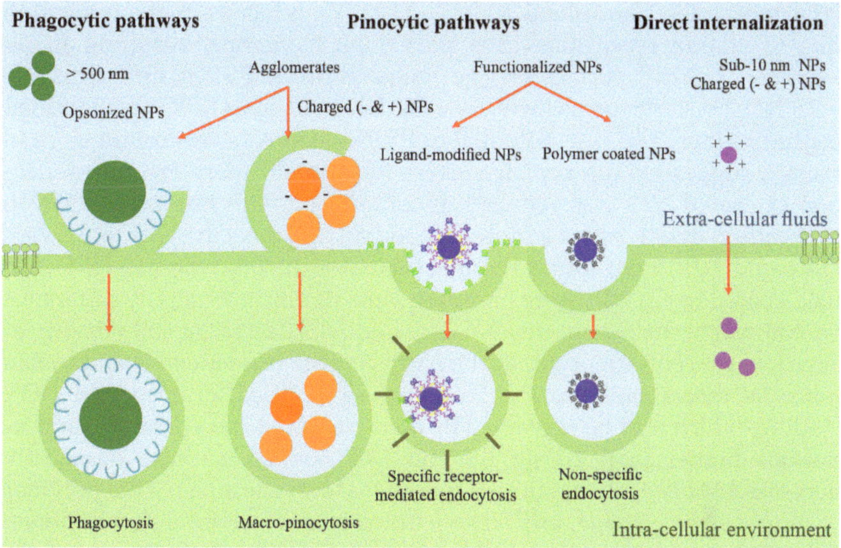

FIG. 5 Overview on the dependency of NPs cellular uptake as a function of their size and surface characteristics.

magnetic resonance imaging (MRI) contrast agents, or thermal seeds, for improving magnetic hyperthermia-based therapies.

For instance, Bekaroğlu et al. [88] performed a comparative study using ionic biopolymers to cover the surface of SPIONs developed for magnetic-assisted targeted drug delivery. The capping with ionic biocompatible polymers allows enhancing NPs' colloidal stability, drug-loading capacity, and cell interaction (adhesion, internalization, and toxicity). Fig. 5 shows an overview on NPs cellular uptake pathways as a function of their size and the overall coating characteristic being discussed in this section. The hydroxyl ethylene cellulose (HEC, cationic), nano-crystalline cellulose.

(NCC, anionic), and polyvinyl pyrrolidone (PVP, neutral) were used to modify the surface charge of SPIONs and to evaluate their biological impact in healthy and cancer cells.

Nanoparticles were coated following a simple adsorption method mediated by ultrasonication in distilled water. DOX was loaded onto the polymer-coated SPION samples using potassium phosphate buffer solutions at pH5. The cytotoxicity trigged by NPs was then evaluated using human fetal osteoblast (hFOB) and MCF-7 cells. Cell viability was determined by MTS assay. In brief, uncoated Fe_3O_4 showed a zeta potential of −18.6 mV and were found as microscopic aggregates by DLS. The cationic HEC coating was achieved by electrostatic interactions. The zeta potential changed to +2 mV, which indicates the coverage of SPIONs with

HEC molecules. The anionic NCC (−39.1 mV) induced either aggregation due to charge repulsions or a partial NCC coating, resulting in an average − 14.5 mV surface charge value. Neutral PVP mainly interacts with NPs by steric forces (Mw_{PVP} = 360,000) and coats SPIONs at defined concentrations. The zeta potential of PVP-coated NPs was found close to −1 mV, suggesting a complete surface functionalization. The magnetization of coated NPs was reduced in comparison with uncoated SPIONs. In healthy hFOB cells, a dose-dependent cytotoxicity was observed. Uncoated NPs were found the most lethal to hFOB, whereas HEC and PVP-coated Fe_3O_4 NPs were harmless (non-significant viability difference vs. control) up to 250 μg/mL. NCC-coated NPs decreased cell viability at even low concentrations due to the partial coverage assumed from the zeta potential measurements. The PVP coating induced an enhanced DOX loading, which may be explained by an increased surface area. At concentrations higher than 200 μg/mL, all DOX-loaded samples drastically decreased MCF-7 cell viability as free DOX does. The effect of the polymer coating, surface charge, and drug loading was observed at lower doses. The HEC-coated SPIONs were found more effective in killing cancer cells, which may be explained by a preferential cellular uptake due to their cationic coverage [88].

Moreover, it has been shown that chitosan (Ch) efficiently coordinates iron, zinc, and other metallic ions, making Ch an excellent candidate to functionalize inorganic NPs [89]. Hobson et al. [90] developed clustering (i.e., aggregated) sub-10 nm SPIONs coated with a Ch-based polymer and determined their suitability for MRI contrast imaging. The authors proposed a simple method to prepare clustering SPIONs in the absence of organic solvents, hydrophobic surfactants, and high temperatures. This method involved a mild alkaline (NH_4OH) nanoprecipitation step under inert atmospheric conditions (N_2 gas) followed by an ultra-sonication NPs clustering performed at room temperature. An amphiphilic Ch-based molecule (GCPQ, 0.1–5 mg/mL) that consists of a glucosamine sugar backbone with amine groups replaced by either long-chain hydrophobic pendant group or a quaternary ammonium cation, was used. SPIONs after GCPQ sonication-mediated functionalization resulted in stable colloidal formulations containing NPs with an average hydrodynamic diameter of about 152 nm (PdI = 0.22) at a 1:0.04 SPION:GCPQ ratio. These formulations contained close to 95% *w*/w of encapsulated SPIONs. The surface charge changed from −4.3 to +22.5 mV at neutral pH when coated with the cationic Ch-based polymer as GCPQ has a pKa of ca. 5.99. The Ch-clustering SPIONs material displayed attractive T2-weighted contrast properties, having a r_2/r_1 ratio of 21.1.

Lachowicz et al. [91] coated Zn-doped SPIONs (ZNF) with a charged polysaccharide and evaluated the influence of the dopant content and coating shell to their magnetic hyperthermia activity in vitro. In brief, NPs were synthesized following an ultrasound-assisted co-precipitation method. Aqueous solutions containing the Fe^{2+}, Fe^{3+}, and Zn^{2+} precursors, in the presence or not of the cationic derivate chitosan (CCh(+), 50–190 kDa, 3 g/L), were purged with argon and sonicated by adding ammonium hydroxide. Complementary spectroscopy and microscopy techniques were used to study the physicochemical properties of the obtained uncoated (hydrophobic) and in situ CCh(+)-coated (hydrophilic) ZNF NPs. The NPs' cytotoxicity was evaluated in murine neuroblastoma cells by performing the MTT tetrazolium salt-based assay. The optimized Zn^{2+} and CCh(+) formulation displayed an average hydrodynamic diameter and a zeta potential of about 198 nm (PdI = 0.191) and +40 mV, respectively, in water at neutral pH. The hydrodynamic size distribution changed as a function of the Zn content, with an observed bimodal distribution at higher Zn content. Hence, the CCh(+)-coated ZNF NPs were found as hydrophilic stable cationic aggregates containing 10-nm inorganic cores. The hyperthermia effect, characterized by an increase of ca. 6°C from the physiological temperature, was more pronounced for Zn-doped SPIONs (0.55 mg Fe/mL). The CCh(+)-coated ZNF NPs did not induce significant cytotoxicity, in terms of decreasing the metabolic activity of murine neuroblastoma cells (from a brain tumor), up to 0.54 mg Fe/mL. The magnetic hyperthermia activation showed very encouraging results killing cancer cells with a cell viability decrease of about 70%. Hence, positively charged chitosan renders ZNF NPs biocompatible, hydrophilic, and highly active for magnetic hyperthermia cell killing.

On the other hand, Somvanshi et al. [92,93] modified the surface of nano-spinel zinc and magnesium ferrite (MGF) NPs, synthesized by sol–gel auto-ignition in the presence of citric acid using OA, the most common monounsaturated fatty acid in nature (18:1 n − 9). The OA molecule is capable of coating inorganic NPs in the perspective of producing biocompatible nano-therapeutic agents. The coating process was performed in liquid ammonia and assisted by ultrasonication. The spinel nanostructured ZNF and MGF NPs, before and after surface modification, were characterized by standard and specialized techniques. Briefly, the FTIR spectra of OA-capped NPs confirmed the presence of carboxylic groups and thus, the OA shell formation over the ZNF and MGF cores. Moreover, their hydrophobic/hydrophobicity character was evaluated by angle contact measurements, which correlate the interfacial energy of the examined media (a fluid and a surface). Even if OA has a strong non-polar character with negligible water solubility, its use as biocompatible capping agent enhanced the wettability of the ZNF and MGF inorganic cores. The passage from hydrophobic to hydrophilic was characterized by a notable decrease in the NPs contact angle, from 97.4 to 47.3° and from 99.7 to 47.9°, respectively, for uncoated and OA-coated ZNF and MGF NPs. The capping agent did not interfere with the superparamagnetic core properties. The isoelectric point of the modified ferrite NPs was found at neutral pH, which indicated that at physiological conditions, the NPs tend to form aggregates that may influence their NP-cell interaction and uptake. Their hydrodynamic size is very similar to the core diameter or even smaller, suggesting an OA monolayer conformation and steric stabilization [92,93].

2.3 Ligands and other molecules to improve nanoparticles' biocompatibility in living organisms

2.3.1 *Polymers to increase biocompatibility of metallic and metallic oxide nanoparticles*

Polymers are large and complex molecules, which can be made up of the same repeating units, or they can contain two or more different building blocks. The former species are called homopolymers, while the last is defined as copolymers [94]. They represent the largest and most versatile class of biomaterials and are used in a wide variety of biomedical applications; this is attributed to the easiest they can be designed and synthesized with a wide variety of structures and suitable physical, chemical, and surface properties [95].

Both synthetic and natural polymers are widely used to functionalize metal and metal oxide NPs, thereby helping to overcome many of the deficiencies that these NPs can present without compromising their original properties, since the proper choice of polymer improves the properties of the NPs [96].

This functionalization has offered control over the size of the NPs, with improved colloidal stability, and the possibility to adjust the solubility to aqueous and organic media and also change their optical, magnetic, and fluorescent properties [97].

For example, Abedin et al. [98] functionalized gold-ferric oxide superparamagnetic NPs (Au-Fe_3O_4_NPs) with poly-L-lysine (PLL) polymer (Au-Fe_3O_4_NPs-PLL) to improve stability. In addition to this, PLL has a positive charge that could favor the internalization of NPs to cells in biomedical applications. The functionalization of Au-Fe_3O_4 NPs with PLL was carried out thanks to Coulomb interactions between the anionic surface of Au and the aliphatic monoamines of PLL. The results showed that functionalization helps improve stability by reducing agglomeration of these NPs in aqueous solutions such as water, phosphate-buffered saline (PBS), PBS with fetal bovine serum, and cell culture medium. The Au-Fe_3O_4-NPs-PLL did not show significant cytotoxicity for breast cancer cells (BT-474 and MDA-MB-231); however, after irradiating with a laser at 808 nm (1 Wcm^{-2}), the growth cell was inhibited by up to 60%. The cellular uptake of Au-Fe_3O_4-NPs-PLL was observed by bright-field microscopy, fluorescence microscopy, and TEM; it showed a significant uptake of Au-Fe_3O_4-NPs-PLL by cells.

NPs designed for biomedical applications are also frequently coated with polymers that contain reactive functional groups; to incorporate drugs or ligands that direct them to their target of interest, the polymers used in these applications must be biocompatible and biodegradable [99]. Also, polymers such as polyethylene glycol (PEG), polyethyleneimine (PEI), and poly (D, L-lactic-co-glycolic acid) (PLGA) have been widely used as biodegradable carriers for drug delivery; sometimes more than one polymer is used in the synthesis of NPs to potentiate the benefits they provide. For example, Encabo-Berzosa et al. [100] synthesized AuNPs using PEI as a reducing agent, stabilizer, and surface functional agent. Furthermore, the NPs were coated with PEG-SH to reduce the cytotoxicity of PEI. Using FTIR, functionalization with PEG was corroborated through two types of interactions: a partially covalent bond of the S—Au type of approximately 35% and another type of electrostatic interaction that is equivalent to 65% of all interactions. These NPs were used as non-viral vectors in the transfection of different cell lines and were compared with commercial lipoplexes (commercial transfection reagent). The results show that PEI helps to protect the transported plasmid from enzymatic degradation. PEGylation of these NPs by covalent coupling reduces the cytotoxicity of PEI and also improves the dispersion of the NPs in culture media. Besides, expression levels equal to or greater than those obtained by lipoplexes were obtained, which makes them potential non-viral vectors to be used as efficient transfection reagents.

On the other hand, the design of NPs functionalized with copolymers and responsive to certain stimuli (temperature, light, and pH) has attracted increasing attention [101], for example, to achieve a controlled release through changes in the pH of DOX and in this way reducing the toxicity of DOX in non-target tissue due to its unwanted interactions and its rapid absorption. Also, in another study by Ebrahimi et al. [102], DOX has been encapsulated in nanocarriers formed by functionalized MNPs by a double emulsion method (water/oil/water) with a PLGA-PEG triblock copolymer (Fe_3O_4-PLGA-PEG). Their results confirm the correct encapsulation of DOX, along with a drug release study, the sensitivity of NPs to changes in the pH was demonstrated, which triggers a controlled release of DOX. Therefore, these NPs could be suitable candidates for drug administration, although they have yet to be evaluated in in vivo assays [102].

2.3.2 Incorporation of polymers (PEG, PAH, etc.) to increase residence time in blood for organic nanoparticles (e.g., lipidic nanoparticles)

Many novel NPs of different chemical compositions and physical properties are currently being developed; however, more chemical modifications are often required for these structures to be compatible with the biological environment [103].

Lipid-based drug delivery systems have been widely used in oncology because they can modify the pharmacokinetics and biodistribution of encapsulated drugs, improving therapeutic efficacy, decreasing drug toxicity, and increasing accumulation in tumor tissues [104,105]. As a result, different types of nanocarriers have been developed such as liposomes, microemulsions, solid lipid NPs, and nanostructured lipid transporters [106]. However, these nanostructures are prone to a rapid elimination from the bloodstream, which limited their therapeutic efficacy, due to the opsonization of plasma components and uptake by macrophages of the reticuloendothelial system (RES) [107].

Many of the proposals to combine liposomes with NPs remain in the preclinical development stages with promising opportunities for designing nanoscale stages with promising opportunities for designing nanoscale theranostic delivery systems. Such NPs have shown promise as diagnostic and therapeutic tools and, in general, offer a large surface area that allows for covalent and non-covalent surface functionalization with hydrophilic polymers, therapeutic moieties, and targeted ligands [103].

Hybrid polymer-lipid formulations that combine the solubilizing ability of lipids with the stabilizing matrix of polymeric systems present a synergistic approach to oral drug delivery with proven potential to overcome the limitations associated with precursor systems [108]. Furthermore, functionalization of liposomes with polymers, or vice versa, is a key strategy to improve circulation times and prevent elimination by RES through steric stabilization [107]. For example, PEG-modified liposomal DOX (Doxil) was the first liposome with anticancer effects to be approved by the US Food and Drug Administration [109].

New nanoformulations have recently been designed that take advantage of the characteristics of both compounds; for example, lipid NPs composed of L-α-phosphatidylcholine (PC) and cholesterol (CHOL), functionalized with Ch (PC-CHOL-NPs-Ch), were loaded with indomethacin, an anti-inflammatory drug used to treat rheumatoid arthritis, which is known to cause certain unwanted effects in the gastrointestinal tract. Thus, these NPs were designed with the intention of reducing these side effects. Functionalization was carried out using simil-microfluidic and drop-wise methods. For both methods, the PC-CHOL-NPs were previously synthesized and the drug was encapsulated. Subsequently, Ch dissolved in an acetic acid solution was added to the PC-CHOL-NPs. By ATR-FTIR, it was shown that the functionalization is carried out through an electrostatic interaction between the phosphate groups of liposomes and the amino groups of chitosan. The results showed stable nanosystems highly charged with indomethacin thanks to the positive charge of Ch; in addition, in vitro tests on gastric and intestinal fluids showed that Ch-coated nanostructures can be used as oral administration systems for the controlled release of indomethacin [110].

On the other hand, Abdou et al. [111] reported the synthesis of hybrid NPs using the poly (D, L-lactide-*co*-glycolide) polymer and the lipid DSPE-PEG-2000 (LPHNPs), with the intention of loading them with β-sitosterol (BSS, a phytosterol with hepatoprotective effect). BSS-LPHNPs were prepared using a single-step nanoprecipitation method. NPs were evaluated in rats previously induced hepatotoxicity by carbon tetrachloride (CCl_4). BSS-LPHNPs exhibited good stability and controlled drug release due to the slow degradation of the polymeric matrix coating the drug. BSS-LPHNPs possess the ability to restore liver enzymes alanine aminotransferase and aspartate aminotransferase, liver lipid peroxidation markers, and total bilirubin and albumin to normal levels. These results make the nanosystem a promising delivery system for hepatoprotective drugs.

Lipid-coated polymeric NPs have also been developed; e.g., Wang et al. [112] synthesized biodegradable polymer nanoparticles (poly (lactic-*co*-glycolic acid) (PLGA)) coated with lipid-polyethylene glycol (PEG2k-DSPE) and loaded with DOX and CUR, with the intention of using them in the treatment of osteosarcoma and thus reduce the side effects caused by drugs. The nanocomposites showed good stability, even in the presence of fetal bovine serum, and in vitro drug release profiles show that drug-loaded NPs have the ability to release DOX and CUR at a sustained rate for 48h. In addition, cytotoxicity tests were performed in osteosarcoma cancer cells (KHOS cells) and in cancer-bearing mice. The cytotoxicity of the drugs was higher when soluble drugs were delivered through the NPs. These results demonstrate that these hybrid NPs are excellent candidates for use in drug delivery.

3 Bioconjugation for single and dual targeting in biomedical applications

Non-specific distribution and uncontrollable release of drugs in conventional approaches have led to the development of intelligent nanocarrier-based systems, which may improve the delivery of drugs to target sites. Exploration of multifunctional nanomaterials from biocompatible components, with integrated imaging and targeted combined therapeutic modalities of tumors in vivo, offers great prospects for clinical theranostic applications mainly for the treatment of cancer [113].

NPs targeting can be carried out passively or actively. The passive targeting allows an efficient accumulation of NPs within the tumor microenvironment and through the enhanced permeability and retention effect (EPR). For this type of targeting, macromolecules ($\geq$ 40kDa), such as biocompatible polymers and proteins, are widely used to functionalize the NPs surface. For instance, Unnikrishnan et al. [114] report the design of

NPs formed by a polysaccharide from seed kernel of Tamarindus (PST001), BSA protein, and gold NPs (PST-BSA-AuNPs). This complex was loaded with DOX with the aim of being used in cancer therapy. The first step was the synthesis of PST-BSA that was prepared by Maillard method [115], which is carried out in distilled water and continuous stirring. The synthesis of PST-BSA was achieved through a possible covalent union of the aldehyde group with the amino group of BSA; then PST-BSA was used as a reducing agent for gold chloride in the synthesis of AuNPs, and functionalization with PST-BSA helped stabilize AuNPs. Finally, DOX was encapsulated within the purified PST-BSA-AuNPs by a reaction of PST-BSA-AuNPs with DOX hydrochloride that allows the BSA gelation, which in turn promotes the encapsulation of drugs into the PST-BSA AuNPs; PST001 acts as a doorkeeper to preserve DOX in the BSA matrix and prevent unpredictable doze from the nanosystem; the controlled release of the drug is carried out through changes in the pH of the medium. In vitro and in vivo tests were carried out using two mammalian cancer cell lines: murine melanoma (B16) and A549 cells to evaluate its performance as alternative to chemotherapy via drug delivery through passive targeting [114].

On the other hand, active targeting facilitates the selective uptake of NPs by conjugation of the NPs surface with specific target ligands and/or receptors [116,117], causing them to internalize in cancer cells through receptor-mediated endocytosis processes, such as phagocytosis, macropinocytosis, clathrin-dependent endocytosis, caveolin-dependent endocytosis, and clathrin- and caveolin-independent endocytosis [118–120] (Fig. 4). Moreover, targeting the NPs to overexpressed cancer cell receptors represents a strategy for a better therapeutic result, with improved efficacy and reduced toxicity [121]. Conjugation of NPs with biomolecules such as antibodies, peptides, vitamins, and aptamers can solve current challenges, including low accumulation, internalization, and retention at the target site in cancer diagnosis and therapy through active targeting [122]. Table 2 shows some of the main characteristics of these biomolecules such as their size, stability, costs, and production time.

In the following sections, we will focus on presenting some recent advances in the bioconjugation of NPs and how the previous functionalization of these NPs facilitates the binding of biomolecules.

3.1 Antibodies

Antibodies and fragments of these have been commercially used in immunotherapy for several years. The design of biofunctional nanomaterials considers antibodies as a source of targeted ligands to bind specific receptors overexpressed in cancer cells [123,124]. In this respect, as

TABLE 2 Main characteristics of the biomolecules used for the conjugation of NPs.

	Antibodies	Peptides	Aptamers
Target	Immunogenetic	Any	Any
Stability	Sensitive to temperature	Aggregation, sensitive to temperature and pH	Stable
Affinity	High	High	High
Denaturation	Irreversible	Irreversible	Reversible
Production process	In vivo, require immune response	In vivo and in vitro	In vitro, reproducible
Time and cost of production	Months, expensive	Months, expensive	Weeks, cost-effective

mentioned before, NPs require modifications on their surface to allow covalent binding of antibodies to NPs. Interactions can be carried out through amide bonds ($-NH_2$), dopamine-mediated bonding, maleimide groups, disulfide bonds (-SH), and the use of an alkyl hydrazide linker [125].

For example, anti-HER2 monoclonal antibody is one of the most widely used to functionalize NPs for active targeting, because HER2 is overexpressed in various breast cancer cell lines. In the search for dual-modal ultrasound/magnetic resonance imaging and photothermal therapy of breast cancer, Dong et al. [126] used HER2 to bioconjugate a nanocomplex made up of gold-nanoshells poly (lactic-*co*-glycolic acid) magnetic hybrid NPs (HER2-GPH-NPs). PLGA NPs were synthesized and co-loaded with perfluorooctyl bromide (PFOB, ultrasound contrast agent) and SPIOs for MRI, using an adapted oil/water emulsion solvent evaporation method, and then, PLGA NPs were encapsulated with gold nanoshells (AuNPs-PLGA NPs). After these steps, AuNPs-PLGA NPs were functionalized with SH-PEG-COOH with the intention of facilitating the subsequent binding of anti-HER2 antibody, using a classical carbodimide technology to activate the carboxylic acid groups of AuNPs-PLGA NPs PEGylated and promote the covalent bonding of amino groups to antibodies. In addition, cell-targeting studies demonstrated a receptor-mediated specific binding of the HER2-GPH-NPs to HER2-positive human breast carcinoma SKBR3 cell line (SKBR3 cells), and its binding rate was significantly higher than that of human breast cancer MDA-MB-231 cell line (MDA-MB-231, HER2-negative cells). HER-2 conjugation showed a substantial improvement on the images obtained through ultrasound/magnetic resonance, and cytotoxicity assays induced by temperature increases confirmed that the nanosystem could potentially be useful for the thermal ablation of tumors, making it a promising theranostic agent.

Aires et al. [127] used anti-CD44 antibody to functionalize MNPs loaded with gemcitabine (GEM) as a potential chemotherapeutic agent. CD44 antibody is overexpressed in several cancer cells. Thus, this nanosystem was aimed to target CD44-positive cancer cells like human pancreatic epithelioid carcinoma (PANC-1) and MDA-MB-231. For this, a modification was made to the anti-CD44 antibody by the reaction between Traut's reagent and the amine groups of the antibodies, with the final result of introducing free thiol groups onto the antibodies. MNPs were previously coated with dimercaptosuccinic acid (DMSA) and functionalized by introducing free thiols on their surface; the functionalization was achieved by the reaction of cysteamine with the carboxylic groups of the DMSA coating in the presence of 1-ethyl-3-(3-dimethylaminopropyl) carbodiimide (EDC) and n-hydroxysuccinimide (NHS). Disulfide bond-based linkers have been employed for the conjugation of MNPs between the reactive thiol of the modified anti-CD44 antibodies and the activated sulfhydryl groups of the MNPs. The authors performed in vitro studies using a CD44 negative non-tumorigenic cell line as a control (Human mammary epithelial cell line, MCF-10A). The results show that the nanosystem specifically binds to the CD44 surface cell receptor, and greater efficacy of the drug was demonstrated when it bounds to the anti-CD44-functionalized nanoparticle compared to the free drug and the NPs without the antibody. The results demonstrate the potential of anti-CD44-MNPs for use in targeted therapy.

On the other hand, AuNPs conjugated with DOX and the anti-PD-L1 antibody (PD-L1-AuNP-DOX) have been explored to enhance the chemo-photothermal therapy of colorectal cancer (CRC), whose poor prognosis is significantly correlated with PD-L1 antibody overexpression. DOX and anti-PD-L1 antibody were attached to the α-terminal end group of polyethylene glycol N-hydroxy succinimide (LA-PEG-NHS), to later attach them covalently to the surface of AuNPs by dithiol-Au covalent bonds. Furthermore, PD-L1-AuNP-DOX were coated with PEG-SH chains to improve their stability. The dual treatment using PD-L1-AuNP-DOX and NIR irradiation (808 nm, 2.5 Wcm^{-2}) significantly suppressed in vitro proliferation of murine colorectal cancer cells (CT-26 cells), demonstrating its potential application in the treatment of localized colorectal cancer [128].

For prostate cancer, the most used tumor marker for its diagnosis is the prostate-specific antigen (PSA), so that most of the studies focus on the development of PSA-conjugated NPs. For example, Du et al. [129] recently designed a manganese oxide–mesoporous silica NPs (Mn-MSNS), conjugated with PSA antibody and Cy7 (near-infrared fluorophore) (PSA-Mn-MSNS-Cy7), to use it in the early detection of prostate cancer. To promote the conjugation of PSA and Cy7, Mn-MSNS were functionalized with DSPE-PEG2000-COOH to produce PEG-Mn-MSNS. Then, this complex was added with 2-(N-morpholino) ethanesulfonic acid buffer, EDC and

NHS solutions. The conjugation of anti-PSA antibody, Cy7, and PEG-Mn-MSNS was performed through an esterification reaction between carboxyl group and an amino group. The efficacy of PSA-Mn-MSNS-Cy7 was evaluated in human prostate adenocarcinoma cells (LNCap), and a prostate cancer mouse model, using human skin fibroblasts cells (HSF cells) as a control. Therefore, optical and MR imaging was evaluated and the results showed that PSA-Mn-MSNS-Cy7 achieved a selective accumulation in cancer cells and cancer tissues in mice. The authors suggested that these results demonstrated the potential of PSA-Mn-MSNS-Cy7 to be used in the diagnosis of prostate cancer and drug delivery applications.

3.2 Peptides

Peptides are excellent candidates for NPs targeting cancer cells or for crossing the blood–brain barrier (BBB), a physiological barrier that avoids the entrance of xenobiotics to the central nervous system. Thus, NPs conjugated with peptides are emerging as potential candidates in theranostic applications [130]. In this sense, tumor homing peptides (THP) are short peptides, which specifically recognize and bind to tumor cells or tumor vasculature. Table 3 shows four examples of the most used THP to conjugate NPs.

For instance, RGD-peptide is a THP that has been widely used to conjugate NPs. Thus, MNPs conjugated with RGD peptide can be directed to cancer cells that express αvβ3 receptor, and subsequently, perform hyperthermia assays. RGD peptides were bound to MNPs by click chemistry. MNPs were functionalized with poly(maleic anhydride-*alt*-1-octadecene (PMAO) to increasing aqueous stability, the surface of the polymer was

TABLE 3 List of the most used peptides for bioconjugation of NPs.

Tumor homing peptide	Type of conjugation	Site of union	Ref
RGD (CDCRGDCFC)	-NH_2, covalent	Various tumor types, αvβ3 and αvβ5 integrins	[131]
NGR (CNGRCVSGCAGRC)	-SH, covalent	Various tumor types, CD13 (a membrane protease expressed on the endothelial cells of tumor blood vessels) and αvβ3 integrins	[132]
CGKRK	MAL-thiol, covalent	Heparan sulfate located near the surface cell and overproduced in glioma cells	[133]
CREKA	MAL-thiol, covalent	Fibrin in microthrombus and stroma of tumors	[134]

hydrolyzed, and free -COOH groups were obtained. To add an alkyne group to the MNPs, the -COOH groups of PMAO were modified with an amino linker (10-(2-cyclooctyn-1-oxy)-3-aza-5,8-dioxa-4-oxodecyl-1-amine) (cyOct). Finally, RGD was modified with azide (RGD—N_3), and in this way, its conjugation with functionalized MNPs was facilitated. MNPs were administered intravascularly in rats with colorectal tumors (WAG/RijCrl). Then, the rats were exposed to a magnetic field to induce hyperthermia, resulting in a significant reduction in tumor viability without inducing a significant damage to healthy liver tissue [135].

Another example is glutathione, a small peptide of glutamate, cysteine, and glycine (GSH), which together with lipoic acid-CUR (LA-CUR) were used to conjugate Fe_3O_4@AuNPs through the ligand exchange technique by an Au—S interaction. In the case of GSH, the conjugation with Fe_3O_4@AuNPs was done through ligand exchange of thiol group of GSH with citrate groups of AuNPs; for the conjugation with LA-CUR, the reaction was by ligand exchange of disulfide group of LA with citrate groups of AuNPs. In this work, the GSH conjugation resulted in a 2-fold increase in cell uptake in GSH receptor-positive astrocyte cells (U87-MG cells), which could indicate the ability of Fe_3O_4@AuNPs-LA-CUR/GSH to cross BBB. The authors suggest that Fe_3O_4@AuNPs-LA-CUR/GSH seems to be a promising candidate for the diagnosis and treatment of brain cancer since treatments resulted in higher cytotoxicity against cancerous U87MG cells than against healthy fetal human astrocyte cells [136].

Kalmodia et al. [137] designed a nanosystem of AuNPs coated with the peptide PHCKRM formed by proline-histidine-cysteine-lysine-arginine-methionine (PepA). AuNPs-PepA were designed to attack reactive oxygen species (ROS) in a retinoblastoma cancer model (Y79 cells) by using PepA that was previously modified with thioctic acid (TA). The main advantages of using TA are its radical scavenging capacity, metal chelation, and its ability to enhance the antioxidant effect. Furthermore, TA has antioxidant linker molecules that contain sulfur, so in the presence of AuNPs, the modified PepA binds to NPs through an Au—S bond (AuNPs-PepA). The results show that coating the AuNPs with PepA reduces the toxicity of the metal ions and in turn, the bioconjugate provides stability to PepA. Also, AuNPs-PepA was observed to target ROS in Y79 cells more effectively and reduce ROS by up to 70% compared to untreated cells.

3.3 Vitamins

Cancer cells require a high concentration of vitamins to maintain their uncontrollable proliferation. Therefore, the receptors involved in the absorption of vitamins are overexpressed [138]. Therefore, various vitamins have served as biomarkers to develop selective nano-agents for diagnosis and cancer treatment [139]. For that reason, they have also been explored to functionalize NPs. For example, folate receptors are overexpressed on the surface of many types of tumors such as breast, prostate, and cervical cancer. This feature could be used to target therapeutic compounds directly to cancerous tissues [140]. In this context, the use of manganese-doped carbon QD (Mn-CQD) with dual conjugation using chlorine e6 (C6, a photodynamic therapy agent) and FA was recently reported. Waste green tea powder was used for the synthesis of Mn-CQD following a one-pot hydrothermal method. Surface Mn-CQD contains abundant carboxyl groups, which promote its solubility in water and also allow direct conjugation of C6 and FA through binding to amine. To activate the carboxyl groups of the CQDs and achieve amine-terminated Mn-CQDs, a reaction was performed between EDC, NHS, and ethylenediamine (EDA) in the presence of the Mn-CQDs. The amine-terminated Mn-CQDs were then conjugated with the activated carboxyl group on the FA and subsequently to the C6 via amide bonds to finally obtain Mn-CQD-FA/C6. The results obtained in HeLa cells show a specific targeting of the Mn-CQD-FA/C6 to HeLa cells; this is attributed to the binding of the NPs to the overexpressed folate receptor on their surface. Additionally, biocompatibility analyses were performed on HeLa, B16, and human liver carcinoma cells (HepG2), which indicated good biocompatibility and high cellular absorption of the nanocomplex, in addition to excellent cytotoxicity after irradiation (671 nm, 1 Wcm^{-2}) for 5 min [141].

Another molecule that has been used for the conjugation of NPs as a targeting agent is pyridoxal 5′-phosphate (PLP) that is the most important cofactor of vitamin B6-dependent enzymes that is involved in the process of metabolism that could be exploited to specifically target highly metabolic cells, such as metastatic tumor cells; Bonvin et al. [142] suggest that this cofactor could be used to specifically target metastatic tumor cells. The authors coated SPIONs with PLP (SPIONs-PLP) and evaluated their biological effect into LnCaP and PC3. Conjugation of SPIONs with PLP was carried out by a one-step coating reaction, through electrostatic interactions. SPIONs-PLP showed not toxicity effect to any of the cell lines. However, using TEM, it was observed that PLP increased the number of NPs absorbed by the cells, suggesting a high potential for PLP as a targeting agent for biomedical applications [142].

3.4 Aptamers

Aptamers are now emerging as promising molecules to attack cancer-specific epitopes in clinical diagnosis and therapy. Furthermore, due to their high affinity and specificity and low toxicity, aptamers could be considered as the compounds of choice for in vivo cell recognition [143,144]. For example, Table 4 shows a small variety of aptamers used in different

TABLE 4 Some examples of reports using aptamers as targeting agents.

Nanoparticle	Multifunctional ligand/active molecule	Aptamer	Application	In vitro model	Ref
Poly(lactide-*co*-glycolic acid) (PLGA) nanobubbles	Paclitaxel	A10–3.2	Imaging, treatment (MRI)	Prostate cancer (LNCaP and PC3)	[145]
Gold nanoparticles	Chlorine6 Banoxantrone (AQ4N)	TLS11a	Multi-modal therapy: Controlled released, imaging, photodynamic /photothermal therapy	Carcinoma Hepatocellular (HepG2)	[146]
BSA nanoparticles	Indocyanine green KALA (cell-penetrating peptide) Doxorubicin Indocyanine green	AS1411	Multi-modal therapy: tumor imaging, photodynamic / photothermal therapy and chemotherapy	MCF-7 cells	[147]
Lipid-polymer hybrid nanoparticles	Doxorubicin	A6	Treatment, delivery of doxorubicin	Cells positive to HER2, Breast cancer (SKBR-3 and 4T1-R)	[148,149]
Fluorescent gold nanoclusters/ superparamagnetic nanoparticles	–	KH1C12	Imaging (MRI)	Leukemia Cells (HL-60)	[150]
Amino- and carboxyl-modified silica-coated terbium (III) thiacalix[4]arenesulfonate luminescent nanoparticles	Terbium (III) complexes with thiacalix[4] arenesulfonate	Sgc8	Imaging, luminescent detection (fluorescence microscopy)	Leukemia Cells (CCL-119, CCL-86 and TIB-152)	[151]

in vitro models as targeting agents, which have also shown great affinity for these cells.

In this context, Zhao et al. [152] propose a NIR photothermal response system directed by binding aptamers (AS1411 and S2.2) on thermosensitive liposomes (TSL) as targeting agents to MCF-7 cells that also contain docetaxel (DTX) capable of being released by increasing temperature in the medium. Nanohybrid consists of TSL loaded with DTX and ammonium bicarbonate (ABC, NH_4HCO_3), coated with gold nanoshells (TSL-AuNPs-DTX/ABC). ABC was used mainly to add intravenous stability. To add the property of directionality, TSL-AuNPs-DTX/ABC were bioconjugated with aptamers previously modified with sulfhydryl groups to facilitate the union with the nanohybrid through an Au—S union. Using fluorescence microscopy, it was shown that functionalization with both ligands significantly increased cell uptake; besides the nanosystem, it has excellent biocompatibility. After irradiation (808 nm, 1 Wcm^{-2}) of cells previously incubated with TSL-AuNPs-DTX/ABC, the DTX was released in a controlled manner and inhibited tumor growth. The authors propose TSL-AuNPs-DTX/ABC as a promising theranostic agent since ultrasound images at the tumor site were achieved.

4 Conclusions and perspectives

Under the principle of «*structure determines function*», enormous efforts have been made to develop functionalized NPs for a wide variety of applications. For instance, the controlled modification of the NMs surface quality enhances their interaction with biological systems on the way to achieve powerful tools for alternative medical and biomedical solutions.

Simply, nanostructured materials (NMs) functionalization may start by modifying their polarity and solubility in biomimetic fluids in order to improve parameters such as colloidal and chemical stability, and biocompatibility, all these in the perspective of creating advanced NMs to be used as imaging contrast agents, drug delivery carriers, or multi-modal nanoagents.

As far as it can be verified in current research, the privileged functionalization methodology consists of post-synthesis coverage or grafting chemical routes, which use anchoring ligands to assure the activation of the surface. These methods occur generally in multi-steps, requiring time and energy consumption to separate and purify the final product. Hence, the remaining challenge is to increase the design and applicability of one-pot synthesis pathways where in situ biocompatible functionalization may be the key selection route parameter.

As perspectives, we can extend here some recommendations regarding wet chemical routes to obtain functionalized inorganic and hybrid NMs for biomedical applications, such as i) the precursors should be low cost salts or complexes with particular safely handing that may form NMs with unique physicochemical characteristics with their own useful ligands for an in situ stable coating, ii) the redox-active molecule should be a natural non-toxic compound soluble in aqueous media, iii) the surfactant or stabilization compound should be carefully selected as a multi-performer capable to control the NMs synthesis steps (i.e., nucleation, coalesce, and growth) at the time it confers biorelevant characteristics (e.g., anchoring groups), and iv) the solvent or reaction medium should be biocompatible and be used at physiological pH. At the same time, soft reaction conditions such as room temperature and atmospheric pressure may be preferentially applied.

On the other hand, the conversion yield may be enhanced using external controlled stimuli (e.g., electromagnetic or mechanical forces). Depending on the NMs' chemical composition and physical state, simple phase separation techniques may be applied without compromising the final product amount, in the understanding that the designed protocol does not require additional purification steps. Hence, by obtaining in situ functionalized and well-activated NMs decreases the operational initial cost for developing potential nano-agents for biomedical applications and allows performing a direct and single bioconjugation gaining specificity for the nanotechnology-based medical tool.

References

[1] V.J. Mohanraj, Y. Chen, Nanoparticles - a review, Trop. J. Pharm. Res. 5 (1) (2007) 561–573, https://doi.org/10.4314/tjpr.v5i1.14634.

[2] I. Khan, K. Saeed, I. Khan, Nanoparticles: properties, applications and toxicities, Arab. J. Chem. 12 (7) (2019) 908–931, https://doi.org/10.1016/j.arabjc.2017.05.011.

[3] J. Jeevanandam, A. Barhoum, Y.S. Chan, A. Dufresne, M.K. Danquah, Review on nanoparticles and nanostructured materials: history, sources, toxicity and regulations, Beilstein J. Nanotechnol. 9 (abril) (2018) 1050–1074, https://doi.org/10.3762/bjnano.9.98.

[4] V.J. Lingayat, N.S. Zarekar, R.S. Shendge, Solid lipid nanoparticles: a review, Nanosci. Nanotechnol. Res. 4 (2) (2017) 67–72, https://doi.org/10.12691/nnr-4-2-5.

[5] C.N.R. Rao, P.J. Thomas, G.U. Kulkarni, Nanocrystals: Synthesis, Properties and Applications. Nanocrystals: Synthesis, Properties and Applications, Springer Berlin Heidelberg, Berlin, Heidelberg, 2007, https://doi.org/10.1007/978-3-540-68752-8.

[6] S. Anu Mary Ealia, M.P. Saravanakumar, A review on the classification, characterisation, synthesis of nanoparticles and their application, IOP Conf. Ser.: Mater. Sci. Eng. 263 (noviembre) (2017) 032019, https://doi.org/10.1088/1757-899X/263/3/032019.

[7] N.R. Jana, Colloidal Nanoparticles: Functionalization for Biomedical Application, Taylor & Francis Group, Boca Raton, FL, 2019. https://books.google.com.mx/books/about/Colloidal_Nanoparticles.html?id=o3G0wwEACAAJ&redir_esc=y.

[8] G.-Z. Jin, A. Chakraborty, J.-H. Lee, J.C. Knowles, H.-W. Kim, Targeting with nanoparticles for the therapeutic treatment of brain diseases, J. Tissue Eng. 11 (enero) (2020), https://doi.org/10.1177/2041731419897460, 204173141989746.

[9] N.D. Thorat, J. Bauer, Functional smart hybrid nanostructures based nanotheranostic approach for advanced cancer treatment, Appl. Surf. Sci. 527 (octubre) (2020), https://doi.org/10.1016/j.apsusc.2020.146809, 146809.
[10] M.R. Zocchi, F. Tosetti, R. Benelli, A. Poggi, Cancer nanomedicine special issue review anticancer drug delivery with nanoparticles: extracellular vesicles or synthetic Nanobeads as therapeutic tools for conventional treatment or immunotherapy, Cancers 12 (7) (2020) 1886, https://doi.org/10.3390/cancers12071886.
[11] L. Chen, J. Liang, An overview of functional nanoparticles as novel emerging antiviral therapeutic agents, Mater. Sci. Eng. C 112 (julio) (2020), https://doi.org/10.1016/j.msec.2020.110924, 110924.
[12] S. Ghafoorianfar, A. Ghorani-Azam, S.A. Mohajeri, D. Farzin, Efficiency of nanoparticles for treatment of ocular infections: systematic literature review, J. Drug Deliv. Sci. Technol. 57 (junio) (2020), https://doi.org/10.1016/j.jddst.2020.101765, 101765.
[13] M. Tavakolian, S.M. Jafari, T.G.M. van de Ven, A review on surface-functionalized cellulosic nanostructures as biocompatible antibacterial materials, Nano-Micro Lett. 12 (1) (2020) 73, https://doi.org/10.1007/s40820-020-0408-4.
[14] Y. Ma, F. Cai, Y. Li, J. Chen, F. Han, W. Lin, A review of the application of nanoparticles in the diagnosis and treatment of chronic kidney disease, Bioact. Mater. 5 (3) (2020) 732–743, https://doi.org/10.1016/j.bioactmat.2020.05.002.
[15] P. Shende, C. Mallick, Nanonutraceuticals: a way towards modern therapeutics in healthcare, J. Drug Deliv. Sci. Technol. 58 (agosto) (2020), https://doi.org/10.1016/j.jddst.2020.101838, 101838.
[16] N. Xin, D. Wei, Y. Zhu, M. Yang, S. Ramakrishna, O. Lee, H. Luo, H. Fan, Upconversion nanomaterials: a platform for biosensing, theranostic and photoregulation, Mater. Today Chem. 17 (septiembre) (2020), https://doi.org/10.1016/j.mtchem.2020.100329, 100329.
[17] D. Zhi, T. Yang, J. O'Hagan, S. Zhang, R.F. Donnelly, Photothermal therapy, J. Control. Release 325 (septiembre) (2020) 52–71, https://doi.org/10.1016/j.jconrel.2020.06.032.
[18] E. Georgilis, M. Abdelghani, J. Pille, E. Aydinlioglu, J.C.M. van Hest, S. Lecommandoux, E. Garanger, Nanoparticles based on natural, engineered or synthetic proteins and polypeptides for drug delivery applications, Int. J. Pharm. 586 (agosto) (2020) 119537, https://doi.org/10.1016/j.ijpharm.2020.119537.
[19] A. Jin, Y. Wang, K. Lin, L. Jiang, Nanoparticles modified by polydopamine: working as 'drug' carriers, Bioact. Mater. 5 (3) (2020) 522–541, https://doi.org/10.1016/j.bioactmat.2020.04.003.
[20] B.C. Cha, J. Kim, Functional mesoporous silica nanoparticles for bio-imaging applications, Wiley Interdiscip. Rev. Nanomed. Nanobiotechnol. 11 (1) (2019), https://doi.org/10.1002/wnan.1515, e1515.
[21] N. Aslan, B. Ceylan, M.M. Koç, F. Findik, Metallic nanoparticles as X-ray computed tomography (CT) contrast agents: a review, J. Mol. Struct. 1219 (noviembre) (2020), https://doi.org/10.1016/j.molstruc.2020.128599, 128599.
[22] A. Avasthi, C. Caro, E. Pozo-Torres, M.P. Leal, M.L. García-Martín, Magnetic nanoparticles as MRI contrast agents, in: Topics in Current Chemistry, Springer, 2020, https://doi.org/10.1007/s41061-020-00302-w.
[23] Y. Liu, J. Guo, L. Huang, Modulation of tumor microenvironment for immunotherapy: focus on nanomaterial-based strategies, Theranostics (2020), https://doi.org/10.7150/thno.42998. Ivyspring International Publisher.
[24] L. Gárate-Vélez, C. Escudero-Lourdes, D. Salado-Leza, A. González-Sánchez, I. Alvarado-Morales, D. Bahena, G.J. Labrada-Delgado, J.L. Rodríguez-López, Anthropogenic Iron oxide nanoparticles induce damage to brain microvascular endothelial cells forming the blood-brain barrier. Editado por Lilian Calderón-Garcidueñas, J. Alzheimers Dis. 76 (4) (2020) 1527–1539, https://doi.org/10.3233/JAD-190929.

[25] R. Gupta, H. Xie, Nanoparticles in daily life: applications, toxicity and regulations, J. Environ. Pathol. Toxicol. Oncol. 37 (3) (2018) 209–230, https://doi.org/10.1615/JEnvironPatholToxicolOncol.2018026009.

[26] B.A. Maher, I.A.M. Ahmed, V. Karloukovski, D.A. MacLaren, P.G. Foulds, D. Allsop, D. M.A. Mann, R. Torres-Jardón, L. Calderon-Garciduenas, Magnetite pollution nanoparticles in the human brain, Proc. Natl. Acad. Sci. 113 (39) (2016) 10797–10801, https://doi.org/10.1073/pnas.1605941113.

[27] L. Yildirimer, N.T.K. Thanh, M. Loizidou, A.M. Seifalian, Toxicology and clinical potential of nanoparticles, Nano Today 6 (6) (2011) 585–607, https://doi.org/10.1016/j.nantod.2011.10.001.

[28] G.R. Tortella, O. Rubilar, N. Durán, M.C. Diez, M. Martínez, J. Parada, A.B. Seabra, Silver nanoparticles: toxicity in model organisms as an overview of its hazard for human health and the environment, J. Hazard. Mater. 390 (mayo) (2020), https://doi.org/10.1016/j.jhazmat.2019.121974, 121974.

[29] R. Raliya, T.S. Chadha, K. Haddad, P. Biswas, Perspective on nanoparticle technology for biomedical use, Curr. Pharm. Des. 22 (17) (2016) 2481–2490, https://doi.org/10.2174/1381612822666160307151409.

[30] S. Kango, S. Kalia, A. Celli, J. Njuguna, Y. Habibi, R. Kumar, Surface modification of inorganic nanoparticles for development of organic–inorganic nanocomposites—a review, Prog. Polym. Sci. 38 (8) (2013) 1232–1261, https://doi.org/10.1016/j.progpolymsci.2013.02.003.

[31] R. Mout, D.F. Moyano, S. Rana, V.M. Rotello, Surface functionalization of nanoparticles for nanomedicine, Chem. Soc. Rev. 41 (7) (2012) 2539–2544, https://doi.org/10.1039/c2cs15294k.

[32] A.A. Mieloch, M. Żurawek, M. Giersig, N. Rozwadowska, J.D. Rybka, Bioevaluation of superparamagnetic iron oxide nanoparticles (SPIONs) functionalized with dihexadecyl phosphate (DHP), Sci. Rep. 10 (1) (2020) 2725, https://doi.org/10.1038/s41598-020-59478-2.

[33] M. Soleymani, S. Khalighfard, S. Khodayari, H. Khodayari, M.R. Kalhori, M.R. Hadjighassem, Z. Shaterabadi, A.M. Alizadeh, Effects of multiple injections on the efficacy and cytotoxicity of folate-targeted magnetite nanoparticles as theranostic agents for MRI detection and magnetic hyperthermia therapy of tumor cells, Sci. Rep. 10 (1) (2020) 1695, https://doi.org/10.1038/s41598-020-58605-3.

[34] Y.A. Selim, M.A. Azb, I. Ragab, M.H.M. Abd, El-Azim., Green synthesis of zinc oxide nanoparticles using aqueous extract of Deverra tortuosa and their cytotoxic activities, Sci. Rep. 10 (1) (2020) 3445, https://doi.org/10.1038/s41598-020-60541-1.

[35] D. Cao, X. Shu, D. Zhu, S. Liang, M. Hasan, S. Gong, Lipid-coated ZnO nanoparticles synthesis, characterization and cytotoxicity studies in cancer cell, Nano Convergence 7 (1) (2020) 14, https://doi.org/10.1186/s40580-020-00224-9.

[36] R.V. Lupusoru, D.A. Pricop, C.M. Uritu, A. Arvinte, A. Coroaba, I. Esanu, M.F. Zaltariov, M. Silion, C. Stefanescu, M. Pinteala, Effect of TAT-DOX-PEG irradiated gold nanoparticles conjugates on human osteosarcoma cells, Sci. Rep. 10 (1) (2020) 6591, https://doi.org/10.1038/s41598-020-63245-8.

[37] S. Shikha, S.R. Chaudhuri, M.S. Bhattacharyya, Facile one pot greener synthesis of Sophorolipid capped gold nanoparticles and its antimicrobial activity having special efficacy against gram negative vibrio cholerae, Sci. Rep. 10 (1) (2020) 1463, https://doi.org/10.1038/s41598-019-57399-3.

[38] X. Zeng, J. Sun, S. Li, J. Shi, H. Gao, W.S. Leong, W. Yiqi, et al., Blood-triggered generation of platinum nanoparticle functions as an anti-cancer agent, Nat. Commun. 11 (1) (2020) 567, https://doi.org/10.1038/s41467-019-14131-z.

[39] Y.N. Yoon, D.-S. Lee, H.J. Park, J.-S. Kim, Barium Titanate nanoparticles sensitise treatment-resistant breast Cancer cells to the antitumor action of tumour-treating fields, Sci. Rep. 10 (1) (2020) 2560, https://doi.org/10.1038/s41598-020-59445-x.

[40] J. Kim, S.H. Chun, L. Amornkitbamrung, C. Song, J.S. Yuk, S.Y. Ahn, B.W. Kim, Y.T. Lim, O. Byung-Keun, S.H. Um, Gold nanoparticle clusters for the investigation of therapeutic efficiency against prostate cancer under near-infrared irradiation, Nano Convergence 7 (1) (2020) 5, https://doi.org/10.1186/s40580-019-0216-z.

[41] M. Vahdati, T.T. Moghadam, Synthesis and characterization of selenium nanoparticles-lysozyme Nanohybrid system with synergistic antibacterial properties, Sci. Rep. 10 (1) (2020) 510, https://doi.org/10.1038/s41598-019-57333-7.

[42] L. Palanikumar, S. Al-Hosani, M. Kalmouni, V.P. Nguyen, L. Ali, R. Pasricha, F.N. Barrera, M. Magzoub, pH-responsive high stability polymeric nanoparticles for targeted delivery of anticancer therapeutics, Commun. Biol. 3 (1) (2020) 95, https://doi.org/10.1038/s42003-020-0817-4.

[43] A.D. Permana, M. Mir, E. Utomo, R.F. Donnelly, Bacterially sensitive nanoparticle-based dissolving microneedles of doxycycline for enhanced treatment of bacterial biofilm skin infection: a proof of concept study, Int. J. Pharm.: X 2 (diciembre) (2020), https://doi.org/10.1016/j.ijpx.2020.100047, 100047.

[44] H. Thai, C.T. Nguyen, L.T. Thach, M.T. Tran, H.D. Mai, T.T.T. Nguyen, G.D. Le, et al., Characterization of chitosan/alginate/lovastatin nanoparticles and investigation of their toxic effects in vitro and in vivo, Sci. Rep. 10 (1) (2020) 909, https://doi.org/10.1038/s41598-020-57666-8.

[45] A.S. Espeseth, P.J. Cejas, M.P. Citron, D. Wang, D.J. DiStefano, C. Callahan, G.O. Donnell, et al., Modified mRNA/lipid nanoparticle-based vaccines expressing respiratory syncytial virus F protein variants are immunogenic and protective in rodent models of RSV infection, npj Vaccines 5 (1) (2020) 16, https://doi.org/10.1038/s41541-020-0163-z.

[46] H. Lee, S. Park, J.E. Kang, H.M. Lee, S.A. Kim, S.J. Rhie, Efficacy and safety of nanoparticle-albumin-bound paclitaxel compared with solvent-based taxanes for metastatic breast cancer: a meta-analysis, Sci. Rep. 10 (1) (2020) 530, https://doi.org/10.1038/s41598-019-57380-0.

[47] X. Yu, Y. Dai, Y. Zhao, S. Qi, L. Liu, L. Lisen, Q. Luo, Z. Zhang, Melittin-lipid nanoparticles target to lymph nodes and elicit a systemic anti-tumor immune response, Nat. Commun. 11 (1) (2020) 1110, https://doi.org/10.1038/s41467-020-14906-9.

[48] A. Heuer-Jungemann, N. Feliu, I. Bakaimi, M. Hamaly, A. Alkilany, I. Chakraborty, A. Masood, et al., The role of ligands in the chemical synthesis and applications of inorganic nanoparticles, Chem. Rev. Am. Chem. Soc. (2019), https://doi.org/10.1021/acs.chemrev.8b00733.

[49] M. Rambukwella, N.A. Sakthivel, J.H. Delcamp, L. Sementa, A. Fortunelli, A. Dass, Ligand structure determines nanoparticles' atomic structure, metal-ligand interface and properties, Front. Chem. (2018), https://doi.org/10.3389/fchem.2018.00330. Frontiers Media S.A.

[50] J. Chen, M. Liu, Q. Huang, R. Jiang, H. Huang, F. Deng, Y. Wen, J. Tian, X. Zhang, Y. Wei, A novel strategy for fabrication of fluorescent hydroxyapatite based polymer composites through the combination of surface ligand exchange and self-catalyzed ATRP, Mater. Sci. Eng. C92 (noviembre) (2018) 518–525, https://doi.org/10.1016/j.msec.2018.06.054.

[51] W. Kong, T. Sun, B. Chen, X. Chen, F. Ai, X. Zhu, M. Li, W. Zhang, G. Zhu, F. Wang, A general strategy for ligand exchange on upconversion nanoparticles, Inorg. Chem. 56 (2) (2017) 872–877, https://doi.org/10.1021/acs.inorgchem.6b02479.

[52] Y. Lee, J. Jang, J. Yoon, J.W. Choi, I. Choi, T. Kang, Phase transfer-driven rapid and complete ligand exchange for molecular assembly of phospholipid bilayers on aqueous gold nanocrystals, Chem. Commun. 55 (22) (2019) 3195–3198, https://doi.org/10.1039/c8cc10037c.

[53] S. Munjal, N. Khare, C. Nehate, V. Koul, Water dispersible CoFe2O4 nanoparticles with improved colloidal stability for biomedical applications, J. Magn. Magn. Mater. 404 (abril) (2016) 166–169, https://doi.org/10.1016/j.jmmm.2015.12.017.

[54] P.B. Santhosh, N. Thomas, S. Sudhakar, A. Chadha, E. Mani, Phospholipid stabilized gold nanorods: towards improved colloidal stability and biocompatibility, Phys. Chem. Chem. Phys. 19 (28) (2017) 18494–18504, https://doi.org/10.1039/c7cp03403b.

[55] W.N. Wenger, F.S. Bates, E.S. Aydil, Functionalization of cadmium selenide quantum dots with poly(ethylene glycol): ligand exchange, surface coverage, and dispersion stability, Langmuir 33 (33) (2017) 8239–8245, https://doi.org/10.1021/acs.langmuir.7b01924.

[56] G. Zeng, M. Liu, C. Heng, Q. Huang, L. Mao, H. Huang, J. Hui, F. Deng, X. Zhang, Y. Wei, Surface polyPEGylation of Eu3+ doped luminescent hydroxyapatite nanorods through the combination of ligand exchange and metal free surface initiated atom transfer radical polymerization, Appl. Surf. Sci. 399 (marzo) (2017) 499–505, https://doi.org/10.1016/j.apsusc.2016.12.107.

[57] H. Terraschke, M. Rothe, P. Lindenberg, In situ monitoring metal-ligand exchange processes by optical spectroscopy and X-ray diffraction analysis: a review, Rev. Anal. Chem. (2018), https://doi.org/10.1515/revac-2017-0003.

[58] R. Dinkel, W. Peukert, B. Braunschweig, In situ spectroscopy of ligand exchange reactions at the surface of colloidal gold and silver nanoparticles, J. Phys. Condens. Matter (2017), https://doi.org/10.1088/1361-648X/aa5a3c.

[59] A. Caragheorgheopol, V. Chechik, Mechanistic aspects of ligand exchange in au nanoparticles, Phys. Chem. Chem. Phys. 10 (33) (2008) 5029–5041, https://doi.org/10.1039/b805551c.

[60] S. Munjal, N. Khare, Transforming single domain magnetic CoFe2O4 nanoparticles from hydrophobic to hydrophilic by novel mechanochemical ligand exchange, J. Nanopart. Res. 19 (1) (2017), https://doi.org/10.1007/s11051-016-3700-y.

[61] H.H. Kyaw, S.H. Al-Harthi, A. Sellai, J. Dutta, Self-organization of gold nanoparticles on silanated surfaces, Beilstein J. Nanotechnol. 6 (1) (2015) 2345–2353, https://doi.org/10.3762/bjnano.6.242.

[62] A. Bekdemir, S. Liao, F. Stellacci, On the effect of ligand shell heterogeneity on nanoparticle/protein binding thermodynamics, Colloids Surf. B: Biointerfaces 174 (febrero) (2019) 367–373, https://doi.org/10.1016/j.colsurfb.2018.11.027.

[63] C.H. Wong, S.C. Zimmerman, Orthogonality in organic, polymer, and supramolecular chemistry: from Merrifield to click chemistry, Chem. Commun. 49 (17) (2013) 1679–1695, https://doi.org/10.1039/c2cc37316e.

[64] H.C. Kolb, M.G. Finn, K. Barry Sharpless, Click chemistry: diverse chemical function from a few good reactions, Angew. Chem. Int. Ed. 40 (11) (2001) 2004–2021, https://doi.org/10.1002/1521-3773(20010601)40:11<2004::aid-anie2004>3.3.co;2-x.

[65] V.V. Rostovtsev, L.G. Green, V.V. Fokin, K. Barry Sharpless, A stepwise huisgen cycloaddition process: copper(I)-catalyzed regioselective 'ligation' of azides and terminal alkynes, Angew. Chem. Int. Ed. 41 (14) (2002) 2596–2599, https://doi.org/10.1002/1521-3773(20020715)41:14<2596::AID-ANIE2596>3.0.CO;2-4.

[66] E. Kim, H. Koo, Biomedical applications of copper-free click chemistry: in vitro, in vivo, and ex vivo, Chem. Sci. 10 (34) (2019) 7835–7851, https://doi.org/10.1039/c9sc03368h.

[67] C. Finetti, L. Sola, M. Pezzullo, D. Prosperi, M. Colombo, B. Riva, S. Avvakumova, C. Morasso, S. Picciolini, M. Chiari, Click chemistry immobilization of antibodies on polymer coated gold nanoparticles, Langmuir 32 (29) (2016) 7435–7441, https://doi.org/10.1021/acs.langmuir.6b01142.

[68] R.M. Fratila, M. Navascuez, J. Idiago-López, M. Eceiza, J.I. Miranda, J.M. Aizpurua, J.M. De La Fuente, Covalent immobilisation of magnetic nanoparticles on surfaces: via strain-promoted azide-alkyne click chemistry, New J. Chem. 41 (19) (2017) 10835–10840, https://doi.org/10.1039/c7nj01822c.

[69] Y. Liu, W. Hou, H. Sun, C. Cui, L. Zhang, Y. Jiang, W. Yongxiang, et al., Thiol–ene click chemistry: a biocompatible way for orthogonal bioconjugation of colloidal nanoparticles, Chem. Sci. 8 (9) (2017) 6182–6187, https://doi.org/10.1039/C7SC01447C.

[70] W.H. Chen, Y. Xu, A. Cecconello, Y.S. Sohn, R. Nechushtai, I. Willner, Stimuli-responsive nucleic acid-functionalized metal-organic framework nanoparticles using pH- and metal-ion-dependent DNAzymes as locks, Chem. Sci. 8 (8) (2017) 5769–5780, https://doi.org/10.1039/c7sc01765k.
[71] Y. Yan, F. Jie, T. Wang, L. Xiuyang, Controlled release of silyl ether camptothecin from thiol-ene click chemistry-functionalized mesoporous silica nanoparticles, Acta Biomater. 51 (marzo) (2017) 471–478, https://doi.org/10.1016/j.actbio.2017.01.062.
[72] J. Yoo, S. Choi, J. Son, G. Yi, E. Kim, H. Koo, Click chemistry-mediated tumor-targeting of SN38-loaded nanoparticles using trastuzumab, Biochem. Biophys. Res. Commun. 515 (1) (2019) 207–213, https://doi.org/10.1016/j.bbrc.2019.05.128.
[73] S. Lee, H. In Yoon, J.H. Na, S. Jeon, S. Lim, H. Koo, S.S. Han, et al., In vivo stem cell tracking with imageable nanoparticles that bind bioorthogonal chemical receptors on the stem cell surface, Biomaterials 139 (septiembre) (2017) 12–29, https://doi.org/10.1016/j.biomaterials.2017.05.050.
[74] S.B. Meer, V. Der, K. Loza, K. Wey, M. Heggen, C. Beuck, P. Bayer, M. Epple, Click chemistry on the surface of Ultrasmall gold nanoparticles (2 nm) for covalent ligand attachment followed by NMR spectroscopy, Langmuir 35 (22) (2019) 7191–7204, https://doi.org/10.1021/acs.langmuir.9b00295.
[75] J.G. Croissant, S. Picard, D. Aggad, M. Klausen, C.M. Jimenez, M. Maynadier, O. Mongin, et al., Fluorescent periodic mesoporous organosilica nanoparticles dual-functionalized: via click chemistry for two-photon photodynamic therapy in cells, J. Mater. Chem. B 4 (33) (2016) 5567–5574, https://doi.org/10.1039/c6tb00638h.
[76] X.Z. Mou, X.Y. Chen, J. Wang, Z. Zhang, Y. Yang, Z.X. Shou, T. Yue Xing, et al., Bacteria-instructed click chemistry between functionalized gold nanoparticles for point-of-care microbial detection, ACS Appl. Mater. Interfaces 11 (26) (2019) 23093–23101, https://doi.org/10.1021/acsami.9b09279.
[77] X.R. Zhang, Y. Zhang, F.T. Chen, Y. Li, S.S. Zhang, Visual detection of single-nucleotide polymorphisms and DNA methyltransferase based on cation-exchange of CuS nanoparticles and click chemistry of functionalized gold nanoparticles, Chem. Commun. 52 (90) (2016) 13261–13264, https://doi.org/10.1039/c6cc06735b.
[78] E.W. Elliott, A.L. Ginzburg, Z.C. Kennedy, Z. Feng, J.E. Hutchison, Single-step synthesis of small, Azide-functionalized gold nanoparticles: versatile, water-dispersible reagents for click chemistry, Langmuir 33 (23) (2017) 5796–5802, https://doi.org/10.1021/acs.langmuir.7b00632.
[79] J. Croissant, X. Cattoën, M.W.C. Man, A. Gallud, L. Raehm, P. Trens, M. Maynadier, J.O. Durand, Biodegradable ethylene-bis(propyl)disulfide-based periodic mesoporous organosilica nanorods and nanospheres for efficient in-vitro drug delivery, Adv. Mater. 26 (35) (2014) 6174–6180, https://doi.org/10.1002/adma.201401931.
[80] S. Bharathiraja, N.Q. Bui, P. Manivasagan, M.S. Moorthy, S. Mondal, H. Seo, N.T. Phuoc, et al., Multimodal tumor-homing chitosan oligosaccharide-coated biocompatible palladium nanoparticles for photo-based imaging and therapy, Sci. Rep. 8 (1) (2018) 1–16, https://doi.org/10.1038/s41598-017-18966-8.
[81] H.C. Hang, Y. Chong, D.L. Kato, C.R. Bertozzi, A metabolic labeling approach toward proteomic analysis of mucin-type O-linked glycosylation, Proc. Natl. Acad. Sci. U. S. A. 100 (25) (2003) 14846–14851, https://doi.org/10.1073/pnas.2335201100.
[82] E.M. Sletten, C.R. Bertozzi, Bioorthogonal chemistry: fishing for selectivity in a sea of functionality, Angew. Chem. Int. Ed. (2009), https://doi.org/10.1002/anie.200900942.
[83] N.K. Devaraj, The future of bioorthogonal chemistry, ACS Cent. Sci. 4 (8) (2018) 952–959, https://doi.org/10.1021/acscentsci.8b00251.
[84] S. Leza, E.P. Daniela, X. Yang, L. Stefancikova, M. Bolsa-Ferruz, F. Savina, D. Dragoe, et al., Green one-step synthesis of medical nanoagents for advanced radiation therapy, Nanotechnol. Sci. Appl. 13 (agosto) (2020) 61–76, https://doi.org/10.2147/nsa.s257392.

[85] T. Sabri, P.D. Pawelek, J.A. Capobianco, Dual activity of rose Bengal functionalized to albumin-coated lanthanide-doped upconverting nanoparticles: targeting and photodynamic therapy, ACS Appl. Mater. Interfaces 10 (32) (2018) 26947–26953, https://doi.org/10.1021/acsami.8b08919.

[86] X.-W. Huang, H. Liang, Z. Li, J. Zhou, X. Chen, S.-M. Bai, H.-H. Yang, Monodisperse phase transfer and surface bioengineering of metal nanoparticles via a silk fibroin protein corona, Nanoscale 9 (8) (2017) 2695–2700, https://doi.org/10.1039/C6NR09581J.

[87] K. Debnath, K. Mandal, N.R. Jana, Phase transfer and surface functionalization of hydrophobic nanoparticle using amphiphilic poly(amino acid), Langmuir 32 (11) (2016) 2798–2807, https://doi.org/10.1021/acs.langmuir.6b00282.

[88] M.G. Bekaroğlu, A. Alemdar, S. İşçi, Comparison of ionic polymers in the targeted drug delivery applications as the coating materials on the Fe3O4 nanoparticles, Mater. Sci. Eng. C 103 (May) (2019) 0–5, https://doi.org/10.1016/j.msec.2019.109838.

[89] V. Bui, D. Park, Y.-C. Lee, Chitosan combined with ZnO, TiO2 and ag nanoparticles for antimicrobial wound healing applications: a mini review of the research trends, Polymers 9 (12) (2017) 21, https://doi.org/10.3390/polym9010021.

[90] N.J. Hobson, X. Weng, M. Ashford, N.T.K. Thanh, A.G. Schätzlein, I.F. Uchegbu, Facile aqueous, room temperature preparation of high transverse relaxivity clustered iron oxide nanoparticles, Colloids Surf. A Physicochem. Eng. Asp. 570 (junio) (2019) 165–171, https://doi.org/10.1016/j.colsurfa.2019.03.023.

[91] D. Lachowicz, W. Górka, A. Kmita, A. Bernasik, J. Zukrowski, W. Szczerba, M. Sikora, C. Kapusta, S. Zapotoczny, Enhanced hyperthermic properties of biocompatible zinc ferrite nanoparticles with a charged polysaccharide coating, J. Mater. Chem. B 7 (18) (2019) 2962–2973, https://doi.org/10.1039/c9tb00029a.

[92] S.B. Somvanshi, P.B. Kharat, M.V. Khedkar, K.M. Jadhav, Hydrophobic to hydrophilic surface transformation of nano-scale zinc ferrite via oleic acid coating: magnetic hyperthermia study towards biomedical applications, Ceram. Int. 46 (6) (2020) 7642–7653, https://doi.org/10.1016/J.CERAMINT.2019.11.265.

[93] S.B. Somvanshi, S.R. Patade, D.D. Andhare, S.A. Jadhav, M.V. Khedkar, P.B. Kharat, P.P. Khirade, K.M. Jadhav, Hyperthermic evaluation of oleic acid coated nano-spinel magnesium ferrite: enhancement via hydrophobic-to-hydrophilic surface transformation, J. Alloys Compd. 835 (septiembre) (2020), https://doi.org/10.1016/j.jallcom.2020.155422, 155422.

[94] V. Causin, Polymers: an overview, in: Polymers on the Crime Scene, Springer International Publishing, Cham, 2015, pp. 9–52, https://doi.org/10.1007/978-3-319-15494-7_2.

[95] A.S. Kulshrestha, A. Mahapatro, Polymers for biomedical applications, ACS Symp. Ser. (2008) 1–7, https://doi.org/10.1021/bk-2008-0977.ch001.

[96] S. Sarkar, E. Guibal, F. Quignard, A.K. SenGupta, Polymer-supported metals and metal oxide nanoparticles: synthesis, characterization, and applications, J. Nanopart. Res. 14 (2) (2012) 715, https://doi.org/10.1007/s11051-011-0715-2.

[97] S. Razzaque, S. Hussain, I. Hussain, B. Tan, Design and utility of metal/metal oxide nanoparticles mediated by Thioether end-functionalized polymeric ligands, Polymers 8 (4) (2016) 156, https://doi.org/10.3390/polym8040156.

[98] M.R. Abedin, S. Umapathi, H. Mahendrakar, T. Laemthong, H. Coleman, D. Muchangi, S. Santra, M. Nath, S. Barua, Polymer coated gold-ferric oxide superparamagnetic nanoparticles for theranostic applications, J. Nanobiotechnol. 16 (1) (2018) 80, https://doi.org/10.1186/s12951-018-0405-7.

[99] N. Tepale, V.V.A. Fernández-Escamilla, C. Carreon-Alvarez, V.J. González-Coronel, A. Luna-Flores, A. Carreon-Alvarez, J. Aguilar, Nanoengineering of gold nanoparticles: green synthesis, characterization, and applications, Crystals 9 (12) (2019) 612, https://doi.org/10.3390/cryst9120612.

[100] M.M. Encabo-Berzosa, M. Sancho-Albero, V. Sebastian, S. Irusta, M. Arruebo, J. Santamaria, P.M. Duque, Polymer functionalized gold nanoparticles as nonviral gene delivery reagents, J. Gene Med. 19 (6–7) (2017), https://doi.org/10.1002/jgm.2964, e2964.

[101] Y. Sun, Q. Wang, J. Chen, L. Liu, L. Ding, M. Shen, J. Li, B. Han, Y. Duan, Temperature-sensitive gold nanoparticle-coated Pluronic-PLL nanoparticles for drug delivery and chemo-Photothermal therapy, Theranostics 7 (18) (2017) 4424–4444, https://doi.org/10.7150/thno.18832.

[102] E. Ebrahimi, A.A. Khandaghi, F. Valipour, S. Babaie, F. Asghari, S. Motaali, E. Abbasi, A. Akbarzadeh, S. Davaran, In vitro study and characterization of doxorubicin-loaded magnetic nanoparticles modified with biodegradable copolymers, Artif. Cells, Nanomed., Biotechnol. 44 (2) (2016) 550–558, https://doi.org/10.3109/21691401.2014.968822.

[103] W.T. Al-Jamal, K. Kostarelos, Liposomes: from a clinically established drug delivery system to a nanoparticle platform for theranostic nanomedicine, Acc. Chem. Res. 44 (10) (2011) 1094–1104, https://doi.org/10.1021/ar200105p.

[104] B. Layek, B. Gidwani, S. Tiwari, V. Joshi, V. Jain, A. Vyas, Recent advances in lipid-based nanodrug delivery systems in cancer therapy, Curr. Pharm. Des. 26 (junio) (2020), https://doi.org/10.2174/1381612826666200622133407.

[105] M.S. Oliveira, L.P. Mendes, V.P. Torchilin, Targeted delivery of anticancer drugs: new trends in lipid nanocarriers, in: Nanostructures for Cancer Therapy, Elsevier, 2017, pp. 455–484, https://doi.org/10.1016/B978-0-323-46144-3.00018-0.

[106] L. Montenegro, F. Lai, A. Offerta, M.G. Sarpietro, L. Micicchè, A.M. Maccioni, D. Valenti, A.M. Fadda, From nanoemulsions to nanostructured lipid carriers: a relevant development in dermal delivery of drugs and cosmetics, J. Drug Deliv. Sci. Technol. 32 (abril) (2016) 100–112, https://doi.org/10.1016/j.jddst.2015.10.003.

[107] L. Sercombe, T. Veerati, F. Moheimani, S.Y. Wu, A.K. Sood, S. Hua, Advances and challenges of liposome assisted drug delivery, Front. Pharmacol. 6 (diciembre) (2015), https://doi.org/10.3389/fphar.2015.00286.

[108] P. Joyce, H.B. Schultz, T.R. Meola, C.A. Prestidge, Polymer lipid hybrid (PLH) formulations, Deliv. Drugs (2020) 1–27. Elsevier https://doi.org/10.1016/B978-0-12-817776-1.00001-8.

[109] M. Alavi, M. Hamidi, Passive and active targeting in cancer therapy by liposomes and lipid nanoparticles, Drug Metab. Pers. Ther. 34 (1) (2019), https://doi.org/10.1515/dmpt-2018-0032.

[110] A. Dalmoro, S. Bochicchio, S.F. Nasibullin, P. Bertoncin, G. Lamberti, A.A. Barba, R.I. Moustafine, Polymer-lipid hybrid nanoparticles as enhanced indomethacin delivery systems, Eur. J. Pharm. Sci. 121 (agosto) (2018) 16–28, https://doi.org/10.1016/j.ejps.2018.05.014.

[111] E.M. Abdou, M.A.A. Fayed, D. Helal, K.A. Ahmed, Assessment of the hepatoprotective effect of developed lipid-polymer hybrid nanoparticles (LPHNPs) encapsulating naturally extracted β-Sitosterol against CCl4 induced hepatotoxicity in rats, Sci. Rep. 9 (1) (2019) 19779, https://doi.org/10.1038/s41598-019-56320-2.

[112] L. Wang, W.G. Wang, Z. Rui, D.S. Zhou, The effective combination therapy against human osteosarcoma: doxorubicin plus curcumin co-encapsulated lipid-coated polymeric nanoparticulate drug delivery system, Drug Deliv. 23 (9) (2016) 3200–3208, https://doi.org/10.3109/10717544.2016.1162875.

[113] L. Ao, C. Wu, K. Liu, W. Wang, L. Fang, L. Huang, W. Su, Polydopamine-derivated hierarchical nanoplatforms for efficient dual-modal imaging-guided combination in vivo cancer therapy, ACS Appl. Mater. Interfaces 10 (15) (2018) 12544–12552, https://doi.org/10.1021/acsami.8b02973.

[114] B.S. Unnikrishnan, S. Maya, G.U. Preethi, K.S. Anusree, P.L. Reshma, M.G. Archana, M.K. Kaustabh, T.T. Sreelekha, Self-assembled drug loaded glycosyl-protein metal nanoconstruct: detailed synthetic procedure and therapeutic effect in solid tumor treatment, Colloids Surf. B: Biointerfaces 193 (septiembre) (2020), https://doi.org/10.1016/j.colsurfb.2020.111082, 111082.
[115] C.M. Oliver, L.D. Melton, R.A. Stanley, Creating proteins with novel functionality via the Maillard reaction: a review, Crit. Rev. Food Sci. Nutr. 46 (4) (2006) 337–350, https://doi.org/10.1080/10408690590957250.
[116] R. Bazak, M. Houri, S. El Achy, S. Kamel, T. Refaat, Cancer active targeting by nanoparticles: a comprehensive review of literature, J. Cancer Res. Clin. Oncol. 141 (5) (2015) 769–784, https://doi.org/10.1007/s00432-014-1767-3.
[117] R.P. Das, V.V. Gandhi, B.G. Singh, A. Kunwar, Passive and active drug targeting: role of nanocarriers in rational design of anticancer formulations, Curr. Pharm. Des. 25 (28) (2019) 3034–3056, https://doi.org/10.2174/1381612825666190830155319.
[118] S. Behzadi, V. Serpooshan, W. Tao, M.A. Hamaly, M.Y. Alkawareek, E.C. Dreaden, D. Brown, A.M. Alkilany, O.C. Farokhzad, M. Mahmoudi, Cellular uptake of nanoparticles: journey inside the cell, Chem. Soc. Rev. 46 (14) (2017) 4218–4244, https://doi.org/10.1039/C6CS00636A.
[119] B. Fortuni, T. Inose, M. Ricci, Y. Fujita, I. Van Zundert, A. Masuhara, E. Fron, et al., Polymeric engineering of nanoparticles for highly efficient multifunctional drug delivery systems, Sci. Rep. 9 (1) (2019) 2666, https://doi.org/10.1038/s41598-019-39107-3.
[120] J. Mosquera, I. García, L.M. Liz-Marzán, Cellular uptake of nanoparticles versus small molecules: a matter of size, Acc. Chem. Res. 51 (9) (2018) 2305–2313, https://doi.org/10.1021/acs.accounts.8b00292.
[121] J.O. Eloy, R. Petrilli,G.L. Raspantini, R.J. Lee, Targeted liposomes for siRNA delivery to cancer, Curr. Pharm. Des. 24 (23) (2018) 2664–2672, https://doi.org/10.2174/1381612824666180807121935.
[122] S. Molavipordanjani, S.J. Hosseinimehr, Strategies for conjugation of biomolecules to nanoparticles as tumor targeting agents, Curr. Pharm. Des. (2019), https://doi.org/10.2174/1381612825666190903154847.
[123] A.C. Marques, P.J. Costa, S. Velho, M.H. Amaral, Functionalizing nanoparticles with cancer-targeting antibodies: a comparison of strategies, J. Control. Release 320 (2020) 180–200, https://doi.org/10.1016/j.jconrel.2020.01.035.
[124] L.M. Weiner, R. Surana, S. Wang, Monoclonal antibodies: versatile platforms for cancer immunotherapy, Nat. Rev. Immunol. 10 (5) (2010) 317–327, https://doi.org/10.1038/nri2744.
[125] J.J. Giner-Casares, M. Henriksen-Lacey, M. Coronado-Puchau, L.M. Liz-Marzán, Inorganic nanoparticles for biomedicine: where materials scientists meet medical research, Mater. Today 19 (1) (2016) 19–28, https://doi.org/10.1016/j.mattod.2015.07.004.
[126] Q. Dong, H. Yang, C. Wan, D. Zheng, Z. Zhou, S. Xie, L. Xu, J. Du, F. Li, Her2-functionalized gold-nanoshelled magnetic hybrid nanoparticles: a theranostic agent for dual-modal imaging and photothermal therapy of breast cancer, Nanoscale Res. Lett. 14 (1) (2019) 235, https://doi.org/10.1186/s11671-019-3053-4.
[127] A. Aires, S.M. Ocampo, B.M. Simoes, M. Josefa Rodriguez, J.F. Cadenas, P. Couleaud, K. Spence, et al., Multifunctionalized iron oxide nanoparticles for selective drug delivery to CD44-positive cancer cells, Nanotechnology 27 (6) (2016) 65103, https://doi.org/10.1088/0957-4484/27/6/065103.
[128] F. Emami, A. Banstola, A. Vatanara, S. Lee, J.O. Kim, J.-H. Jeong, S. Yook, Doxorubicin and anti-PD-L1 antibody conjugated gold nanoparticles for colorectal cancer photochemotherapy, Mol. Pharm. 16 (3) (2019) 1184–1199, https://doi.org/10.1021/acs.molpharmaceut.8b01157.

[129] D. Du, F. Hui-Jun, W.-w. Ren, X.-L. Li, L.-H. Guo, PSA targeted dual-modality manganese oxide–mesoporous silica nanoparticles for prostate cancer imaging, Biomed. Pharmacother. 121 (enero) (2020), https://doi.org/10.1016/j.biopha.2019.109614, 109614.

[130] C. Arcangeli, C. Lico, S. Baschieri, M. Mancuso, Characterization of blood-brain barrier crossing and tumor homing peptides by molecular dynamics simulations, Int. J. Nanomedicine 14 (2019) 10123–10136, https://doi.org/10.2147/IJN.S225793.

[131] A.S. Yadav, N.N.V. Radharani, M. Gorain, A. Bulbule, D. Shetti, G. Roy, T. Baby, G.C. Kundu, RGD functionalized chitosan nanoparticle mediated targeted delivery of raloxifene selectively suppresses angiogenesis and tumor growth in breast cancer, Nanoscale 12 (19) (2020) 10664–10684, https://doi.org/10.1039/C9NR10673A.

[132] M. Wu, Y. Zhang, Y. Zhang, W. Mingjie, W. Menglin, W. Hongyi, L. Cao, L. Li, X. Li, X. Zhang, Tumor angiogenesis targeting and imaging using gold nanoparticle probe with directly conjugated cyclic NGR, RSC Adv. 8 (3) (2018) 1706–1716, https://doi.org/10.1039/C7RA10155D.

[133] L. Lv, X. Li, W. Qian, S. Li, Y. Jiang, Y. Xiong, J. Xu, et al., Enhanced anti-glioma efficacy by borneol combined with CGKRK-modified paclitaxel self-assembled redox-sensitive nanoparticles, Front. Pharmacol. 11 (abril) (2020), https://doi.org/10.3389/fphar.2020.00558.

[134] Z. Huang, Y. Song, Z. Pang, B. Zhang, H. Yang, H. Shi, J. Chen, H. Gong, J. Qian, J. Ge, Targeted delivery of thymosin beta 4 to the injured myocardium using CREKA-conjugated nanoparticles, Int. J. Nanomedicine 12 (abril) (2017) 3023–3036, https://doi.org/10.2147/IJN.S131949.

[135] O.K. Arriortua, E. Garaio, B. Herrero, M. de la Parte, L.L. Insausti, F. Plazaola, J.A. Garcia, et al., Antitumor magnetic hyperthermia induced by RGD-functionalized Fe3O4 nanoparticles, in an experimental model of colorectal liver metastases, Beilstein J Nanotechnol. 7 (2016) 1532–1542, https://doi.org/10.3762/bjnano.7.147.

[136] M. Ghorbani, B. Bigdeli, L. Jalili-Baleh, H. Baharifar, M. Akrami, S. Dehghani, B. Goliaei, et al., Curcumin-lipoic acid conjugate as a promising anticancer agent on the surface of goldiron oxide nanocomposites: a pH-sensitive targeted drug delivery system for brain cancer theranostics, Eur. J. Pharm. Sci. 114 (2018) 175–188, https://doi.org/10.1016/j.ejps.2017.12.008.

[137] S. Kalmodia, S. Vandhana, B.R. Tejaswini Rama, B. Jayashree, T. Sreenivasan Seethalakshmi, V. Umashankar, W. Yang, C.J. Barrow, S. Krishnakumar, S.V. Elchuri, Bio-conjugation of antioxidant peptide on surface-modified gold nanoparticles: a novel approach to enhance the radical scavenging property in cancer cell, Cancer Nanotechnol. 7 (2016) 1, https://doi.org/10.1186/s12645-016-0013-x.

[138] G. Russell-Jones, K. McTavish, J. McEwan, J. Rice, D. Nowotnik, Vitamin-mediated targeting as a potential mechanism to increase drug uptake by tumours, J. Inorg. Biochem. 98 (10) (2004) 1625–1633, https://doi.org/10.1016/j.jinorgbio.2004.07.009.

[139] S. Chen, X. Zhao, J. Chen, J. Chen, L. Kuznetsova, S.S. Wong, I. Ojima, Mechanism-based tumor-targeting drug delivery system. Validation of efficient vitamin receptor-mediated endocytosis and drug release, Bioconjug. Chem. 21 (5) (2010) 979–987, https://doi.org/10.1021/bc9005656.

[140] G.L. Zwicke, G. Ali Mansoori, C.J. Jeffery, Utilizing the folate receptor for active targeting of cancer nanotherapeutics, Nano Rev. 3 (1) (2012) 18496, https://doi.org/10.3402/nano.v3i0.18496.

[141] N. Irmania, K. Dehvari, G. Gedda, P.J. Tseng, J.Y. Chang, Manganese-doped green tea-derived carbon quantum dots as a targeted dual imaging and photodynamic therapy platform, J Biomed Mater Res B Appl Biomater 108 (4) (2020) 1616–1625, https://doi.org/10.1002/jbm.b.34508.

[142] D. Bonvin, U.J. Aschauer, J.A.M. Bastiaansen, M. Stuber, H. Hofmann, M.M. Ebersold, Versatility of pyridoxal phosphate as a coating of iron oxide nanoparticles, Nanomaterials (Basel) 7 (8) (2017), https://doi.org/10.3390/nano7080202.

[143] L. Cerchia, V. de Franciscis, Targeting cancer cells with nucleic acid aptamers, Trends Biotechnol. 28 (10) (2010) 517–525, https://doi.org/10.1016/j.tibtech.2010.07.005.

[144] O.C. Farokhzad, S. Jon, A. Khademhosseini, T.-N.T. Tran, D.A. LaVan, R. Langer, Nanoparticle-aptamer bioconjugates, Cancer Res. 64 (21) (2004) 7668–7672, https://doi.org/10.1158/0008-5472.CAN-04-2550.

[145] M. Wu, Y. Wang, Y. Wang, M. Zhang, Y. Luo, J. Tang, Z. Wang, W. Dong, L. Hao, Z. Wang, Paclitaxel-loaded and A10-3.2 aptamer-targeted poly(lactide-co-glycolic acid) nanobubbles for ultrasound imaging and therapy of prostate cancer, Int. J. Nanomedicine 12 (julio) (2017) 5313–5330, https://doi.org/10.2147/IJN.S136032.

[146] D. Zhang, A. Zheng, J. Li, W. Ming, W. Lingjie, Z. Wei, N. Liao, et al., Smart Cu(II)-aptamer complexes based gold nanoplatform for tumor micro-environment triggered programmable intracellular prodrug release, photodynamic treatment and aggregation induced photothermal therapy of hepatocellular carcinoma, Theranostics 7 (1) (2017) 164–179, https://doi.org/10.7150/thno.17099.

[147] L. Xu, S.-B. Wang, X. Chang, D. Han, X.-H. Ren, X.-Z. Zhang, S.-X. Cheng, Multifunctional albumin-based delivery system generated by programmed assembly for tumor-targeted multimodal therapy and imaging, ACS Appl. Mater. Interfaces 11 (42) (2019) 38385–38394, https://doi.org/10.1021/acsami.9b11263.

[148] S. Chandra, H. Nguyen, K. Wiltz, N. Hall, S. Chaudhry, G. Olverson, T. Mandal, S. Dash, A. Kundu, Aptamer-functionalized hybrid nanoparticles to enhance the delivery of doxorubicin into breast cancer cells by silencing P-glycoprotein, J. Cancer Treat. Diagn. 4 (1) (2020) 1–13, https://doi.org/10.29245/2578-2967/2020/1.1176.

[149] D. Powell, S. Chandra, K. Dodson, F. Shaheen, K. Wiltz, S. Ireland, M. Syed, et al., Aptamer-functionalized hybrid nanoparticle for the treatment of breast cancer, Eur. J. Pharm. Biopharm. 114 (mayo) (2017) 108–118, https://doi.org/10.1016/j.ejpb.2017.01.011.

[150] F.H. Haghighi, R. Binaymotlagh, S.Z. Mirahmadi-Zare, H. Hadadzadeh, Aptamer/magnetic nanoparticles decorated with fluorescent gold nanoclusters for selective detection and collection of human promyelocytic leukemia (HL-60) cells from a mixture, Nanotechnology 31 (2) (2020), https://doi.org/10.1088/1361-6528/ab484a, 025605.

[151] Y.A. Grechkin, S.L. Grechkina, E.A. Zaripov, S.V. Fedorenko, A.R. Mustafina, M.V. Berezovski, Aptamer-conjugated Tb(III)-doped silica nanoparticles for luminescent detection of leukemia cells, Biomedicine 8 (1) (2020) 14, https://doi.org/10.3390/biomedicines8010014.

[152] F. Zhao, J. Zhou, S. Xiangjie, Y. Wang, X. Yan, S. Jia, D. Bin, A smart responsive dual aptamers-targeted bubble-generating Nanosystem for cancer triplex therapy and ultrasound imaging, Small 13 (20) (2017) 1603990, https://doi.org/10.1002/smll.201603990.

CHAPTER

8

Toxicity of nanoparticles

Iliana E. Medina-Ramirez[a], *Arturo Jimenez-Chavez*[b], *and Andrea De Vizcaya-Ruiz*[b,*]

[a]Departamento de Química, Universidad Autonoma de Aguascalientes, Aguascalientes, Mexico, [b]Departamento de Toxicología. CINVESTAV-IPN, Mexico City, Mexico

Abbreviations

AAS	atomic absorption spectroscopy
Ag NPs	silver nanoparticles
Au NPs	gold nanoparticles
DMEM	Dulbecco's Modified Eagle's Medium
DNCs	drug nanocrystals
ENMs	engineered nanomaterials
HCPT	10-hydroxy-camptothecin
i.v.	intravenous
IDs	infectious diseases
MDR	multidrug resistant
MOs	microorganisms
MPS	mononuclear phagocytic system
NMs	nanomaterials
NPs	nanoparticles
PC	protein corona
PEG	poly (ethylene glycol)
RPMI	Roswell Park Memorial Institute Medium

1 Introduction

Evidence indicates that new chemical or technological developments might cause damage or health impairment to live organisms or the environment if they enter the commercial market before a complete

*Current address: Department of Environmental and Occupational Health, Program in Public Health, University of California Irvine, Irvine, CA, United States

https://doi.org/10.1016/B978-0-12-821637-8.00006-7
Copyright © 2023 Elsevier Inc. All rights reserved.

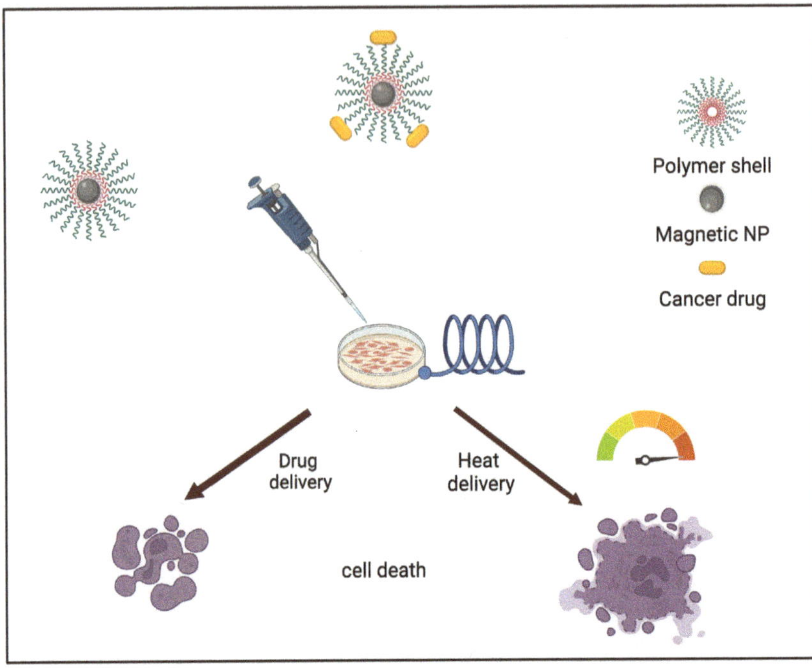

FIG. 1 Development of multifunctional platforms for cancer treatment.

biocompatibility evaluation is performed. In particular, nano-based technologies might cause adverse effects at different trophic levels [1,2]. However, nanotechnology has enormous potential to aid in solving the current challenges of the health segment [3–6]. For example, the development of multifunctional platforms that performs several activities against nonhealthy tissues (Fig. 1). The adhesion of antitumor pharmaceuticals to magnetic NMs as carriers results in an increased amount of a therapeutic agent delivered directly to a cancerous tumor, as compared with the conventional infusion. Besides, the NMs respond to an alternating magnetic field, increasing their surroundings' temperature and leading to nonhealthy cell death [7].

Nowadays, a considerable number of medical-related products that contain NMs are commercially available [3]. In particular, NMs with antibacterial activity, drug delivery agents, and regenerative medical products are widely used. Furthermore, remarkable progress is observed in cancer treatment with the use of nanotechnology. In this respect, NMs can be employed themselves or act as drug delivery agents. Upon magnetic activation, NMs exert anticancer activity by different mechanisms [4]. However, concerns have appeared regarding the use of nanotechnologies in medicine. It is well known that NMs have unique physicochemical properties that regulate their activity. This behavior makes it burdensome to evaluate the possible adverse effects and fate of NMs in the human body. Several techniques need to be used to determine the transformation and final destination of NMs in living organisms. Due to their size, electron microscopy techniques need to be used to investigate the location of the NMs. Due to their composition (e.g., carbon-based NMs), some NMs will not scatter electrons very differently from the biological specimen. Thus labeling is necessary for proper identification. For soluble metal-containing NMs, atomic absorption spectroscopy (AAS) aids in the evaluation of the amount of metal ion lixiviation.

The understanding of NM properties came with the development of quantum mechanics. Also, the implementation of instruments to measure the properties of NMs contributed to reproducibility in their fabrication. Although a scientific justification of NMs' behavior evolved in the last decades, exposures to these materials date back in history [8]. NMs can enter the environment due to numerous natural phenomena, also named incidental NMs. For example, all combustion processes, volcanic activity, pollen from flowers, etc. Also, man unconsciously manipulated NMs before. Romans used metallic colloids to taint glass, whereas other societies fabricated gold colloids for medicinal purposes. Nowadays, manufacturers or scientists can control NM properties to fabricate engineered NMs (ENMs) with enhanced properties for a specific application. The production of these new ENMs grew exponentially in the last decade and is believed to have increased up to 58,000 tons by 2020 [9]. The mass production of nano-based products has changed the abundance of NMs in everyday activities, increasing the probability of human exposure.

Today it is a great challenge to assess the risks associated with the use of ENMs. The difficulties emerge from the uncertainties around the interactions of ENMs with living organisms and the environment. The lack of data and material safety data sheets for ENMs impede a complete and objective hazard evaluation. An ideal scheme for conducting reproducible and accurate toxicology studies must include the following elements: (a) rigorous control to NM characterization, (b) consideration of potential routes of exposure, (c) justification of NM doses, and (d) inclusion of standard controls [10–12].

2 The physicochemical properties of nanoparticles (NPs) of biomedical importance

As previously mentioned, nanotechnology has experienced exponential growth in its field of applications and the commercialization of nano-based products [1]. This advancement has set the establishment of nanomedicine, a discipline that aims to exploit the unique properties of NMs for the diagnostic and treatment of diseases at the molecular level. Nanomedicine-based therapy intends to either kill target cells or repair them. NMs become active by a specific stimulus to exert their action [13]. Nanotherapy may minimize adverse effects by preventing the non-specific uptake of therapeutic agents by healthy tissues [5,6,14–16].

On the other hand, the fate of the ENMs in the organism once its intended activity finishes is still an issue of concern that needs to be investigated. A better understanding of ENMs and their interactions with living organisms is needed. The toxicological behavior of bulk inorganic and organic chemicals cannot be directly correlated to ENMs, since for the latter, their physical and biological properties depend on the novel relationships between their size, composition, crystalline structure, morphology, and surface composition (Fig. 2) [17]. The following paragraphs will describe the influence of different physicochemical properties of NMs with their bioactivity and/or toxicity.

2.1 Size and polydispersity

A general consideration is that as the size of the NMs decreases, their bioactivity and potential toxicity increase. This expected behavior is supported since a reduction in the size of the NMs increases their surface area and reactivity. Also, the reduced size of NMs increases the opportunities

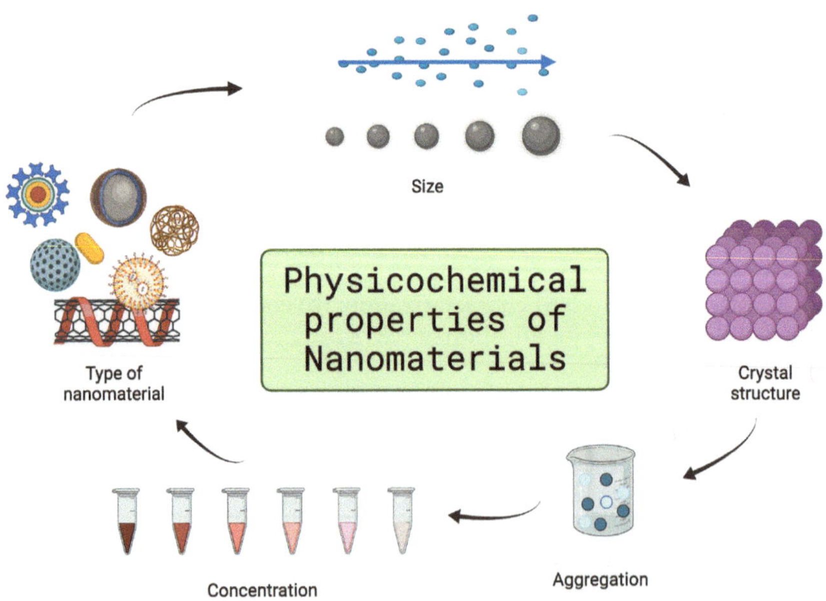

FIG. 2 Physicochemical properties of ENMs that must be considered to evaluate their bioactivity and/or biocompatibility.

for interaction and uptake by living cells [18]. This biological behavior could result in adverse effects that would not be possible with the same material in the macroscale. For example, decreasing the size of ZnO NPs favors its dissolution in the exposure media. Also, reduced size facilitates the movement of particles across cellular and intracellular barriers [9].

However, the properties of the NMs in the solid state do not dictate their behavior in different exposure media. For the accurate evaluation of the biocompatibility of specific NMs, a systematic characterization of the materials must be conducted, considering their properties in the solid state, suspended in different media, and interacting with biological samples [19,20]. Despite the remarkable progress in the emerging field of nanotoxicology, some misconceptions still prevail in this discipline [21]. For instance, many studies only focus on the chemical composition of NMs to evaluate their toxicity. However, other properties influence their bioactivity (or toxicity) such as their chemical composition, shape, and size. Additionally, the properties of the biomicroenvironment modulate NM-living organism's interactions (Fig. 3).

Nevertheless, some nanotoxicity studies correlate the composition of NMs to their bulk counterparts since considerable information on health and environmental impact already exists for them. Besides, the absence of data and material safety data sheets for NMs promotes considering

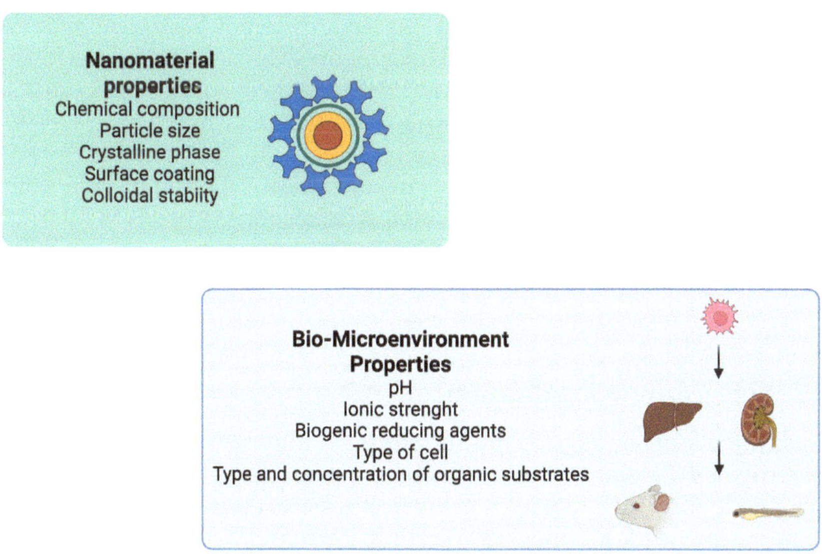

FIG. 3 Evaluation of the bioactivity of NMs. The physicochemical properties of the NMs and the exposure microenvironment must be considered for a better understanding of NM bioactivity or toxicity.

the NMs like their bulk counterparts, despite their unique properties. However, numerous reports indicate the influence of particle size on the bioactivity of NMs. The size of NMs and their possible agglomeration can influence their internalization in the cells. For example, NPs in the range of 10–100 nm achieve higher cellular uptake, NPs higher than 100 nm used the phagocytic pathways, and also NPs less than 5 nm are internalized by diffusion [22]. These differences in the behavior of the NMs can influence their accumulation, interaction, and response of the biological system.

Although most NMs are spherical, recently, more sophisticated synthetic approaches allow for the fabrication of NMs with intricate structures (rods, tubes, cubes, 2D prisms, and branched structures) [10]. As previously mentioned, size and composition are not the only properties of NMs that determine their bioactivity. More recent studies emphasize the role of NM morphology in their biological action. Nanotoxicology discoveries discuss the relation of shape-dependent toxicity. For instance, a recent study shows the shape regulated toxicity of drug nanocrystals (DNCs) of 10-hydroxy-camptothecin (HCPT). Rods are more toxic than spheres due to their enhanced cell internalization [23]. In a different report, gold NPs exhibit contrary behavior. Spherical Au NPs exert more toxicity than their rod-shaped analogous [24]. The shape of the NMs can contribute to their accumulation in the different tissues. The inhalation exposure to differently shaped TiO_2 NMs (nanospheres or nanobelts) showed a faster clearance of the spherical NMs from the lungs [25].

Thus there is a need for more systematic nanotoxicity studies since slight changes in NMs properties render a different biological activity. Understanding the bioactivity and toxicity of NMs is critical for the future applications of these emerging NMs. For biomedical applications, particle size affects cellular uptake, biodistribution, and pharmacokinetics. However, the NMs upon entry into the cell interact with the biomolecules and form NP-biomolecule complexes [10,26]. These NP-biomolecule complexes vary depending on the host biological response upon NM internalization. Thus the toxicity of NMs significantly depends on cell type and NP physicochemical properties [27].

2.2 Surface functionalization

At the nanoscale regime the surface is an important moiety, since the majority of the atoms are located there. Atoms at the surface are very reactive due to their incomplete coordination sphere, thus proper passivation of the surface leads to controlled reactivity (activity). The functionalization of the NMs' surface dictates their behavior in different exposure media and their interactions with biological specimens. The influence of the surface properties of NMs on NP-cell interactions is of considerable importance [28]. For example, hydrophobicity and hydrophilicity affect most of the biointeractions to these materials.

The hydrophobicity of the NMs is one of the properties that can influence their stability and biodistribution. Hydrophobicity has an important role in biological processes such as protein adsorption, interaction with membranes, cellular uptake, immune response, among others [29]. For instance, differently coated Au NPs with different charges and hydrophobicity showed a differential pattern of cellular uptake and toxicity in A549 cells [30]. Differences in the uptake of NMs in the gastrointestinal tract result due to the extent of hydrophilic/hydrophobic surface properties. Surface hydrophilicity favors the adsorption of the NMs to the cell membrane. After their adsorption, impairment of the membrane wrapping process occurs. The NMs enter the cells via energy-independent pathways and exert toxicity via ROS generation [31].

The chemical nature of NM coatings is diverse. Inorganic or organic compounds, polymers, or surfactants are examples of compounds that contribute to the stability of the NMs as a colloidal suspension. In aqueous suspension, the surface properties of NMs are strongly dependent on the composition of these coatings [1]. Specific functional groups at the surface of NMs such as amine, thiol, hydroxyl, or carboxylic acid may enhance their ability to penetrate cell membranes and facilitate their circulation in the body [32]. Sometimes the modification of the surface of NMs is intended to prevent the interaction between the particle's surface and the cells. For example, NMs intended for drug delivery can be coated with poly(ethylene glycol) PEG to increase their half-life in the blood circulation preventing opsonizing proteins from adhering to the NM surface and reducing rapid metabolisms and clearance [33].

Furthermore, the surface properties of naked metal oxide NPs depend on their acidity constants and point of zero charge. For example, titania nanoparticles are positively charged at pH6 and negatively charged at pH7. Silica nanoparticles are generally negatively charged, as their zero point of charge is around pH2 [34]. Surface charge is a determinant of NM colloidal behavior, which influences the biological response by changing the shape and size of NMs through aggregate/agglomerate formation. Particle aggregation and deposition are related phenomena and play a vital role in metal oxide bioactivity. Ionic charges affect the interactions between cells and NMs. At physiological pH, biomolecules are negatively charged, thus favoring their interactions to positively charged NMs. Positively charged NMs are considered more toxic than negatively charged with the same composition [17]. Additionally, positively charged NMs are more suitable as drug delivery vehicles for anticancer drugs since they can stay longer in the bloodstream [35].

For clinical applications, PEG is commonly used to passivate NM surface. PEG functionalization reduces the clearance of the NMs by immune system components and proteins, increasing blood circulation time [13]. We have emphasized that slight changes in NM properties result in different properties and bioactivity. Although most of the studies discussed in this section highlight the influence of the surface coating to regulate cellular response, recent studies demonstrate the impact of core materials. Core materials influence cellular ROS generation and cytotoxicity similar to the surface ligand. For example, Au, Pt, and Pd NPs of the same size and functionality show different toxicity due to core composition [28].

Surface functionalization not only affects the colloidal stability of the NM suspension but also specific biofunctionalization results in active targeting that renders reduced toxicity and improved stability in biological fluids. Specific surface functionalization is the foundation for the applications of NMs to nanomedicine. For example, the functionalization with human albumin reduces toxicity and favors active targeting at the same time. Besides, zwitterionic coatings facilitate the specific targeting of NMs. Also, effective drug administration is achieved by the binding of therapeutic agents to NM surface. NM surface functionalization modulates their activity for applications such as drug delivery, imaging, and cancer treatment [9]. Surface properties are important in determining in vivo biocompatibility and toxic potential of NMs. If done correctly, it will allow a balanced approach to NM medicine design and implementation [36].

2.3 Stability in the exposition media or biological fluids

A reliable hazard assessment of manufactured NMs is complicated because these materials considerably change their intrinsic physicochemical properties upon dispersion in biological fluids [37]. NMs do not behave in solution as inert objects or soluble compounds. They are labile entities that undergo aggregation and/or agglomeration processes, leading to the formation of new chemical species with different properties. The changes in the properties of the NMs affect their biological interaction. On the other hand, the properties of the exposure media also contribute to determining NM-cell interactions.

Some studies indicate that NMs are more stable in the protein-rich medium due to protein corona (PC) formation. For instance, there is a report on the stability of citrate-Au NPs in different exposure media. NMs are more stable in Roswell Park Memorial Institute Medium (RPMI) in comparison to their suspension in Dulbecco's Modified Eagle's Medium (DMEM) due to PC formation in the RPMI medium. On the other hand, surface functionalization also dictates the behavior of the NMs in different exposure media [38,39]. For example, polymer-coated iron oxide NMs exhibit differences in their behavior (colloidal stability, cytotoxicity, and cellular uptake) which is dependent on both NMs coating and exposure media. The optimal medium is different and dependent on coating composition [40].

In biological environments, NMs associate with proteins forming a protein corona (PC). The PC influences the NM pharmacokinetics and pharmacodynamics, thus affecting their toxicity. PC might generate complexes that are more active and can enter otherwise inaccessible tissue sites. The corona on NMs is complex and variable (dependent on NM properties and microenvironment). PC determines the internalization, accumulation, and elimination of NMs [41].

3 Interactions of NMs and living organisms

The developments in nanoscience allow scientists and engineers the design and fabrication of NMs with optimal properties for a specific application. Because of its improved performance, numerous NMs are currently commercially available for a range of industrial and consumer products. Up to now, the Nanodatabase has registered 4494 products containing different kinds of NMs [42]. The fabrication, use, and disposal of manufactured NMs will increase the probability to become environmental pollutants and interact with living organisms. Human exposure to NMs takes place most likely during the manufacturing or use of these NMs [43]. For example, the continuous use of sunscreens and personal care products that contain NMs will result in dermal exposure of human beings to NMs. Moreover, human exposure to NMs might result from inhalation of NMs released into the atmosphere. Oral ingestion is also likely due to the consumption of water or food (fruits, fish) that has accumulated NMs.

Because of the intended use of NMs in medicine, they will interact closely with humans, and the interaction will depend on the pharmaceutical form of the NMs. NMs can be adsorbed on the skin, inhaled into the lungs, ingested into the gastrointestinal tract, injected into the blood, and then internalized into cells of specific tissues through various endocytosis pathways. In the following paragraphs, we will describe the effect of the route of entry of NMs in their biodistribution and toxicity.

3.1 Exposure routes

The portal of entry of NMs to living organisms is an important variable to be considered to estimate their biodistribution and biokinetics. The route of administration is generally chosen based on the therapeutic application of the NMs [44]. However, there is a preference for the intravenous

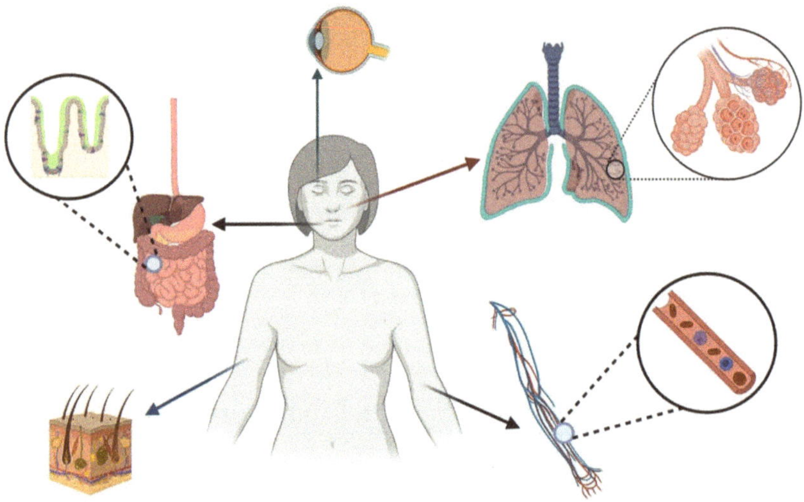

FIG. 4 The main routes of NM entry into the human body and their subsequent biodistribution.

administration due to fast response, increased bioavailability of drugs, and its suitability for the delivery of materials that cannot be absorbed efficiently or that can undergo proteolytic or pH disruption.

NMs can potentially enter the human body through a range of exposure routes: intravenous injection, inhalation, and ingestion via the digestive tract (Fig. 4). They can then translocate to the blood from where they can reach most organs and possibly accumulate before being eliminated through processes that are not yet clearly understood. Soluble or metal ion releasing NMs show systemic distribution. Exposure to these NMs induces inflammation and oxidative stress at the site of entry. The translocation of NMs in blood capillaries is easy for particles with a diameter of less than 100 nm. After translocation, they can biodistribute to all organs [35].

3.1.1 *Intravenous route (i.v.)*

The administration of NMs intravenously is directly related to nanomedicine applications [45]. Given the characteristics of this route, NMs quickly spread throughout the circulatory system reaching different organs [46]. Currently, the main applications of NMs in medicine include diagnosis, drug delivery, and therapy. Within diagnosis, medical imaging strongly relies on the use of iron-based magnetic nanoparticles. Recent studies indicate the biocompatibility of commercially available magnetically responsive contrast agents. The report points out that although ~55% of the injected iron accumulates in the liver at 6 h after administration, such accumulation does not result in any notable or long-term alteration of serum biomarker levels. Although iron nanoparticles are considered biocompatible due to their composition, *high doses* of these materials result in oxidative damage [47].

The time of permanence of the NMs in the body is an influencing variable that dictates their bioactivity and toxicity. The use of Au and iron oxide NMs for cancer treatment requires their accumulation in the tumor. Their fast clearance from circulation causes reduced target accumulation. Careful consideration of the route of administration is needed to increase the therapeutic activity of the NMs [48].

3.1.2 *Oral route*

Oral administration of NMs follows their fast elimination in feces. Nevertheless, the ingestion of some NMs can induce adverse effects on humans due to their absorption in the gastrointestinal tract and translocation into the circulatory system. However, the rate of absorption depends on the physicochemical properties of the NMs. If absorption is not favored, the NMs are degraded or eliminated before absorption [9]. The critical information required for hazard assessment of NMs after oral administration is the extent of absorption into the blood. Following absorption, similar principles to those described for *i.v.* administration of NMs can be observed [11]. After the oral administration of colloidal silver, the NMs distribute throughout the body. However, some organs are more prone to silver deposition. Following oral exposure to colloidal silver results in particle deposition in tissues such as the skin epidermis, the glomeruli, and the intestines [49].

3.1.3 *Dermal route*

The main entry of NMs in humans due to environmental or occupational exposure is through inhalation. There are numerous studies regarding the nanotoxicity of inhaled NMs. Dermal exposure to NMs present in the environment is not likely. However, nanomedicine, cosmetics, textiles, and personal care products are good sources for dermal exposure to NMs. Healthy skin is a natural barrier to exogenous substances. Recent studies show that healthy skin can be prone to suffer alterations (skin sensitization) due to NM exposure. When there is an alteration to skin barrier integrity, the NM penetration is possible [50,51].

ZnO and TiO_2 find applications in sunscreen formulations. Because of its antibacterial properties and odor destroying capacity, Ag NPs are applied as inner coatings in socks, clothes, medical masks, and underwear. Ag NMs also find application for the fabrication of topical antiseptic formulations and wound dressings. TiO_2 is chemically stable, thus is not considered to cause damages to humans due to dermal exposure. ZnO and Ag NMs lixiviate ions under different conditions [52]. These metal ions are responsible for the adverse effects due to dermal exposure to these NMs. The mechanism of their toxicity is via ROS generation that induces DNA damage, mitochondrial dysfunction, and lipid peroxidation in the membrane, which could eventually lead to cell death [10].

Wound-dressing products and Ag topical disinfectants are in direct contact with injured tissues, facilitating the entry of NMs into the human body. Ag NMs can be transported in the bloodstream and distribute to different organs, including the liver, spleen, kidneys, lungs, and brain. Besides, the use of Ag sprays might result in pulmonary exposure [53].

3.1.4 *Respiratory route*

As previously mentioned, inhalation is the main route of entry of NMs to living organisms under ambient or occupational NM exposures. After inhalation of NMs, clearance from the lungs is an important factor influencing long-term effects. The biodistribution of the NMs depends on their physical properties, in particular, their chemical stability. For poorly soluble particles, clearance is slow, thus causing sustained inflammation and eventually resulting in carcinogenesis. Even if a particle has poor solubility and low toxicity, administration of high doses results in bioaccumulation. Chemical reactive NMs are cleared promptly from the lungs [54].

As previously stated, the use of silver sprays can induce pulmonary exposure. After inhalation, silver shows a systemic biodistribution. Contrarily, the subchronic inhalation of Au Np in Sprague Dawley rats (males and females) results in a dose-dependent accumulation of gold in *only* lungs (port of entry) and kidneys (clearance) with gender-related differences in Au NP content in kidneys [55]. The differences in the biodistribution of the Ag and Au NPs are due to the chemical instability of silver NPs.

3.2 Interaction of NMs with MOs

Infectious diseases (IDs) are a challenge to the health sector because of the increasing number of multidrug-resistant (MDR) MOs (bacteria, fungi, viruses, protozoa) as well as the lack of potent, specific, and effective antimicrobial agents [38,39,56]. IDs affect not only humans but also animals and plants are susceptible to the attack of MOs, increasing their economic burden. Nowadays, therapies and vaccines based on NMs to control IDs are currently being developed (Fig. 5). The main advantages of NMs for the control of IDs are the multifunctionality of NMs to exert antimicrobial action (multiple antibacterial mechanisms), specificity, reduced toxicity, and low cost. NMs inhibit MO growth by a single or multistep mechanism. Of utmost importance, nanotechnology developments can contribute

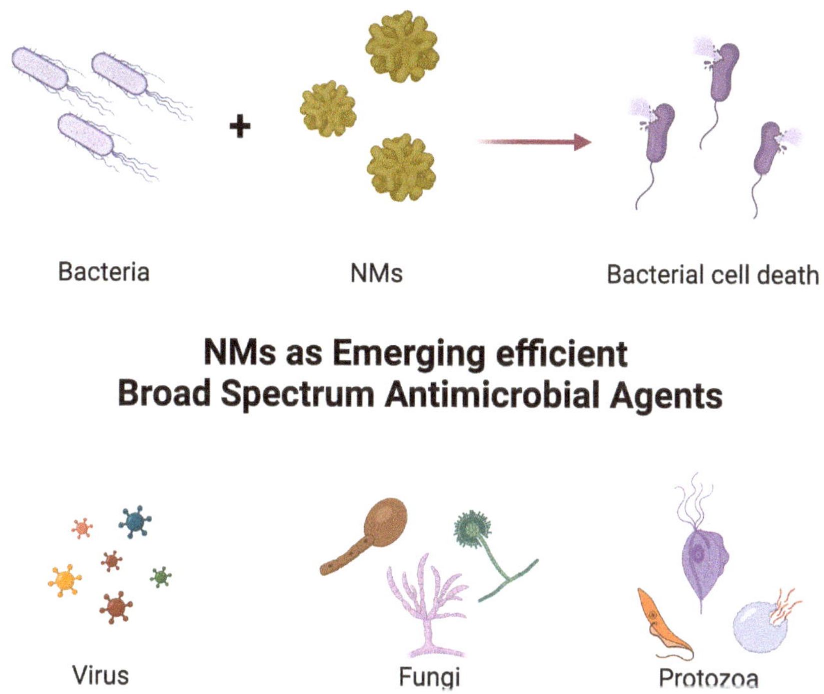

FIG. 5 NMs have become versatile, broad-spectrum agents for the treatment of diverse infectious diseases. NMs act by multiple mechanisms simultaneously to combat MOs.

significantly to the fight against COVID-19 [57–59]. NMs are versatile tools to develop diagnostic devices, therapeutic agents, vaccines, and personnel protective equipment (PPE). For example, Ag NPs exert broad-spectrum, potent antimicrobial activity. A report on the fabrication of Ag-coated condoms indicates the antiviral (against HIV-1 and HSV-1/2), antibacterial (against *E. coli*, *S. aureus*, *M. luteus*, *K. pneumonia*), and antifungi (against Candida spp.) properties of the material [60].

NMs exhibit antimicrobial activity through different mechanisms (Fig. 6). They can also act as drug carriers or drug enhancers (Table 1). The activity of NMs toward MOs includes the destruction (lysis) or alteration of the cell wall or cell membranes, oxidative stress, metal ion release, and ATP depletion [61]. Oxidative stress impedes bacterial metabolism and continuously damage essential cellular functions, causing bacterial inactivation and eventually killing bacteria. NMs can cross the cell wall and release metal ions inside the cell. Metal ion lixiviation enhances NM antimicrobial activity due to their redox activity, interaction with proteins, or damage to the cell wall, leading to pathogen death [62].

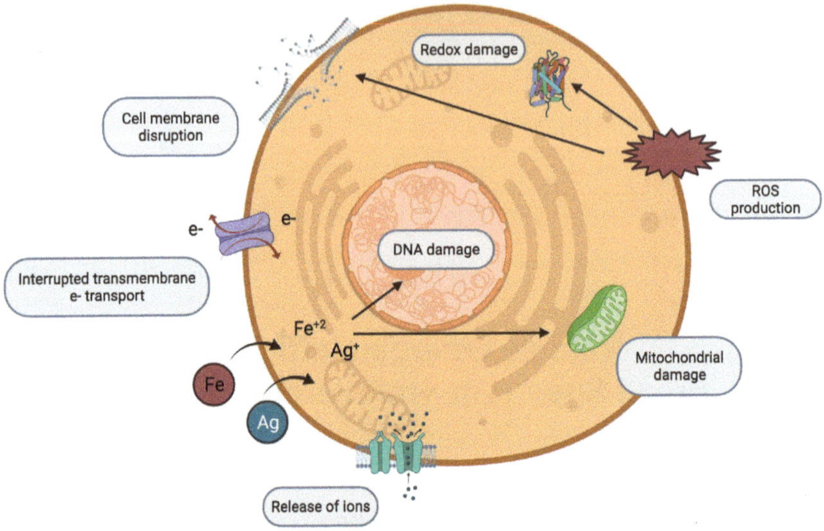

FIG. 6 Activity of NMs to inhibit the growth or to kill MOs.

TABLE 1 Representative examples on the application of NMs for the treatment of infectious diseases.

Type of NMs	Mode of activity	Ref.
Antimicrobial NMs		
Silver (Ag) NPs	Disruption of cell membrane, alteration of the respiratory chain in the bacterial mitochondria, disruption of ATP production and DNA replication, reactive oxygen species (ROS) generation	[63,64]
Copper (Cu), Copper oxide (CuO) NPs	Fenton-type reactions for ROS generation, cell wall damage, lipid peroxidation, protein oxidation, DNA degradation	[65–68]
Gold (Au) NPs	Gold itself exhibits low or null antibacterial activity. However antibiotic surface-modified Au (Au@Van, Au@AM) exerts excellent antibacterial activity. Au NPs are effective for the photothermal treatment of MOs	[67,69]
Iron (Fe), Fe_xO_y (Iron oxide) NPs	Fenton-type reactions for ROS generation, mild hyperthermia, decomposition of functional groups in membrane proteins and lipopolysaccharides	[38,39,70]
Zinc Oxide (ZnO) NPs	ROS generation (damage/inhibition of proteins, damage to mRNA, damage to peptidoglycan and/or cell membrane)	[67]

TABLE 1 Representative examples on the application of NMs for the treatment of infectious diseases—cont'd

Type of NMs	Mode of activity	Ref.
Titanium Dioxide (TiO_2) NPs	Photocatalytic production of reactive oxygen species, ROS (damage/inhibition of proteins, damage to mRNA, damage to peptidoglycan and/or cell membrane), unknown antimicrobial mechanisms	[61,67]
Fullerenes (C_{60}) and fullerene-derivatives	Physical rupture of bacterial cell membranes	[71]
Carbon nanotubes (CNTs)	Oxidative stress, physical rupture of bacterial cell membranes, oxidation of cell membrane proteins and lipids	[18,63]
Antimicrobial peptides and chitosan	Bacterial membrane lysis, damage to DNA or inhibition of transcription of DNA, metal chelation, thereby decreasing metalloprotein activity	[67]
Nanocomposites	Exert multiple mechanisms simultaneously to inhibit or kill microbes	[72]
NMs for efficient drug delivery		
Dendrimers, polymeric NPs, solid lipid NPs, liposomes	Drugs that have poor pharmacodynamics can be delivered using NPs. This approach improves their half-life, stability, and bioavailability	[73]
NMs as drug enhancers		
Magnetic oxides	Mild hyperthermia for drug delivery, physical disruption of bacteria attachment, ROs	[38,39,74]

3.2.1 *Efficiency as antibacterial agents*

Different metal and metal oxide NPs have been used as antibacterial agents given their characteristics. Au NPs exhibit excellent antibacterial activity against *Corynebacterium pseudotuberculosis.* Au NPs penetrate the cell wall and accumulate inside the bacteria [75]. Single metallic NPs as Ag or Cu NPs have shown their effectiveness as broad antibacterial agents [76]. Enhanced antibacterial activity results from the combination of Cu, Ag, and tungsten carbide (WC) NPs in different concentrations. The mesoporous SiO_2 NPs have been used as the carrier for Parmetol S15, an antibacterial and antifouling agent. The controlled delivery of the Parmetol S15 by SiO_2 NPs results in complete inhibition of *Escherichia coli* and *S. aureus* [77].

Despite the combined effect of the different physicochemical properties of NMs to exert antibacterial activity, the dose is an important variable to modulate. Low doses of NMs are needed if their antibacterial activity is optimal. Also, by using small amounts of NMs, the probability of adverse effects on nontarget organisms is scarce. Thus, lately, composite materials have been designed. Their composition includes more than one element (or compound) with antibacterial activity. The composites exhibit several mechanisms for their bactericidal effects. Since different compounds exert antimicrobial activity, smaller amounts are included in the composites, reducing their potential toxicity. For example, the composite, Ag@TiO_2-Cu^{2+} is used at low doses for air disinfection [61]. Some other reports indicate the enhanced antibacterial activity of bimetallic graphene composites (Ag-Cu-Graphene) or Cu-Chitosan-Iodide. The composites inhibit microbial growth at low doses [78,79].

3.2.2 *Efficiency as antifungal agents*

Compared with antibacterial agents, the number of antifungals is limited. Besides, the similarities of fungi with the host cells increase the complexity for the development of new drugs. There are limited options of antifungal agents for the treatment of systemic fungal infections. Additionally, these therapeutic agents are expensive and toxic. Recent studies show promising results on the use of different NMs for the treatment of fungal diseases [80].

Recent studies demonstrate the potential antifungal activity of a variety of NMs. Currently, biodegradable polymers, copolymers, metallic NPs, metallic nanocomposites, and lipid-based nanosystems have shown the capacity for the development of efficient and specific antifungal agents. Magnetic platforms exhibit numerous advantages as therapeutic agents against mycotic infections. Literature reports indicate that using magnetic NMs as drug delivery vehicles results in reduced toxicity of conventional antimycotic agents. Recent reviews include more examples of the application of NMs for the cure of fungal diseases [80–82].

3.2.3 *Efficiency as antiviral agents*

NMs can interact with the viruses outside or inside the host cells. NMs can engage directly with the viral genome or bind to the viral particle outside the host cells. NMs can also compete with the viruses by the binding sites, block the virus-host binding or penetration [83]. Carbon dots with different surface passivation molecules were synthesized and evaluated as antivirals against human norovirus, the most common cause of nonbacterial acute gastroenteritis outbreaks. The carbon dots inhibit the binding between the norovirus particles and specific receptors without affecting the virus integrity [84].

Green synthesized Au NPs in *Allium sativa* (garlic) extract showed their capability as antiviral against the measles virus in infected Vero cells. The Au NPs act at low concentrations to inhibit the replication of the measles virus [85]. The antiviral activity of silver NPs was evaluated using VeroE6/TMPRSS2 and Calu-3 cell line infected with SARS-CoV-2 virus. Ag NPs disrupted the integrity of the SARS-CoV-2 virus avoiding the virus entrance to the exposed cells proving their antiviral effectiveness [86]. PEGylated and bare ZnO NPs were used against the H1N1 influenza virus in infected Madin-Darby canine kidney (MDCK)-SIAT1 cells. Naked and functionalized ZnO NPs act effectively to inhibit the growth of the H1N1 influenza virus. Functionalized ZnO-PEG NPs showed a higher antiviral efficiency compared with bare ZnO NPs [87].

There are many advantages to the use of NMs to control infectious diseases. For example, in comparison to conventional treatment, the use of NMs results in reduced acute toxicity, capacity to overcome microbial resistance, and lower treatment costs. However, despite the advantages of their application, it is necessary to emphasize the unspecific activity of NMs. Thus, to avoid unexpected adverse effects, their development must include a complete toxicity evaluation on different living organisms (plants, humans, and aquatic organisms).

4 Toxicity of NPs

The toxicity of all new consumer products must be evaluated before their large-scale production. Nanotoxicity evaluation is not an exception. As previously observed with other pharmaceuticals (auranofin) or chemical compounds (DDT), the same properties that are desirable for a specific application can also give rise to unexpected and hazardous toxicities [88]. Thus a complete toxicity assessment of NMs is necessary to elucidate possible adverse effects and their fate inside the human body. This assessment must comprise an exhaustive characterization of NMs and inclusion of different toxicity assays (Table 2).

4.1 In vitro toxicity of NMs

In vitro studies generate information about the potential risks of NMs and provide mechanistic information at the cellular level. In vitro studies are simple and allow the researcher to investigate numerous variables at the same time in an experimental unit. Although in vitro studies can generate important information useful to predict the behavior of NMs in living organisms, this correlation is not always totally accurate. Some studies demonstrate that there is no correlation between in vivo and in vitro data [89]. Discrepancies result mainly because of the differences in the

TABLE 2 Methodological approaches to estimate the toxicity of NMs [17].

In vitro methods	In vivo methods
Dye exclusion assays	Quantitation and biodistribution of NPs from tissues
Colorimetric assays	Electron microscopy and detection of NP accumulation
Fluorescence-based assays	Liquid scintillation counting
Luminometric methods	In vivo optical imaging
Cell viability test in real time	Computed tomography
Apoptosis based assays	Magnetic resonance
Genotoxicity assays	

complexity of the biological system and experimental conditions. For this reason more data needs to be systematically generated to harmonize the information obtained [90].

In vitro studies indicate that NMs affect living organisms at different levels. The interaction of NMs and living beings can cause several adverse effects inside and outside the cell (Fig. 7). Oxidative stress is one of the main contributors to NM toxicity. NMs can generate ROS by different mechanisms depending on their optical or physicochemical properties [91–93]. ROS are very reactive and unspecific chemical species. They can interact with numerous biomolecules and alter their activity. Oxidative stress can lead to inflammation; activation of adaptive immune responses; initiation of apoptotic/necrotic pathways; and oxidative damage to lipids, proteins, and DNA.

The presence of some NMs results in the modification of the membrane fluidity [94]. Also, the surface charge of the NMs can contribute to membrane damage. The particle can bind to and damage membranes when its zeta-potential has a high positive value [43]. If NMs are strongly attracted to the cell membrane, their binding results in physical and chemical damage. For example, the interaction of NMs with the cell membrane results in outer membrane lipid peroxidation; increased cell permeability, leading to the uncontrolled passage of NMs from the extracellular environment into the cytoplasm [95].

NM exposure damages the cytoskeleton, leading to changes in cell structure and activity [96]. It is also responsible for impaired transport and organelle dysfunction [17]. The possible mechanisms for cytoskeletal dysfunction induced by NMs include ROS induction, direct physical interaction, and intracellular energetic imbalances [97].

The available methods for the assessment of nanotoxicity using 2D cell cultures and fluorescence-based assays show limited sensitivity and a lack of correspondence with in vivo studies. However, recent developments

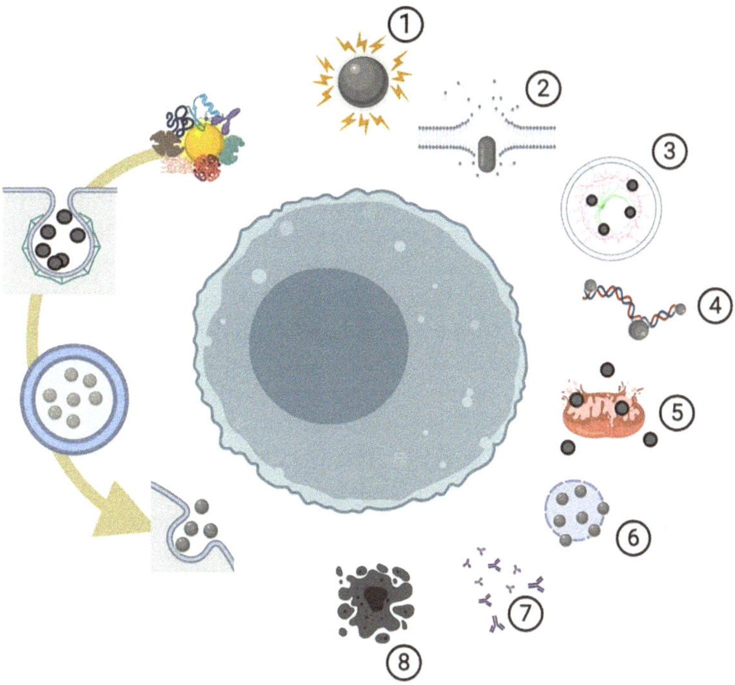

FIG. 7 Nanoparticles interaction and toxicity through the cell. The nano-bio interaction with the different biological media can lead to the NP "corona" formation, the possible NP cell internalization, their vesicular transportation, and the NP elimination/translocation. During their journey through the biological systems NPs can cause different toxicity effects in the cells: ROS generation (1), perforation of cell membranes (2), damage of the cytoskeleton (3), damage in the DNA (4), injury of the mitochondria (5), interfere in the formation of lysosomes (6), activate the synthesis of inflammatory mediators (7), cell death (8), among others.

make available 3D cell cultures and advanced cell imaging techniques, such as electron microscopy and holotomography [98,99]. Holotomographic imaging allows characterizing the intracellular biodistribution of NMs and their effects in real time. These new tools facilitate the understanding of the mechanisms of NMs in cells.

4.2 In vivo toxicity of NMs

There are several studies regarding NM toxicity in unicellular and multicellular organisms. As the complexity of the organism increases, the complexity of toxicity evaluation also increases. However, in vivo studies aim to investigate the biodistribution and toxic kinetics of NMs in animal models. Ideally, these results could be correlated with the possible effects of NMs in the human body. Numerous animal models can be used to investigate the in vivo toxicity of NMs. Zebrafish (*Danio rerio*) is an excellent model for the evaluation of NM toxicity. Zebrafish have rapid development and significant homology to humans [100]. Some other animal models are also in use to study the organ-specific toxicity of NMs.

Depending on the route of administration (intravenous, oral, dermal, inhalation, or ocular) the biodistribution and biological effects of NMs can be determined. Once the NMs enter the bloodstream they can be distributed through the organism. Different NM characteristics can contribute to biodistribution and bioaccumulation after intravenous exposure. The NPs tend to bioaccumulate mainly in organs like the liver and spleen. However, NMs can be present in the bladder, kidney, lungs, heart, or testis. The main effect of NMs in the organs and tissues is the inflammatory response. The NMs after intravenous exposure can be eliminated mainly by the urine [101–103].

Some reports indicate damage to the gastrointestinal tract due to NM interactions. NMs can disturb digestion and absorption of nutrients, causing deficiencies in the body [104]. The particle uptake, ROS generation, oxidative stress, inflammation, and hyperplasia are the main effects observed after in vivo exposure to the NMs by the gastrointestinal tract [105]. The oral ingested NMs can be translocated and distributed to different organs like the liver, spleen, kidneys, and heart. The translocation of NPs can be particle charge dependent. The NPs can be eliminated mainly by feces and in less percentage by the urine [55,106].

The skin is the largest organ in the body and a dynamic system with different functions, as protection against external agents and a selectively permeable barrier [35]. Between the effects on the skin, NMs can cause irritation, sensitization, inflammation, and accelerate skin aging depending on the exposure time (subchronic or chronic exposure) [51]. Some studies observed the impossibility of NMs crossing the skin, but other studies showed the penetration of NMs through hair follicles reaching the basal and spinous layers; generally, this penetration depends on particle size [44]. NMs cannot penetrate the circulatory system or other organs. Only in the case of damaged skin or the use of different techniques that modify the skin structure the NMs can be translocated [107].

Inhalation of NMs can lead to alveoli exposure depending on the NM characteristics such as size, surface reactivity, or biopersistence. The NMs can cause oxidative stress, inflammation, infiltration of the immune cells, and damage of the alveoli [43]. Some diseases like pulmonary fibrosis, lung cancer, or emphysema have been enhanced or developed by NP exposure [108]. Different mechanisms (the proximity of the alveolar cells with the blood vessels, the uptake of the NPs by immune cells) can lead to the translocation of the NPs. The NPs can translocate to organs like the brain, heart, liver, spleen, kidneys, and gastrointestinal tract. After, NMs are eliminated mainly in the urine or feces [109,110].

The deposition and biointeraction of NPs in the eyes can cause different effects. The inflammation of the eyes by the anthropogenic or synthesized

NPs can lead to eye disorders; in the same manner, NPs can lead to oxidative stress, DNA damage, inhibition of antioxidant pathways, or modify the extracellular matrix of the cells in the eyes [111,112]. The NP translocation through the eyes depends on their size; only particles with a size less than 5 nm can translocate the whole organ to the systemic circulation, making difficult the translocation of the NPs through this organ [113].

The central nervous system (CNS) is not directly associated with the NP exposure routes; however, the CNS is considered one of the main affected systems by NP toxicity. NPs can cross the blood-brain barrier (BBB); the inhalation of NPs can produce their translocation mainly in the olfactory bulb and the hippocampus through the olfactory bulb [104]. NPs can cause different effects in the brain like oxidative stress induction, DNA damage, proinflammatory cytokines release, changes in the synthesis and metabolism of the neurotransmitter, dysfunction of transmembrane protein complexes, intracellular calcium ions disturbance, among others [114]. In the same manner, in utero exposure studies have shown some adverse effects as thinning of the cerebral and cerebellar cortex, decrease in neuron density in the cerebrum, damage in hippocampal pyramidal cells, and a deficiency in learning and memory in the neurodevelopment of the offspring after NP exposure [115].

The immune system is in direct contact with the NMs in all the exposure routes in the body. The immune system protects the body against NM damage. The NM protein corona can contribute to this response. The NM protein corona has a critical role in the innate immune response, the definition of the immunological response, and the NM elimination [44,116]. The presence of immune cells in the lungs, the skin, the gastrointestinal tract, or the circulatory system leads to the nano-bio interaction [51]. The immune cells phagocyte the NMs to eliminate them. However, immune cells cannot destroy the biopersistent NMs, leading to their translocation to other organs inside the immune cells and the biodistribution of the NPs in the bloodstream [117]. As mentioned before, this NM biodistribution can lead to systemic damage.

Until now, many studies report the effects and mechanisms of NMs in living organisms. However, many of those studies are very particular (case-by-case studies) and difficult to draw general conclusions. Thus there is a need to obtain more information about the possible cellular and systemic damages of the NMs to group such observations and systematize the study of the toxicity of NMs.

5 Toxicity effects of biomedical relevant NPs

NMs are a heterogeneous and vast group of materials with different physical, chemical, and biological properties. They belong to this group

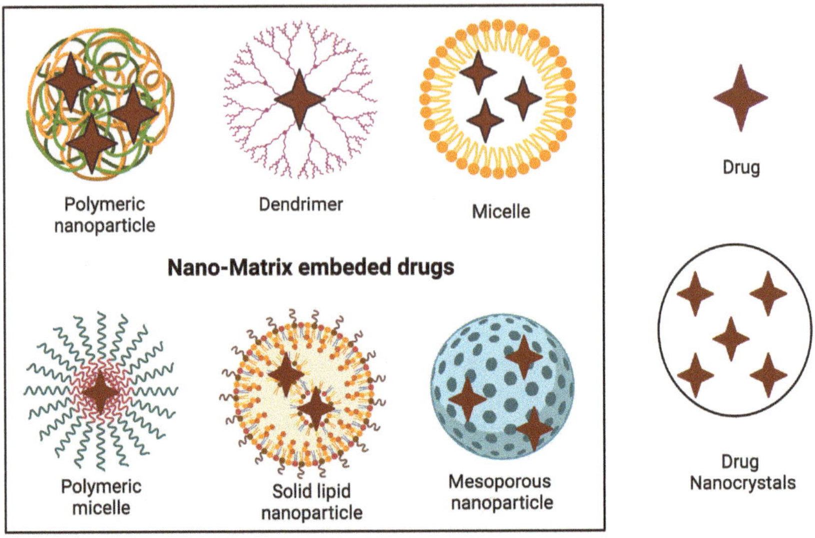

FIG. 8 NMs can be used as drug delivery agents aiming to the controlled release of the drug in the organ of interest. However, drug loading can be low *(left side)*. Drug nanocrystals aimed to increase the therapeutic load and efficacy of poorly soluble drugs *(right)*.

because of their reduced size (at least one of their dimensions less than 100nm). Their composition is diverse. The NMs can be organic (polymers, fullerenes, and carbon nanotubes) or inorganic (metal oxides, metals, and quantum dots). NMs can be used as therapeutic agents or as drug delivery systems, depending on their properties and composition. Lately, the use of drug nanocrystals (DNCs) enhances loading capacity and therapeutic efficacy (Fig. 8) [118].

Nowadays, there is an increased interest in the development of DNCs. DNCs contribute to increased bioavailability of poorly water-soluble drugs. Surface modification of DNCs results in improved efficacy. For example, surface modification with hydrophilic polymers allows DNCs to evade phagocytosis, increasing circulation and therapeutic effect. Recently, improved DNCs result by formulating hybrid drug nanocrystals [119]. For example, drug nanocrystals that integrate fluorophores can perform both treatment and bioimaging in real time.

Currently, there are concerns regarding the toxicity of DNCs. In this regard, their reduced size facilitates the internalization of DNCs into cells. Besides, their long retention times in the body might induce harmful effects. Taking into consideration biopersistency and size-related risks, a toxicity related classification of DNCs is as follows: class I (nanoparticle size >100nm and biodegradable), class II (nanoparticle size >100nm and nonbiodegradable), class III (nanoparticle size <100nm and biodegradable), and class IV (nanoparticle size <100nm and nonbiodegradable). Class I DNCs are the safest, but in general, the toxicity of DNCs is limited.

5.1 Metallic NPs

Silver NMs are among the most widely investigated materials for the development of efficient antimicrobial agents for environmental or biomedical applications. These materials are also popular in the academy due to their catalytic, thermo-plasmonic, and surface Raman properties. Lately, their use for photothermal therapy to inhibit bacterial growth or cancer treatment has become popular [120].

The use of silver for disinfection purposes goes back in history. Romans stored drinking water in containers made of silver. Later, physicians explored the treatment of syphilis with silver-salvarsan. Nowadays, silver is still the focus of numerous researches since Ag-based antiseptics show broad-spectrum activity. Also, it was postulated that microorganisms exhibit a low propensity to induce resistance to Ag treatment. However, the exacerbated use of these materials results in the chronic exposition of MOs to them, favoring the development of microbial resistance. Recent studies report on the development of metal resistance in MOs. Six metal resistance mechanisms exist: exclusion by permeability barrier, intra- and extracellular sequestration, active transport efflux pumps, enzymatic detoxification, and reduction in the sensitivity of cellular targets to metal ions [121].

There is no doubt in the benefits of Ag NPs for the control of infectious diseases. In nanomedicine, the use of Ag NMs is mainly to fabricate topical broad-spectrum bactericides. In this regard, the main route of the entrance of Ag NPs to the body is dermal contact. Also, several medical products are coated with Ag NPs, for example, surgical instruments, bare prostheses, dental alloys, and contraceptive devices. In vitro cell studies demonstrate the dose-dependent toxicity of Ag NMs. Some other studies indicate the surface coating-related toxicity of Ag NPs. For example, citrate coated-colloidal Ag NPs are not toxic at doses of 100 μg/mL to human skin keratinocytes [122]. On the other hand, powdered citrate-coated Ag NPs are toxic under the same exposure conditions. Coating of Ag NPs with a polymer prevents the toxicity of the powder.

In vitro toxicity studies report that the cellular uptake of Ag NPs is both NP properties and cell type-dependent process. Different mechanisms are responsible for the Ag NP cellular uptake: micropinocytosis, lipid raft-dependent endocytosis (caveolin dependent, dynamin dependent, dynamin independent), clathrin-dependent endocytosis, and phagocytosis. Numerous in vitro studies using different cell culture models suggest that increasing the concentrations of Ag NPs leads to decreased viability and proliferation and induction of apoptosis [123].

In vivo studies show that Ag NP toxicity depends on the dose, size, route of the exposure, and exposition time (chronic or acute). The toxic effects of Ag NPs affect several organs and tissues. Reports of Ag NP toxicity on humans are related to colloidal silver consumption. The effects are similar to the ones reported for animal models [62,123].

Gold NMs have applications in medicine, biomedical diagnosis, cancer therapy, and drug delivery. Aurimmune, which is a nanocomposite (NCs) of Au NPs coated with recombinant human tumor necrosis factor-alpha (TNF-a) and PEG, is an efficient antineoplastic drug. The therapeutic use of this material favors the accumulation of its components in the tumor, avoiding its uptake into healthy cells [124]. Recently, a highly sensitive Au NP-based immunoassay to detect SARS-Cov-2 was fabricated [125].

Gold and silver NMs are among the few NMs with FDA approval for medical applications. Regarding the toxicity of these NMs, due to their differences in reactivity, silver NMs induce ROS related toxicity mechanisms. Contrarily, gold is a biopersistent material that accumulates in different organs resulting in long-term toxicity. Regarding Au NP toxicity, literature reports indicate that small particles (below 4 nm) induce damage and oxidative stress in cultured cells. Bigger particles are considered nontoxic and safe for their use in medicine [53].

Therapeutic or delivery agents should not be cleared quickly from the body. They should be able to cross natural barriers to reach the target organ to exert activity. Sequestration of NMs by the mononuclear phagocytic system (MPS) can affect their ability to deliver therapeutics to target cells. An in vivo study regarding the chronic exposition of BALB/mice to Au NMs demonstrates that chronic exposure to 10 mg/kg Au NPs in BALB/c mice did not establish a steady state in the MPS. Additionally, Au NPs accumulated the most in the liver, followed by the spleen, causing tissue discoloration. The carcass showed high amounts of Au NPs due to possible accumulation in lymph node [48].

5.2 Metal oxide NPs

Iron oxide NMs are among the few examples of NMs that are FDA approved for clinical practice. Some examples of commercially available products are Feridex, Resovist, Combidex, and NanoTherm. The first three find application in MRI imaging and the last one for cancer treatment. Some of the adverse effects that accompany the use of these NMs are back pain, vasodilatation, acute urinary retention, and fever [124]. Some other examples are feraheme, venofer, ferrlecit, INFeD, and Dex-Iron. The use of these compounds is as supplements for iron deficiency in chronic kidney disease [3].

The solubility of the metal oxide NMs can be a crucial characteristic for their persistence, interaction, and accumulation inside the cell but also a determinant in the metal oxide NM toxicity. The study of 11 different metal oxide NMs showed the solubility of the ZnO, CuO, Sb_2O_3, Mn_3O_4, and Co_3O_4 NMs in different biological systems (exposure to A549 cells, Caco2 cells, or Balb/c 3T3 cells), which was associated with the toxicity of the metal oxide NMs. Only the soluble NMs caused the death of almost 50% of all the cell lines tested after 24h of exposure [126].

The solubility of the NMs can influence the accumulation and the persistence of the particles in the different organs. For example, the single administration to 0.3mL of a fluorescent ZnO: Eu (Europium-doped zinc oxide) NM suspension (10mg/mL) by gavage to Balb-c mice to determine the possible absorption and biodistribution of the NMs. The results showed the presence of the NMs in the gastrointestinal tract, brain, liver, spleen, and kidneys mainly during the first 24h. Results suggest that the liver and kidney were responsible for NM elimination. However, there was an accumulation in the spleen and adipose tissues [127].

On the other hand, the interaction, biodistribution, and accumulation of persistent NMs can lead to a long-term effect in the biological systems. The administration of CeO_2 NMs (naked and SiO_2 coated) to male Wistar rats showed a different biodistribution in the body due to the surface properties of NMs. Besides, the NM route of exposure affects their biokinetics. IT administration results in the bioaccumulation of the naked and SiO_2-coated CeO_2 NMs in high concentrations in the lungs. The materials were present in secondary organs after 28 days' postexposure. The gavage exposure results in the fast excretion of both CeO_2 NMs. The coating of CeO_2 alters their biodistribution likely due to modifications on PC formation after IT and IV administration [128].

5.3 Carbon NPs

Carbon-based NMs find use in medicine as coatings of inorganic NMs or drug delivery agents. Some representative examples of this group are polymer-based NPs, liposomes, carbon nanotubes, graphene, and fullerenes. Polymers and liposomes have properties suitable for biomedical applications, such as self-assembly and biodegradability. Several liposome-drug formulations are commercially available. For example, DOXIL, AmBisome, Epaxal, and Inflexal V. Polymer-based therapeutics are commercially available. Some examples include Zinostatin Stimalmer, Oncospar PEG-L-asparaginase, and Neulasta PEG-GCSF. Despite the advantages of the use of these materials, they exhibit some drawbacks. For example, NMs are unstable in physiological media, limiting the controlled release of pharmaceutics due to the burst release of their therapeutic cargo [53].

Fullerenes (C60) are carbon spheres that exhibit unique electronic properties for biomedical applications. Fullerenes have many potential applications based on their unique free radical chemistry and antioxidant properties. These NMs find applications for fullerene-based therapeutics as antioxidants for inflammatory diseases, antimicrobial agents, drug delivery vehicles, and MRI contrast agents [129]. Fullerenes induce ROS-mediated cellular damage; however, modification of the fullerene surface yields NMs with biologically useful antioxidant activity. Even at high doses, animal studies have demonstrated minimal dermal and oral toxicity [33].

Certainly, NMs are very active and susceptible to damage to different cellular components; however, in many literature reports, their findings are conflicting, inconclusive, and the subject of much debate. For example, the use of hydrophobic nonderivatized C60 to conduct *in vivo* studies about their toxicity in fish. Due to the hydrophobic behavior of the materials, they can form agglomerates in the aqueous environment, limiting the interaction with biological specimens [130]. Similar misleading examples are found in the literature regarding the toxicity of TiO_2 NMs (chemically stable material), which is generally administered at high doses, thus prone to form aggregates and physical damages (obstruction), not specific to the NM nature.

Numerous investigations are currently active on the biomedical applications of graphene and graphene oxide. Current applications include cell growth control, cancer treatment, biosensing, and mass spectrometry. Lately, concerns regarding the harmful effects of graphene have appeared in the literature. Recent studies demonstrate that surface functionalization is critical to improving the biocompatibility of graphene. Currently, several studies seek on how to abolish the toxicity of graphene and accelerate its excretion in biological systems [131].Thus aiming to fabricate efficient and safe NMs for application in the biomedical sector, before the synthesis, all the possible variables that can influence the biointeractions and the biokinetics of the NMs must be considered.

6 Standardization of nanotoxicity evaluation: Future and challenges

The use of different NMs to improve human life (medicine, technologies, and consumer products) has led to their spread in all the ecosystems. Because nanotechnology is an emerging technology, limited information is available, in particular, regarding the toxicity and health effects of NMs. The fate of the NMs is uncertain. However, their biointeractions might result in damages to humans and nature [132]. A big challenge is to establish standardized assay protocols to evaluate the safety of NMs. There is

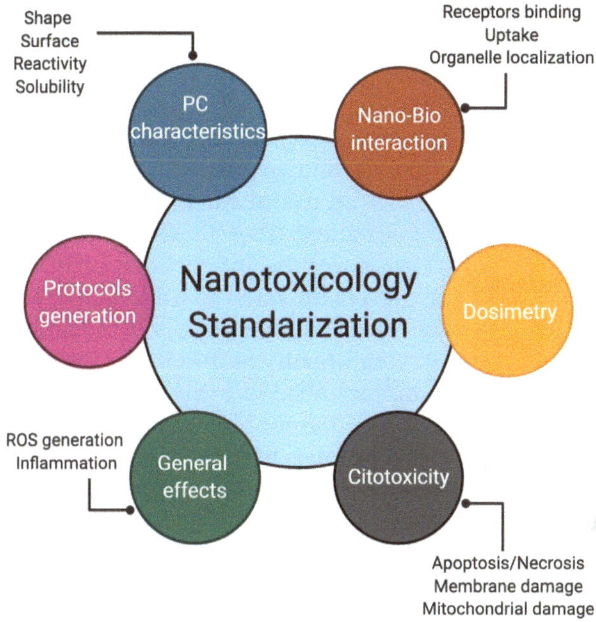

FIG. 9 A scheme of the current challenges for the standardization of nanotoxicity evaluation.

an urgent need for a methodological scheme that allows the classification of NMs into different classes according to physicochemical properties, biological interactions, exposure threats, or endpoint toxicity.

Some of the limitations for the standardization of NM toxicity evaluation are as follows: the unavailability of defined reference material to validate the outcome of the tested material, lack of consensus on predictive endpoints for human toxicity, technical limitations of the tests for NMs (e.g., interference), and batch-to-batch differences due to insufficient control over the manufacturing process. To address this challenge, a multidisciplinary and systematic approach is required (Fig. 9). First, an exhaust characterization (physicochemical properties and biointeractions) of the NMs before their use is mandatory for sustainable development. Also, universal, objective, systematic, and reproducible protocols for the toxicological evaluation of NMs are needed. Much more effort is required to develop NMs with improved activity and biocompatible engineered properties to achieve minimal toxicity for its safe use in medicine.

There are some reports about the categorization of NMs and systematic approaches for their hazard assessment. Gebel and coworkers propose the classification of NMs into three categories according to the route of exposure and mode of action. Following their study, NM classification is as follows. Class 1 NMs, the toxicity of these NMs is due to chemical activity (i.e., metal ion lixiviation) or functional groups on the surface. Class 2 NMs, rigid biopersistent respirable fibrous materials. Class 3 NMs, respirable granular biodegradable particles (GBP) [11].

Currently, there is information regarding the physicochemical properties of NMs (surface, shape, reactivity), biological response to interaction [133,134], specific types of use and exposure, or uptake and kinetics [11,135]. This information can be useful for the categorization of NMs and facilitate their hazard assessment. There is also a need to develop different standard strategies to study the interactions of NMs and living organisms. These studies will set the basis to understand and avoid the possible toxicity of the NMs in all the organisms. A big challenge is to elucidate the fate of NMs in living organisms and the environment (Fig. 9).

Furthermore, more studies are needed to standardize doses and relevant concentrations of NMs in different biological models. Also, there is a need to develop accurate techniques to determine the exposure concentrations of NMs in the environment and consumer products. A better understanding of the entrance of NMs to living organisms by different exposure routes is necessary to achieve a real measurement of NM exposure concentrations and establish the dose-effect relation of NM exposure. The dose-dependent effects observed by in vitro assays cannot be extrapolated to in vivo studies due to the differences of the NM properties on the different exposure media. So, validated high-throughput standardized methods and NMs are needed to overcome these problems.

The unique properties of nanomaterials (NMs) can cause interferences during their toxicity evaluation. In many instances, classical toxicological assays are used to evaluate NMs toxicity. NMs may interact with molecules (chemical or biological), leading to a false result during toxicity evaluation. For example, NMs can interact with the proteins in the exposure media, changing NMs´ bioactivity. They might also react with dye molecules or chemical indicators of the toxicity assay, affecting the outcome of the study. Besides, the optical and chemical reactivity of NMs can also influence toxicity evaluation [38,39,136]. However, the use of controls during the determination of some assays can correct or eliminate the NM interferences [137]; it is necessary to implement standard operating procedures (SOPs) and the appropriate controls for these assays.

Despite the amount of information about the toxicity of the NMs in different biological models, there is a need to investigate all the possible NM toxicological effects. A precise understanding of how and why NMs interact with biological systems is needed. The use of the called "omics" as genomics, transcriptomics, proteomics, and metabolomics can help to identify the possible toxicological effects caused by NMs [138]. New simple biological models (i.e., zebrafish, *Drosophila*) can facilitate the study of NM toxicity [139]. Besides, the application of in silico studies can contribute to obtaining more information about the relation between the NMs and the living organisms in a costless, safer, and efficient manner.

7 Concluding remarks

Nanomedicine is a strong strategy to alleviate the current challenges in the biomedical health segment. One of the major defects of conventional pharmaceutics is the lack of selectivity and severe side effects in healthy organs or tissues. Another problem of conventional therapy is the need for high doses to achieve appropriate therapeutic effects. However, high doses of pharmaceutical agents result in harmful effects, while a reduced drug concentration is not high enough to eradicate illness. The use of nanotechnology to face these challenges is promissory. A complete understanding of the interactions of NMs and living organisms is needed. The development of this knowledge will contribute to the tailored interactions of NMs with biological systems to achieve an improved health system. There are numerous challenges to the standardization of nanotoxicology evaluation. To produce reproducible findings, nanotoxicology studies need to use clear definitions and investigate the toxicity of unambiguously characterized nanotechnology products. Most in vitro studies demonstrate the size-, dose-, and coating-dependent cellular uptake of NMs. Independent of the differences encountered regarding the biointeractions of NMs in cell cultures and in vivo, it is necessary to test nanotoxicity in multiple cell types to increase the power of prediction of the outcomes of NM bioactivity. As part of the National Nanotechnology Initiative and other federal agency collaborations, large-scale research efforts are underway to characterize nanoscale materials and quantify their impact for purposes of developing toxicological assessment and testing tools.

References

[1] Q. Mu, et al., Chemical basis of interactions between engineered nanoparticles and biological systems, Chem. Rev. 114 (15) (2014) 7740–7781, https://doi.org/10.1021/cr400295a.

[2] G. Oberdörster, Safety assessment for nanotechnology and nanomedicine: concepts of nanotoxicology, J. Intern. Med. 267 (1) (2010) 89–105, https://doi.org/10.1111/j.1365-2796.2009.02187.x.

[3] D. Bobo, et al., Nanoparticle-based medicines: a review of FDA-approved materials and clinical trials to date, Pharm. Res. 33 (10) (2016) 2373–2387, https://doi.org/10.1007/s11095-016-1958-5.

[4] M. Cordani, Á. Somoza, Targeting autophagy using metallic nanoparticles: a promising strategy for cancer treatment, Cell. Mol. Life Sci. 76 (7) (2019) 1215–1242, https://doi.org/10.1007/s00018-018-2973-y. Springer International Publishing.

[5] D. Ling, N. Lee, T. Hyeon, Chemical synthesis and assembly of uniformly sized iron oxide nanoparticles for medical applications, Acc. Chem. Res. 48 (5) (2015) 1276–1285, https://doi.org/10.1021/acs.accounts.5b00038.

[6] X. Ma, Y. Zhao, X.J. Liang, Theranostic nanoparticles engineered for clinic and pharmaceutics, Acc. Chem. Res. 44 (10) (2011) 1114–1122, https://doi.org/10.1021/ar2000056.

[7] O. Taratula, et al., Multifunctional nanomedicine platform for concurrent delivery of chemotherapeutic drugs and mild hyperthermia to ovarian cancer cells, Int. J. Pharm. 458 (1) (2013) 169–180, https://doi.org/10.1016/j.ijpharm.2013.09.032. Elsevier B.V.

[8] S. Bayda, et al., The history of nanoscience and nanotechnology: from chemical-physical applications to nanomedicine, Molecules 25 (1) (2020) 1–15, https://doi.org/10.3390/molecules25010112.
[9] P. Ganguly, A. Breen, S.C. Pillai, Toxicity of nanomaterials: exposure, pathways, assessment, and recent advances, ACS Biomater Sci. Eng. 4 (7) (2018) 2237–2275, https://doi.org/10.1021/acsbiomaterials.8b00068.
[10] M. Ajdary, et al., Health concerns of various nanoparticles: a review of their in vitro and in vivo toxicity, Nano 8 (9) (2018) 1–28, https://doi.org/10.3390/nano8090634.
[11] T. Gebel, et al., Manufactured nanomaterials: categorization and approaches to hazard assessment, Arch. Toxicol. 88 (12) (2014) 2191–2211, https://doi.org/10.1007/s00204-014-1383-7.
[12] P.A. Holden, et al., Ecological nanotoxicology: integrating nanomaterial hazard considerations across the subcellular, population, community, and ecosystems levels, Acc. Chem. Res. 46 (3) (2013), https://doi.org/10.1021/ar300069t.
[13] K. Peynshaert, et al., Exploiting intrinsic nanoparticle toxicity: the pros and cons of nanoparticle-induced autophagy in biomedical research, Chem. Rev. 114 (15) (2014) 7581–7609, https://doi.org/10.1021/cr400372p.
[14] V.E. Kagan, H. Bayir, A.A. Shvedova, Nanomedicine and nanotoxicology: two sides of the same coin, Nanomed.: Nanotechnol., Biol. Med. 1 (4) (2005) 313–316, https://doi.org/10.1016/j.nano.2005.10.003.
[15] I. Linkov, F.K. Satterstrom, L.M. Corey, Nanotoxicology and nanomedicine: making hard decisions, Nanomedicine 4 (2008) 167–171, https://doi.org/10.1016/j.nano.2008.01.001.
[16] P. Patel, et al., Doxorubicin-conjugated platinum theranostic nanoparticles induce apoptosis via inhibition of a cell survival (PI3K/AKT) signaling pathway in human breast cancer cells, ACS Appl. Nano Mater. (2020), https://doi.org/10.1021/acsanm.0c02521.
[17] S. Attarilar, et al., The toxicity phenomenon and the related occurrence in metal and metal oxide nanoparticles: a brief review from the biomedical perspective, Front. Bioeng. Biotechnol. 8 (July) (2020), https://doi.org/10.3389/fbioe.2020.00822.
[18] S. Kang, et al., Antibacterial effects of carbon nanotubes: size does matter! Langmuir 24 (13) (2008) 6409–6413, https://doi.org/10.1021/la800951v.
[19] H.C. Fischer, W.C. Chan, Nanotoxicity: the growing need for in vivo study, Curr. Opin. Biotechnol. (2007) 565–571, https://doi.org/10.1016/j.copbio.2007.11.008. Elsevier.
[20] X. Hu, et al., Knowledge gaps between nanotoxicological research and nanomaterial safety, Environ. Int. (2016) 8–23, https://doi.org/10.1016/j.envint.2016.05.001.
[21] D.B. Warheit, Debunking some misconceptions about nanotoxicology, Nano Lett. 10 (12) (2010) 4777–4782, https://doi.org/10.1021/nl103432w.
[22] C. Auría-Soro, et al., Interactions of nanoparticles and biosystems: microenvironment of nanoparticles and biomolecules in nanomedicine, Nano 9 (10) (2019), https://doi.org/10.3390/nano9101365.
[23] M. Zhou, et al., Shape regulated anticancer activities and systematic toxicities of drug nanocrystals in vivo, Nanomed.: Nanotechnol., Biol. Med. 12 (1) (2016) 181–189, https://doi.org/10.1016/j.nano.2015.09.006. Elsevier Inc.
[24] B.D. Chithrani, A.A. Ghazani, W.C.W. Chan, Determining the size and shape dependence of gold nanoparticle uptake into mammalian cells, Nano Lett. 6 (4) (2006) 662–668, https://doi.org/10.1021/nl052396o.
[25] D.W. Porter, et al., Differential mouse pulmonary dose and time course responses to titanium dioxide nanospheres and nanobelts, Toxicol. Sci. 131 (1) (2013) 179–193, https://doi.org/10.1093/toxsci/kfs261.
[26] Y.L. Wu, et al., Biophysical responses upon the interaction of nanomaterials with cellular interfaces, Acc. Chem. Res. (2013), https://doi.org/10.1021/ar300046u.
[27] M.J. Akhtar, M. Ahamed, H.A. Alhadlaq, Challenges facing nanotoxicology and nanomedicine due to cellular diversity, Clin. Chim. Acta 487 (2018) 186–196, https://doi.org/10.1016/j.cca.2018.10.004.

[28] X. Bai, et al., Regulation of cell uptake and cytotoxicity by nanoparticle core under the controlled shape, size, and surface chemistries regulation of cell uptake and cytotoxicity by nanoparticle core under the controlled shape, size, and surface chemistries, ACS Nano 14 (1) (2019) 289–302, https://doi.org/10.1021/acsnano.9b04407.
[29] A. Valsesia, et al., Direct quantification of nanoparticle surface hydrophobicity, Commun. Chem. 1 (1) (2018), https://doi.org/10.1038/s42004-018-0054-7.
[30] H. Sun, et al., Induction of oxidative stress and sensitization of cancer cells to paclitaxel by gold nanoparticles with different charge densities and hydrophobicities, J. Mater. Chem. B 6 (11) (2018) 1633–1639, https://doi.org/10.1039/c7tb03153j.
[31] B.J. Teubl, et al., Interactions between nano-TiO2 and the oral cavity: impact of nanomaterial surface hydrophilicity/hydrophobicity, J. Hazard. Mater. 286 (2015) 298–305, https://doi.org/10.1016/j.jhazmat.2014.12.064. Elsevier B.V.
[32] A. Gnach, et al., Upconverting nanoparticles : assessing the toxicity based on survey of existing studies, low nanotoxicity of lanthanide doped upconverting nanoparticles holds promise for their safety and suitability for biomedical detection and imaging, Chem. Soc. Rev. 44 (6) (2014) 1561–1584.
[33] A. Nel, et al., Toxic potential of materials at the nanolevel, Science 311 (5761) (2006) 622–627, https://doi.org/10.1126/science.1114397.
[34] M. Kosmulski, The pH dependent surface charging and points of zero charge. VII. Update, Adv. Colloid Interface Sci. 251 (2018) 115–138, https://doi.org/10.1016/j.cis.2017.10.005. Elsevier B.V.
[35] Z. Ferdous, A. Nemmar, Health impact of silver nanoparticles: a review of the biodistribution and toxicity following various routes of exposure, Int. J. Mol. Sci. 21 (2020) 1–31, https://doi.org/10.3390/ijms21072375.
[36] G. Liu, et al., Applications and potential toxicity of magnetic iron oxide nanoparticles, Small (2013), https://doi.org/10.1002/smll.201201531.
[37] G. Maiorano, et al., Effects of cell culture media on the dynamic formation of protein-nanoparticle complexes and influence on the cellular response, ACS Nano 4 (12) (2010) 7481–7491, https://doi.org/10.1021/nn101557e.
[38] I.E. Medina-Ramírez, M.A. Díaz de León Olmos, et al., Development and assessment of nano-technologies for cancer treatment: cytotoxicity and hyperthermia laboratory studies, Cancer Invest. 38 (1) (2020) 61–84, https://doi.org/10.1080/07357907.2019.1698593. Taylor and Francis Ltd.
[39] I.E. Medina-Ramírez, C.E. Díaz de León-Macias, et al., Evaluation of the biocompatibility and growth inhibition of bacterial biofilms by ZnO, Fe3O4 and ZnO@Fe3O4 photocatalytic magnetic materials, Ceram. Int. (2020), https://doi.org/10.1016/j.ceramint.2019.12.145. Elsevier Ltd.
[40] A. Petri-Fink, et al., Effect of cell media on polymer coated superparamagnetic iron oxide nanoparticles (SPIONs): colloidal stability, cytotoxicity, and cellular uptake studies, Eur. J. Pharm. Biopharm. 68 (1) (2008) 129–137, https://doi.org/10.1016/j.ejpb.2007.02.024.
[41] L. Marichal, et al., From protein Corona to colloidal self-assembly: the importance of protein size in protein-nanoparticle interactions, Langmuir 36 (28) (2020) 8218–8230, https://doi.org/10.1021/acs.langmuir.0c01334.
[42] The Nanodatabase (2020).
[43] H.M. Braakhuis, et al., Physicochemical characteristics of nanomaterials that affect pulmonary inflammation, Part. Fibre Toxicol. 11 (1) (2014) 1–25.
[44] V. De Matteis, Exposure to inorganic nanoparticles: routes of entry, immune response, biodistribution and in vitro/in vivo toxicity evaluation, Toxics 5 (4) (2017), https://doi.org/10.3390/toxics5040029.
[45] V. Escamilla-Rivera, et al., Protein corona acts as a protective shield against Fe3O4-PEG inflammation and ROS-induced toxicity in human macrophages, Toxicol. Lett. 240 (1) (2016) 172–184. Elsevier Ireland Ltd.

[46] G. Crisponi, et al., Toxicity of nanoparticles: etiology and mechanisms, in: Antimicrobial Nanoarchitectonics: From Synthesis to Applications, 2017, https://doi.org/10.1016/B978-0-323-52733-0.00018-5.

[47] L.H. Reddy, et al., Magnetic nanoparticles: design and characterization, toxicity and biocompatibility, pharmaceutical and biomedical applications, Chem. Rev. 112 (11) (2012) 5818–5878, https://doi.org/10.1021/cr300068p.

[48] P. Natarajan, J.M. Tomich, Understanding the influence of experimental factors on bio-interactions of nanoparticles: towards improving correlation between in vitro and in vivo studies, Arch. Biochem. Biophys. 694 (June) (2020) 108592, https://doi.org/10.1016/j.abb.2020.108592. Elsevier Inc.

[49] N. Hadrup, H.R. Lam, Oral toxicity of silver ions, silver nanoparticles and colloidal silver—a review, Regul. Toxicol. Pharmacol. 68 (1) (2014) 1–7, https://doi.org/10.1016/j.yrtph.2013.11.002. Elsevier Inc.

[50] D.H. Brouwer, et al., Occupational dermal exposure to nanoparticles and nano-enabled products: part 2, exploration of exposure processes and methods of assessment, Int. J. Hyg. Environ. Health 219 (6) (2016) 503–512, https://doi.org/10.1016/j.ijheh.2016.05.003. Elsevier GmbH.

[51] F. Larese Filon, et al., Occupational dermal exposure to nanoparticles and nano-enabled products: part I—factors affecting skin absorption, Int. J. Hyg. Environ. Health 219 (6) (2016) 536–544, https://doi.org/10.1016/j.ijheh.2016.05.009. Elsevier GmbH.

[52] A. Mackevica, S. Foss Hansen, Release of nanomaterials from solid nanocomposites and consumer exposure assessment—a forward-looking review, Nanotoxicology 10 (6) (2016) 641–653, https://doi.org/10.3109/17435390.2015.1132346.

[53] V. Mirshafiee, et al., Facilitating translational nanomedicine via predictive safety assessment, Mol. Ther. 25 (7) (2017) 1522–1530, https://doi.org/10.1016/j.ymthe.2017.03.011.

[54] K. McNamara, S.A.M. Tofail, Nanoparticles in biomedical applications, Adv. Phys.: X. (2017), https://doi.org/10.1080/23746149.2016.1254570.

[55] I.C. Lee, et al., Comparative toxicity and biodistribution assessments in rats following subchronic oral exposure to copper nanoparticles and microparticles, Part. Fibre Toxicol. 13 (56) (2016) 1–16, https://doi.org/10.1186/s12989-016-0169-x.

[56] C.S. Yah, G.S. Simate, Nanoparticles as potential new generation broad spectrum antimicrobial agents, DARU J. Pharm. Sci. 23 (1) (2015), https://doi.org/10.1186/s40199-015-0125-6.

[57] W.C.W. Chan, Nano research for COVID-19, ACS Nano (2020), https://doi.org/10.1021/acsnano.0c02540.

[58] H. Huang, et al., COVID-19: a call for physical scientists and engineers, ACS Nano (2020), https://doi.org/10.1021/acsnano.0c02618.

[59] R. Vazquez-Munoz, J.L. Lopez-Ribot, Nanotechnology as an alternative to reduce the spread of COVID-19, Challenges (2020) 1–14, https://doi.org/10.20944/PREPRINTS202006.0301.V1.

[60] A. Mohammed Fayaz, et al., Inactivation of microbial infectiousness by silver nanoparticles-coated condom: a new approach to inhibit HIV- and HSV-transmitted infection, Int. J. Nanomedicine 7 (2012) 5007–5018, https://doi.org/10.2147/IJN.S34973.

[61] J.H. Martínez-Montelongo, et al., Development of a sustainable photocatalytic process for air purification, Chemosphere 257 (2020), https://doi.org/10.1016/j.chemosphere.2020.127236.

[62] X. Chen, H.J. Schluesener, Nanosilver: a nanoproduct in medical application, Toxicol Lett. 176 (2008) 1–12, https://doi.org/10.1016/j.toxlet.2007.10.004.

[63] A.J. Huh, Y.J. Kwon, '"Nanoantibiotics": a new paradigm for treating infectious diseases using nanomaterials in the antibiotics resistant era, J. Control. Release 156 (2) (2011) 128–145, https://doi.org/10.1016/j.jconrel.2011.07.002. Elsevier B.V.

[64] B. Wang, et al., Physicochemical origin for free radical generation of iron oxide nanoparticles in biomicroenvironment: catalytic activities mediated by surface chemical states, J. Phys. Chem. C 117 (1) (2013) 383–392, https://doi.org/10.1021/jp3101392.
[65] J.M. Boyce, Modern technologies for improving cleaning and disinfection of environmental surfaces in hospitals, Antimicrob. Resist. Infect. Control 5 (1) (2016) 1–10, https://doi.org/10.1186/s13756-016-0111-x.
[66] A.K. Chatterjee, R. Chakraborty, T. Basu, Mechanism of antibacterial activity of copper nanoparticles, Nanotechnology 25 (13) (2014) 1–12, https://doi.org/10.1088/0957-4484/25/13/135101.
[67] R.Y. Pelgrift, A.J. Friedman, Nanotechnology as a therapeutic tool to combat microbial resistance, Adv. Drug Deliv. Rev. 65 (13–14) (2013) 1803–1815, https://doi.org/10.1016/j.addr.2013.07.011. Elsevier B.V.
[68] J. Xia, et al., Improvement of antibacterial activity of copper nanoclusters for selective inhibition on the growth of gram-positive bacteria, Chin. Chem. Lett. 30 (2) (2019) 421–424, https://doi.org/10.1016/j.cclet.2018.07.008. Chinese Chemical Society.
[69] H. Gu, et al., Presenting vancomycin on nanoparticles to enhance antimicrobial activities, Nano Lett. 3 (9) (2003) 1261–1263, https://doi.org/10.1021/nl034396z.
[70] S. Parveen, et al., Preparation, characterization and antifungal activity of iron oxide nanoparticles, Microb. Pathog. 115 (June) (2018) 287–292, https://doi.org/10.1016/j.micpath.2017.12.068. Elsevier Ltd.
[71] Y. Wang, et al., Antibiotic-free antibacterial strategies enabled by nanomaterials: progress and perspectives, Adv. Mater. 32 (18) (2020) 1–21, https://doi.org/10.1002/adma.201904106.
[72] C. Zhang, et al., Pd@Ag nanosheets in combination with amphotericin B exert a potent anti-cryptococcal fungicidal effect, PLoS One 11 (6) (2016) 1–18, https://doi.org/10.1371/journal.pone.0157000.
[73] M.C. Teixeira, C. Carbone, E.B. Souto, Beyond liposomes: recent advances on lipid based nanostructures for poorly soluble/poorly permeable drug delivery, Prog. Lipid Res. 68 (July) (2017) 1–11, https://doi.org/10.1016/j.plipres.2017.07.001. Elsevier.
[74] K.L. Ribeiro, et al., Clavanin A-bioconjugated Fe3O4/Silane core-shell nanoparticles for thermal ablation of bacterial biofilms, Colloids Surf. B Biointerfaces 169 (2018) 72–81, https://doi.org/10.1016/j.colsurfb.2018.04.055. Elsevier B.V.
[75] M.M. Mohamed, et al., Antibacterial effect of gold nanoparticles against Corynebacterium pseudotuberculosis, Int. J. Vet. Sci. Med. 5 (1) (2017) 23–29, https://doi.org/10.1016/j.ijvsm.2017.02.003. Faculty of Veterinary Medicine, Cairo University.
[76] C. Bankier, et al., Synergistic antibacterial effects of metallic nanoparticle combinations, Sci. Rep. 9 (1) (2019) 3–10, https://doi.org/10.1038/s41598-019-52473-2.
[77] M. Michailidis, et al., Modified mesoporous silica nanoparticles with a dual synergetic antibacterial effect, ACS Appl. Mater. Interfaces 9 (44) (2017) 38364–38372, https://doi.org/10.1021/acsami.7b14642.
[78] S. Mallick, et al., Iodine-stabilized Cu nanoparticle chitosan composite for antibacterial applications, ACS Appl. Mater. Interfaces 4 (3) (2012) 1313–1323, https://doi.org/10.1021/am201586w.
[79] A. Perdikaki, et al., Ag and Cu monometallic and Ag/Cu bimetallic nanoparticle-graphene composites with enhanced antibacterial performance, ACS Appl. Mater. Interfaces 8 (41) (2016) 27498–27510, https://doi.org/10.1021/acsami.6b08403.
[80] B. Kischkel, et al., Therapies and vaccines based on nanoparticles for the treatment of systemic fungal infections, Front. Cell. Infect. Microbiol. 10 (September) (2020), https://doi.org/10.3389/fcimb.2020.00463.
[81] A.F. Costa, et al., Development, characterization, and in vitro-in vivo evaluation of polymeric nanoparticles containing miconazole and farnesol for treatment of vulvovaginal candidiasis, Med. Mycol. 57 (1) (2019) 52–62, https://doi.org/10.1093/mmy/myx155.

[82] K. Niemirowicz, et al., Development of antifungal therapies using nanomaterials, Nanomedicine 12 (15) (2017) 1891–1905, https://doi.org/10.2217/nnm-2017-0052.
[83] K. Maduray, R. Parboosing, Metal nanoparticles: a promising treatment for viral and arboviral infections, Biol. Trace Elem. Res. (2020), https://doi.org/10.1007/s12011-020-02414-2.
[84] X. Dong, et al., 'Carbon dots' antiviral functions against noroviruses, Sci. Rep. 7 (1) (2017) 1–10, https://doi.org/10.1038/s41598-017-00675-x. Springer US.
[85] M.A. Meléndez-Villanueva, et al., Virucidal activity of gold nanoparticles synthesized by green chemistry using garlic extract, Viruses 11 (12) (2019) 1–13, https://doi.org/10.3390/v11121111.
[86] S.S. Jeremiah, K. Miyakawa, T. Morita, Y. Yamaoka, Potent antiviral effect of silver nanoparticles on SARS-CoV-2, Biochem. Biophys. Res. Commun. 533 (1) (2020) 195–200.
[87] H. Ghaffari, A. Tavakoli, A. Moradi, A. Tabarraei, F. Bokharaei-Salim, M. Zahmatkeshan, M. Farahmand, et al., Inhibition of H1N1 influenza virus infection by zinc oxide nanoparticles: another emerging application of nanomedicine, J. Biomed. Sci. 26 (1) (2019) 1–10.
[88] B.M. Sutton, Gold compounds for rheumatoid arthritis, Gold Bull. 19 (1) (1986) 15–16, https://doi.org/10.1007/BF03214639.
[89] Y. Zhang, et al., A systematic comparison of in vitro cell uptake and in vivo biodistribution for three classes of gold nanoparticles with saturated PEG coatings, PLoS One 15 (7) (2020) 1–16, https://doi.org/10.1371/journal.pone.0234916.
[90] S. Vranic, I. Gosens, N.R. Jacobsen, K.A. Jensen, B. Bokkers, A. Kermanizadeh, V. Stone, et al., Impact of serum as a dispersion agent for in vitro and in vivo toxicological assessments of TiO_2 nanoparticles, Arch. Toxicol. 91 (1) (2017) 353–363.
[91] B. Fubini, M. Ghiazza, I. Fenoglio, Physico-chemical features of engineered nanoparticles relevant to their toxicity, Nanotoxicology 4 (4) (2010) 347–363, https://doi.org/10.3109/17435390.2010.509519.
[92] M.C. Power, et al., Exposure to air pollution as a potential contributor to cognitive function, cognitive decline, brain imaging, and dementia: a systematic review of epidemiologic research, Neurotoxicology (2016), https://doi.org/10.1016/j.neuro.2016.06.004.
[93] K. Unfried, et al., Cellular responses to nanoparticles: target structures and mechanisms, Nanotoxicology (2007), https://doi.org/10.1080/00222930701314932.
[94] A. Sukhanova, et al., Dependence of nanoparticle toxicity on their physical and chemical properties, Nanoscale Res. Lett. 13 (2018), https://doi.org/10.1186/s11671-018-2457-x.
[95] Z. Yu, et al., Reactive oxygen species-related nanoparticle toxicity in the biomedical field, Nanoscale Res. Lett. 15 (1) (2020), https://doi.org/10.1186/s11671-020-03344-7.
[96] R. Santacruz-Marquez, et al., Comparative effects of TiO2 and ZnO nanoparticles on growth and ultrastructure of ovarian antral follicles, Reprod. Toxicol. (2020).
[97] Y. Kang, et al., Potential links between cytoskeletal disturbances and electroneurophysiological dysfunctions induced in the central nervous system by inorganic nanoparticles, Cell. Physiol. Biochem. 40 (6) (2016) 1487–1505, https://doi.org/10.1159/000453200.
[98] A. Lambert, Live cell imaging with holotomography and fluorescence, Microscopy Today 28 (1) (2020) 18–23, https://doi.org/10.1017/s1551929519001032.
[99] M. Ravi, et al., 3D cell culture systems: advantages and applications, J. Cell. Physiol. 230 (1) (2015) 16–26, https://doi.org/10.1002/jcp.24683.
[100] W. Liu, et al., Zebrafish: a promising model for evaluating the toxicity of carbon dot-based nanomaterials, ACS Appl. Mater. Interfaces 12 (43) (2020) 49012–49020, https://doi.org/10.1021/acsami.0c17492.
[101] H. Guo, et al., Intravenous administration of silver nanoparticles causes organ toxicity through intracellular ros-related loss of interendothelial junction, Part. Fibre Toxicol. 13 (21) (2016) 1–13, https://doi.org/10.1186/s12989-016-0133-9.

[102] N. Waegeneers, et al., Short-term biodistribution and clearance of intravenously administered silica nanoparticles, Toxicol. Rep. 5 (May) (2018) 632–638, https://doi.org/10.1016/j.toxrep.2018.05.004. Elsevier.

[103] L. Yang, et al., Comparisons of the biodistribution and toxicological examinations after repeated intravenous administration of silver and gold nanoparticles in mice, Sci. Rep. 7 (1) (2017) 1–12, https://doi.org/10.1038/s41598-017-03015-1. Springer US.

[104] E. Baranowska-wójcik, et al., Effects of titanium dioxide nanoparticles exposure on human health—a review, Biol. Trace Elem. Res. 193 (2020) 118–129.

[105] H.M. Braakhuis, et al., Mechanism of action of TiO 2 : recommendations to reduce uncertainties related to carcinogenic potential, Annu. Rev. Pharmacol. Toxicol. 61 (2020) 1–21.

[106] A.P. Walczak, et al., Bioavailability and biodistribution of differently charged polystyrene nanoparticles upon oral exposure in rats, J. Nanopart. Res. 17 (5) (2015) 1–13, https://doi.org/10.1007/s11051-015-3029-y. Springer Netherlands.

[107] S. Barua, S. Mitragotri, Challenges associated with penetration of nanoparticles across cell and tissue barriers: a review of current status and future prospects, Nano Today 9 (2) (2014) 223–243, https://doi.org/10.1016/j.nantod.2014.04.008. Elsevier Ltd.

[108] R. Nho, Pathological effects of nano-sized particles on the respiratory system, Nanomed.: Nanotechnol., Biol. Med. 29 (2020) 102242, https://doi.org/10.1016/j.nano.2020.102242. Elsevier Inc.

[109] M. Guttenberg, et al., Biodistribution of inhaled metal oxide nanoparticles mimicking occupational exposure: a preliminary investigation using enhanced darkfield microscopy, J. Biophotonics 9 (10) (2016) 987–993, https://doi.org/10.1002/jbio.201600125.

[110] D. Li, et al., In vivo biodistribution and physiologically based pharmacokinetic modeling of inhaled fresh and aged cerium oxide nanoparticles in rats, Part. Fibre Toxicol. 13 (45) (2016) 1–20, https://doi.org/10.1186/s12989-016-0156-2.

[111] X. Chen, et al., Toxicity and mechanism of mesoporous silica nanoparticles in eyes, Nanoscale 12 (25) (2020) 13637–13653, https://doi.org/10.1039/d0nr03208e.

[112] S. Zhu, et al., Safety assessment of nanomaterials to eyes: an important but neglected issue, Adv. Sci. 6 (2019) 1–16, https://doi.org/10.1002/advs.201802289.

[113] N.K. Mehra, et al., Safety and toxicity of nanomaterials for ocular drug delivery applications, Nanotoxicology 10 (7) (2016) 836–860, https://doi.org/10.3109/17435390.2016.1153165.

[114] K. Sawicki, et al., Toxicity of metallic nanoparticles in the central nervous system, Nanotechnol. Rev. 8 (1) (2019) 175–200, https://doi.org/10.1515/ntrev-2019-0017.

[115] F. Hong, et al., Nano-TiO2 inhibits development of the central nervous system and its mechanism in offspring mice, J. Agric. Food Chem. 66 (44) (2018) 11767–11774, https://doi.org/10.1021/acs.jafc.8b02952.

[116] R. Rampado, et al., Recent advances in understanding the protein Corona of nanoparticles and in the formulation of "stealthy" nanomaterials, Front. Bioeng. Biotechnol. 8 (April) (2020) 1–19, https://doi.org/10.3389/fbioe.2020.00166.

[117] J.G. Cronin, et al., Nanomaterials and innate immunity: a perspective of the current status in nanosafety, Chem. Res. Toxicol. 33 (5) (2020) 1061–1073, https://doi.org/10.1021/acs.chemrestox.0c00051.

[118] R.H. Müller, S. Gohla, C.M. Keck, State of the art of nanocrystals—special features, production, nanotoxicology aspects and intracellular delivery, Eur. J. Pharm. Biopharm. 78 (1) (2011) 1–9, https://doi.org/10.1016/j.ejpb.2011.01.007.

[119] Y. Lu, Y. Lv, T. Li, Hybrid drug nanocrystals, Adv. Drug Deliv. Rev. 143 (2019) 115–133, https://doi.org/10.1016/j.addr.2019.06.006.

[120] D.E. Marx, D.J. Barillo, Silver in medicine: the basic science, Burns 40 (S1) (2014) S9–S18, https://doi.org/10.1016/j.burns.2014.09.010. Elsevier Ltd and International Society of Burns Injuries.

[121] J.S. Kim, et al., Antimicrobial effects of silver nanoparticles, Nanomedicine 3 (1) (2007) 95–101, https://doi.org/10.1016/j.nano.2006.12.001.
[122] W. Lu, et al., Effect of surface coating on the toxicity of silver nanomaterials on human skin keratinocytes, Chem. Phys. Lett. 487 (1–3) (2010) 92–96, https://doi.org/10.1016/j.cplett.2010.01.027. Elsevier B.V.
[123] M. Kruszewski, et al., Toxicity of silver nanomaterials in higher eukaryotes, Adv. Mol. Toxicol. (2011), https://doi.org/10.1016/B978-0-444-53864-2.00005-0.
[124] B.Y.S. Kim, J.T. Rutka, W.C.W. Chan, Nanomedicine, N. Engl. J. Med. 365 (25) (2010) 2434–2443, https://doi.org/10.5005/jp/books/11254_1.
[125] Y. Pan, et al., Serological immunochromatographic approach in diagnosis with SARS-CoV-2 infected COVID-19 patients, J. Infect. 81 (1) (2020) e28–e32, https://doi.org/10.1016/j.jinf.2020.03.051. Elsevier Ltd.
[126] A. Ivask, et al., Toxicity of 11 metal oxide nanoparticles to three mammalian cell types in Vitro, Curr. Top. Med. Chem. 15 (18) (2015) 1914–1929.
[127] P. Kielbik, et al., Biodegradation of the ZnO:Eu nanoparticles in the tissues of adult mouse after alimentary application, Nanomed.: Nanotechnol., Biol. Med. 13 (3) (2017) 843–852, https://doi.org/10.1016/j.nano.2016.11.002.
[128] N.V. Konduru, et al., Silica coating influences the corona and biokinetics of cerium oxide nanoparticles, Part. Fibre Toxicol. 12 (1) (2015) 1–15, https://doi.org/10.1186/s12989-015-0106-4.
[129] A. Dellinger, et al., Application of fullerenes in nanomedicine: an update, Nanomedicine 8 (7) (2013) 1191–1208, https://doi.org/10.2217/nnm.13.99.
[130] E. Oberdörster, Manufactured nanomaterials (fullerenes, C60) induce oxidative stress in the brain of juvenile largemouth bass, Environ. Health Perspect. 112 (10) (2004) 1058–1062, https://doi.org/10.1289/ehp.7021.
[131] C. Chung, et al., Biomedical applications of graphene and graphene oxide, Acc. Chem. Res. 46 (10) (2013) 2211–2224, https://doi.org/10.1021/ar300159f.
[132] S. Sabella, et al., A general mechanism for intracellular toxicity of metal-containing nanoparticles, Nanoscale 6 (12) (2014) 7052–7061, https://doi.org/10.1039/c4nr01234h.
[133] X. Gao, G.V. Lowry, Progress towards standardized and validated characterizations for measuring physicochemical properties of manufactured nanomaterials relevant to nano health and safety risks, NanoImpact 9 (September 2017) (2018) 14–30, https://doi.org/10.1016/j.impact.2017.09.002. Elsevier.
[134] L. Lamon, et al., Grouping of nanomaterials to read-across hazard endpoints: a review, Nanotoxicology 13 (1) (2019) 100–118, https://doi.org/10.1080/17435390.2018.1506060. Taylor & Francis.
[135] J.H.E. Arts, et al., A critical appraisal of existing concepts for the grouping of nanomaterials, Regul. Toxicol. Pharmacol. 70 (2) (2015) 492–506, https://doi.org/10.1016/j.yrtph.2014.07.025. Elsevier Inc.
[136] L. Liang, et al., Nanoparticles' interference in the evaluation of in vitro toxicity of silver nanoparticles, RSC Adv. 5 (82) (2015) 67327–67334, https://doi.org/10.1039/c5ra05863e. Royal Society of Chemistry.
[137] J.P. Piret, et al., Pan-European inter-laboratory studies on a panel of in vitro cytotoxicity and pro-inflammation assays for nanoparticles, Arch. Toxicol. 91 (6) (2017) 2315–2330, https://doi.org/10.1007/s00204-016-1897-2. Springer Berlin Heidelberg.
[138] A. Bannuscher, et al., Metabolomics profiling to investigate nanomaterial toxicity in vitro and invivo, Nanotoxicology 14 (6) (2020) 807–826, https://doi.org/10.1080/17435390.2020.1764123. Taylor & Francis.
[139] M. Riediker, et al., Particle toxicology and health—where are we? Part. Fibre Toxicol. (2019), https://doi.org/10.1186/s12989-019-0302-8.

Index

Note: Page numbers followed by *f* indicate figures and *t* indicate tables.

FSC
www.fsc.org
MIX
Papier | Fördert
gute Waldnutzung
FSC® C083411

Kolibri 360 GmbH
Ferdinand-Jühlke-Straße 7
99095 Erfurt, Deutschland
produktsicherheit@kolibri360.de